SOLID WASTE ANALYSIS AND MINIMIZATION

A SYSTEMS APPROACH

MATTHEW J. FRANCHETTI, Ph.D., P.E.

New York   Chicago   San Francisco   Lisbon   London   Madrid
Mexico City   Milan   New Delhi   San Juan   Seoul
Singapore   Sydney   Toronto

The McGraw-Hill Companies

Library of Congress Cataloging-in-Publication Data

Franchetti, Matthew J.
   Solid waste analysis and minimization : a systems approach /
   Matthew J. Franchetti.
          p.   cm.
   Includes bibliographical references and index.
   ISBN 978-0-07-160524-3 (alk. paper)
   1. Waste minimization.   2. Factory and trade waste.   I. Title.
   TD793.9.F73   2009
   363.72′85—dc22   2009010507

1 2 3 4 5 6 7 8 9 0   DOC/DOC   0 1 5 4 3 2 1 0 9

ISBN 978-0-07-160524-3
MHID 0-07-160524-X

Sponsoring Editor: Taisuke Soda
Editing Supervisor: Stephen M. Smith
Production Supervisor: Pamela A. Pelton
Project Manager: Smita Rajan, International Typesetting and Composition
Copy Editor: Priyanka Sinha, International Typesetting and Composition
Proofreader: Lucia Read
Indexer: WordCo Indexing Services, Inc.
Art Director, Cover: Jeff Weeks
Composition: International Typesetting and Composition

Printed and bound by RR Donnelley.

Dedication

This book is dedicated to my children Jack and Kate so that the world may be a cleaner and healthier place for them and their children.

About the Author

Matthew J. Franchetti, Ph.D., P.E., is an Assistant Professor of Mechanical, Industrial, and Manufacturing Engineering and Director of Undergraduate Studies for the Mechanical and Industrial Engineering Programs at The University of Toledo.

CONTENTS

Part 2   Solid Waste Assessment Strategies

PREFACE

"Green" initiatives have increased in popularity in recent years as many organizations attempt to analyze and reduce environmental impacts. This gain in attention has been brought on by consumer demand for cleaner production, dwindling natural resources, and population growth around the world. Solid waste minimization and management is often an overlooked component of an overall environmental protection plan for organizations. The focusing on sustainability, waste reduction, and green initiatives by individual firms can lead to enhanced economic performance, environmental protection, and health benefits for the worldwide community.

Solid Waste Analysis and Minimization: A Systems Approach provides an up-to-date source of technical information relating to current and potential solid waste minimization practices. This book gives a detailed framework and reference material and is broken into four parts:

- Part 1 gives definitions, benefits, and background information regarding solid waste analysis and minimization.
- Part 2 discusses solid waste assessment strategies and provides the general framework for conducting a solid waste audit.
- Part 3 discusses solid waste modeling, prediction, and evaluation for various industries based on standard industry classifications.
- Part 4 features 21 case studies in various industries, including two illustrating international businesses.

This book serves as both a practical and a technical guide to minimizing solid waste for any organization, from manufacturing to service facilities. It is based on the solid waste minimization hierarchy of reduce, reuse, and recycle. The solid waste auditing procedure detailed in this book has been applied and refined over the past 12 years on more than 80 solid waste assessment projects around the globe. I am confident that this book will contribute to the field of solid waste management and emphasize the need for continued research as we strive for sustainable solutions to solid waste issues.

Acknowledgments

I wish to acknowledge the Lucas County Solid Waste Management District and the Lucas County Board of Commissioners for their continued support and commitment to prevention of pollution and sustainability. Their funding of solid waste reduction research since 1996 has led to many of the examples and case studies discussed in this book. Specifically, I would like to mention Mr. James Walters of the Lucas County Solid Waste Management District and Dr. Robert Bennett of The University of Toledo for their guidance and support related to solid waste minimization. I also wish to thank the numerous undergraduate and graduate students who have conducted research in the Environmentally Conscious Design and Manufacturing Laboratory at The University of Toledo.

I wish to acknowledge Laura Franchetti for her editing and critiquing skills and moral support during the writing this book. Also, I thank the many unnamed people who assisted in the preparation of this book or who worked on some of the projects used as examples.

Matthew J. Franchetti, Ph.D., P.E.

Part 1

WHY SOLID WASTE REDUCTION?

1

DEFINITION OF SOLID WASTE

ANALYSIS AND MINIMIZATION

1.1 Introduction

In the extremely competitive world of business and manufacturing, solid waste disposal is often overlooked. In many companies, few stakeholders are aware of the destinations of the generated wastes and associated costs. There are associated costs with excess raw materials, scrap parts, poor use of resources (including labor), and outdated materials. All of these contribute to a company's solid waste stream, and must be addressed when calculating the costs of disposal. In addition, it is not an ecologically or economically sound decision to allow recyclables to enter landfills. Thus, destinations and the costs associated with them must be addressed.

Solid waste disposal typically accounts for 3 percent of a manufacturing organization's expenses. In 2007, Ford Motor Company (Ford) reported net revenue of $172.5 billion and expenses of $153.6 billion, with a profit of $18.9 billion, and General Motors (GM) reported net revenue of $181.1 billion and, expenses of $166.6 billion, with a profit of $14.5 billion. Solid waste disposal costs are approximately $4.6 billion and $5.0 billion for Ford and GM, respectively. Considering the multibillion dollar expenses associated with solid waste disposal for these companies, significant cost savings can be realized by analyzing and minimizing solid waste generation.

As indicated in the previous example, waste analysis and minimization can be financially very advantageous to companies and should be evaluated. However, many companies do not have any personnel capable of performing the evaluation due to time constraints and lack of knowledge and experience in the field. The fulfillment of this need for an experienced evaluator with available time has led to research opportunities and a competitive disposal and recycling market, and has opened new fields of recycling.

It is easy to get wrapped up in the emotions of environmental protection, waste minimization, and recycling often associated with environmental activists or "tree huggers." The real challenge is to have these initiatives make business sense for corporations around the world. In other words, how will recycling and environmental protection improve the profitability or operating budgets for organizations? That is the focus of this book, to translate solid waste minimization and recycling into terms that appeal to a CEO and that will motivate him or her to take actions based on the economic benefits as well as the environmental benefits. A 2007 survey conducted in Toledo, Ohio of companies that had implemented solid waste minimization and recycling strategies indicated an average annual cost savings of $125,000 per company. These are concepts that motivate CEOs. The central concept is that a company can help the environment and its economics at the same time.

As the world moves toward a global economy and competition gets tougher, there is a greater need for organizations to enhance corporate images and improve financial performance. Waste minimization is very important for companies and organizations of all types. Effective waste minimization can increase profits by reducing waste disposal costs, raw material purchases, and other operating costs. This is primarily accomplished through use of the three Rs, which is to *reduce*, *reuse*, and *recycle* any material in the organization's waste stream.

Many analytical models have been developed to help companies achieve the goal of waste minimization. One such method is presented in this book. This method has been applied to over 75 companies and has achieved demonstrated environmental and economic results. The model hinges on the business maxim "If you can't measure it, you can't manage it." The objective of the model is to quantify the solid waste generation and achieve the following goals:

- Increased manufacturing competitiveness through reduced solid waste disposal costs, reduced energy costs, and optimized use of raw materials, packaging, and floor space.
- Improved corporate image as companies become more green.
- Reduced pollution through reduced energy usage, and the application of clean and renewable energy sources.
- Decreased reliance on landfills for disposal.

The primary goal of this type of research is to empower companies and organizations to reduce operating costs by limiting the amount of raw material and energy used in the facility. This includes minimizing waste, transportation, and storage, reducing disposal fees, generating revenue for the sale of recyclables, and limiting pollution to improve the quality of the environment. Many organizations are able to increase profit by reducing waste disposal costs, raw material purchases, and other operating costs. However, many companies do not have the capability to perform a waste evaluation due to time constraints and lack of knowledge in the field.

This book provides a detailed framework and reference material for solid waste analysis and minimization and it is broken into three parts

- Part 1 provides the definition, benefits, and background information regarding solid waste analysis and minimization.
- Part 2 discusses solid waste assessment strategies and provides the general framework to conduct a solid waste audit.
- Part 3 provides 27 case studies in various businesses, including two international case studies.

1.2   Definitions

In order to study waste analysis and minimization, it is first necessary to understand the terms and definitions that relate to the process. In the field of recycling, people often have differing expectations upon hearing many of the common terms. To compound the problem, finding universal definitions for these terms can be challenging as many companies and government agencies create their own designations, often using combinations of technical and operational components. This section discusses these key terms and definitions as they relate to the topics covered in this book.

First of all, it is important to define solid waste analysis and minimization. *Waste analysis* is the detection of waste streams, their origins, their composition and their destinations. Waste analysis is often accomplished through a waste audit or assessment procedure. This involves a walk through of the facility, as well as tracking and quantification. *Waste minimization* is the process of reducing waste streams through source reduction, reuse and recycling of materials, and the reallocation of resources. Typically, the waste audit involves

■ Research of the company's process and overall solid waste generation
■ Recommendations that are designed to maximize process efficiency and reduce solid waste disposal costs
■ A detailed reference list of vendors that complement the recommendations

The waste minimization procedure is segmented into four phases as displayed in Fig. 1.1, which provides an overview of the waste minimization procedure.

1.2.1   SOURCE REDUCTION

Source reduction is the reduction of materials coming into the system. The Environmental Protection Agency (EPA) defines source reduction as "activities designed to reduce the volume or toxicity of waste generated, including the design and manufacture of products with minimum toxic content, minimum volume of material, and/or a longer useful life." The two interesting components of this definition are *volume* and *toxicity*. This means that an organization does not have to solely focus on reducing volumes for source reduction initiatives, but could focus efforts on reducing the negative impacts of impacts on the environment of those same volumes. This may mean a reduction of pallets entering a building or of flash being generated from a mold.

An example of source reduction includes the environmental efforts of Proctor and Gamble, the makers of Tide laundry detergent. Tide switched from a plastic bottle to a flexible pouch. The flexible pouch weighs 85 percent less than a plastic bottle and the amount of packaging is 84 percent less than the plastic bottle. This source reduction also resulted in lower transportation costs for the company.

1.2.2   REUSE

Reuse is the actual reuse of a material in its present form. Some examples are printing draft copies on the reverse side of previously used paper, using incoming pallets as outgoing pallets, or using incoming boxes as collection containers for recyclables.

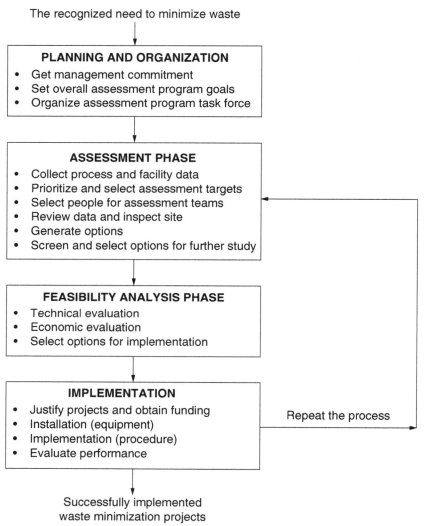

The recognized need to minimize waste

PLANNING AND ORGANIZATION
- Get management commitment
- Set overall assessment program goals
- Organize assessment program task force

ASSESSMENT PHASE
- Collect process and facility data
- Prioritize and select assessment targets
- Select people for assessment teams
- Review data and inspect site
- Generate options
- Screen and select options for further study

FEASIBILITY ANALYSIS PHASE
- Technical evaluation
- Economic evaluation
- Select options for implementation

IMPLEMENTATION
- Justify projects and obtain funding
- Installation (equipment)
- Implementation (procedure)
- Evaluate performance

Repeat the process

Successfully implemented
waste minimization projects

Figure 1.1   The waste minimization assessment procedure.

1.2.3 RECYCLING

Recycling is a type of reuse which involves changing the composition or properties of the material in one way or another. For example, this can be accomplished by melting it down, chipping or grinding the materials. A broad definition of recycling is taking a product or material at the end of its useful life and turning it into a usable raw material to make another product. There are three types of recycling, as discussed in the hierarchy of waste management:

1 In-process recycling
2 On-site recycling
3 Off-site recycling

1.2.4 POLLUTION PREVENTION

Pollution prevention is the broadest and most difficult term to concisely define. In essence it is the overall process of reducing waste and preventing pollution from entering the environment through the air, water, or ground. It encompasses both the aspects of source reduction and waste reduction. The EPA has defined pollution prevention as

> "Pollution prevention means source reduction, as defined under the Pollution Prevention Act, and other practices that reduce or eliminate the creation of pollutants through:
>
> ■ Increased efficiency in the use of raw materials, energy, water, or other resources, or
> ■ Protection of natural resources under conservation"

Based on this definition, pollution prevention covers only the first two elements of the waste management hierarchy: source reduction and in-process recycling.

1.2.5 ACRONYMS AND SYMBOLISM

It is now necessary to clarify abbreviations and symbols for materials that are related to solid waste minimization and recycling. To begin, it is necessary to note the differences in paper. There are several types of papers according to recyclers, and the breakdown is as follows:

■ *White office paper (WOP).* This is one of the highest grades of paper and is also known as white ledger. WOP is clean white sheets from laser printers and copy machines. Colored, contaminated, or lower-grade paper is not acceptable. The wrappers the paper comes in are of lower grade, and not WOP.
■ *Mixed office paper (MOP).* This is a catch-all for types of paper not specifically mentioned above. Everything—magazines to packaging—is acceptable. The paper must still be clean, dry, and free of food, most plastic, wax, and other contamination.
■ *Corrugated containers (OCC).* This is a paper-based construction material consisting of a fluted corrugated sheet and one or two flat linerboards. It is widely used in the manufacture of corrugated boxes and shipping containers. The corrugated medium and linerboard are made of paperboard, a paper-like material usually over 10 mils (0.010 in, or 0.25 mm) thick. Paperboard and corrugated fiberboard are sometimes called cardboard by nonspecialists, although cardboard might be any heavy paper-pulp-based board.
■ *Newspapers (ONP).* This is widely available and of uniform consistency, which makes it valuable. The entire newspaper, including inserts is acceptable, except for things like plastic, product samples, and rubber bands.

■ *Phone books.*   These are made with special glue that breaks down in water. (Some phone books use glue that interferes with recycling.) Printed in the phone book should be information on the source and type of paper used, the nature of the binding, and where phone books can be recycled locally (old Corrugated containers[OCC]).

■ *Waxed cartons.*   These, often used for milk or juice, are plastic laminated on the inside, even if they don't have a plastic spout.

The next set of acronyms and abbreviations that is important to understand regards plastics. The breakdown is as follows:

■ *Type 1—PETE (polyethylene terephthalate).*   Soda and water containers, and some waterproof packaging.

■ *Type 2—HDPE (high-density polyethylene).*   Milk, detergent, and oil bottles, also toys and plastic bags.

■ *Type 3—PVC (vinyl/polyvinyl chloride).*   Food wrap, vegetable oil bottles, blister packages.

■ *Type 4—LDPE (low-density polyethylene).*   Many plastic bags, shrink wrap, and garment bags.

■ *Type 5—PP (polypropylene).*   Refrigerated containers, some bags, most bottle tops, some carpets, some food wrap.

■ *Type 6—PS (polystyrene).*   Throwaway utensils, meat packing, protective packing.

■ *Type 7—Other.*   Usually layered or mixed plastic. This generally has limited recycling potential and is disposed at the landfill.

With a little bit of care most plastics can be recycled, and collection of plastics for recycling is increasing rapidly. Plastic recycling faces one huge problem: plastic types must not be mixed for recycling, yet it is extremely difficult to distinguish one type from another by sight or touch. It is usually very important that plastics are separated prior to recycling. This is due to the fact that plastic recyclers use different processes for different types of plastics, and not all plastics have the same properties, such as melting points. Thus, it is imperative that they can be separated prior to recycling, either by the consumer, the hauler, the processor, or the recycler.

Plastic types are generally easy to distinguish, because they are imprinted with a recycling symbol and a number. The recycling symbol verifies that the material can be recycled, and the number corresponds to the type of plastic used.

1.2.6 UNITS

In the field of solid waste minimization and recycling, there are two conventions for quantifying the waste and materials. Quantification is either done by weight or volume. If by weight, numbers are given in pounds (lb) or more commonly tons (2000 lb = 1 ton). If by volume, then the numbers are given in cubic yards (yd^3). This book incorporates both conventions, because a scale is not used for all waste audits. Most observations are made by volume and then, using density figures, the approximate weight is calculated.

1.3 The Hierarchy of Solid Waste Management

When approaching solid waste analysis and minimization, there is an accepted hierarchy, or order of solution approaches, that should be deployed when analyzing an organization's waste streams. This hierarchy is given before a discussion of definitions of solid waste management terms because it makes the definitions more meaningful and easier to understand. This hierarchy has been defined in the Pollution Prevention Act of 1990 as follows:

"The Congress hereby declares it to be national policy of the United States that pollution should be prevented or reduced at the source whenever feasible; pollution that cannot be prevented should be recycled in an environmentally safe manner, whenever feasible; pollution that cannot be prevented or recycled should be treated in an environmentally safe manner whenever feasible; and disposal or release into the environment should be employed only as a last resort and should be conducted in an environmentally safe manner"

Based on that definition, the solid waste management solutions, in order of preference are

1 Source reduction
2 In-process recycling
3 On-site recycling
4 Off-site recycling
5 Waste treatment to render the waste less hazardous
6 Secure disposal
7 Direct release to the environment

These terms may be easier to understand with an example. A very straightforward and simple example involves a company that receives incoming raw material in cardboard packaging. Suppose the company is a metal stamping plant for the automobile industry. Below are situations where the company could apply the hierarchy:

1 *Source reduction.* Eliminate the waste through engineering process modifications. For example, the company could work with the vendor to eliminate the use of cardboard containers and switch to returnable plastic containers. This is the most preferred method and completely eliminates the problem of waste management and need for recycling.
2 *In-process recycling.* If waste is generated, develop a separation method to use the waste as a raw material for the same process. For example, the company could collect the metal scrap from the stamping process and attempt to process it again in the same process.
3 *On-site recycling.* If waste is generated, develop a separation method to use the waste as a raw material for another in-house process. For example, the company could use the cardboard packages to ship their own final products.

4 *Off-site recycling.* If waste is generated, develop a separation method and transport the waste to another organization so that another company could use the waste as a raw material. For example, the company could transport the cardboard to a third-party processor for recycling.

5 *Waste treatment to render the waste less hazardous.* If waste is generated, develop a separation method and treat the waste so it is less harmful before releasing it to the environment. This solution applies mostly to chemical processes, such as treating waste water and releasing it to the public sewer.

6 *Secure disposal.* Dispose of the waste at a secure landfill. The company could use its waste hauler to transport the cardboard to a local landfill.

7 *Direct release to the environment.* If waste is generated, develop a separation method and release the waste directly to the environment. In this final scenario, the company could stage the cardboard in an outside area to allow it to biodegrade.

1.4 The Three Rs and the Two Es

If this book were to be reduced to one key idea, it would be the application of the three Rs and the two Es as they apply to solid waste minimization. From a technical and regulatory standpoint, the hierarchy mentioned in the previous section is excellent, but it does not hit home with the leadership of many organizations and does not promote the full benefits of solid waste minimization. In terms of communicating and promoting solid waste minimization the three Rs and the two Es have served as very effective tools. The three Rs are

- Reduce
- Reuse
- Recycle

The three Rs are a summary of the hierarchy discussed in the previous section. In terms of the hierarchy, the first item discussed, source reduction, has been separated into two components, reduce and reuse. Reuse has been added to emphasize the fact that many items that are being disposed of at a landfill by organizations could have been reused, such as cardboard containers, plastic caps, or rubber bands. Finally, all recycling methods, in-process, on-site, and off-site, have been lumped into one category for simplicity. And the two Es are

- Environment
- Economics

The concept is to apply the three Rs at an organization to help the two Es. The three Rs provide the solutions to the solid waste problem and are based on the hierarchy of solid waste management. The two Es communicate the goals of these efforts, to lessen the environmental impact of an organization and improve an organization's economics or bottom line. This simple phrase is easy to understand and has served as a great tag line or catch phrase to promote solid waste reduction. By emphasizing economic benefits, solid waste reduction and minimization benefits are much easier to sell to the decision makers because the efforts are placing attention on the financial health of the organization as well. A plant manager of a battery manufacturing firm located in Toledo, Ohio bluntly summarized this concept when he stated that his company "is in the business of producing and selling batteries, not recycling." It is easy to get caught up in the good feelings associated with helping the environment, but unless there is a financial incentive many organizations may give environmental concerns only lip service. A buzzword that is now emerging is "green washing," which is a situation in which a company publicly and verbally promotes their environmental efforts to bolster corporate images, but falls short of the actions associated with the public statements. The goal of emphasizing the three Rs and two Es is to bridge the gap between public statements and corporate actions by demonstrating that environmental concerns makes business sense.

A common misconception of organizations is that they do not have funds in their budgets for recycling or environmental initiatives. The fallacy in this thinking is that virtually all companies have a starting point or a budget for recycling and waste reduction: the annual expenses for trash removal and janitorial efforts. The paradigm shift requires considering the amount of money expended per year as the environmental or recycling budget and devising creative methods and processes to minimize impacts to the environment. Many organizational leaders are surprised to learn the potential cost avoidance or revenue generated from becoming more environmentally conscious. A financial case study is discussed later in this section highlighting some of the typical benefits.

Economic benefits from solid waste analysis and minimization can be achieved a variety of ways. The most common economic benefits derived from solid waste minimization are listed below:

- *Cost avoidance.*   Organizations can save money by diverting solid waste streams from the landfill back to the company through reuse or to an off-site recycler. The monetary savings are derived from no longer paying a waste hauler to remove the trash and dispose of it at the landfill.
- *Recycling revenue.*   Substantial additional revenue can be earned by selling recyclables to third-party processors or recycling commodity brokers. For example, 1 ton of baled cardboard sells for $100 to $180 on the market.
- *Reduced raw material costs.*   When an organization is able to utilize in-process or on-site recycling they reduce their raw material needs directly by replacing virgin material purchases with in-house scrap, rework, or process by-products.
- *Reduced energy costs.*   By reducing the amount of materials within a facility through reduction and reuse, material-handling costs can be minimized.
- *Increased sales.*   Many consumers and businesses look favorably on organizations that are environmentally conscious and purchase products or services from them.
- *Increased productivity.*   As workers are engaged in efforts that they see as meaningful, many of them take pride and put additional efforts into their work. In addition, absenteeism may reduce as well.

The following case study highlights several typical financial benefits associated with waste minimization. This case study involves an automobile battery manufacturer that employs 430 employees and produces approximately 20,000 batteries per day. At the time of the study the company was recycling over 76 percent of its solid waste. The solid waste minimization audit found that an additional 221 tons of solid waste could be recycled per year, saving the company over $62,000 annually. Table 1.1 highlights the existing recycling and the savings by waste stream.

1.5   The Systems Approach for Waste Minimization

The waste minimization and analysis process presented in this book is based on the systems approach. The systems approach is a problem-solving philosophy that focuses on a holistic view of an organization by analyzing the linkages and interactions between the elements that comprise the entire system. A system is defined as a group of interacting, interrelated, or interdependent elements forming a complex whole coordinated to achieve a stated purpose or goal. The systems approach is a framework that is based on the belief that the component parts of a system will act differently when the system's relationships are removed and it is viewed in isolation. The only way to fully understand why a problem or element occurs and persists is to understand the part in relation to the whole. From a macro view, a system is comprised of inputs, processes, and outputs all revolving around accomplishing a given goal or goals. The definition and clear understanding of this goal are critical to defining the system in terms of its processes, required inputs, and desired outputs. For example, there will be very different systems for an organization that produces automobiles versus an organization that provides heath-care services. The key benefit of the systems approach to solid waste management is that it addresses the solid waste problem from a business standpoint, consistently focusing on the organization's goals, and confronting the problem at every stage of the supply chain. Traditional approaches tend to only address the issue of solid waste at the end of the process, when determining how to cost-effectively remove the waste from the facility. Many organizations also manage trash removal and recycling as compartmentalized problems that are managed separately from their core processes and often handled primarily though the accounting departments. The central issue with this traditional approach is that by focusing on these individual outcomes, overall system optimization cannot be achieved. The systems approach confronts the issue at all phases of the supply chain, from procuring raw materials to designing environmentally friendly processes that reduce solid waste. To that end, the complete life cycle of the product and process is analyzed for potential environmental improvements, not just the waste left over at the end of the day. Areas such as raw material wastes, scrap rates, material-handling wastes (cardboard), and end-use disposal are examined and will be covered in this book. Figure 1.2 provides an overview of the system as it relates to business processes and waste reduction.

Defining the terms of a system will provide additional insights into the interactions and relationships as they relate to solid waste analysis and minimization.

TABLE 1.1   SAMPLE RECYCLING ACTIVITY SUMMARY AND COST BENEFITS

MATERIAL	ANNUAL GENERATION (LB.)	QUANTITY CURRENTLY RECYCLED	CURRENT RECYCLING ACTIVITY	POTENTIAL RECYCLING ACTIVITY	ADDITIONAL RECYCLING ACTIVITY (LB.)	DISPOSAL COST SAVINGS	POTENTIAL RECYCLING REVENUE	TOTAL WASTE RECYCLED
Uncoated Lead*	437,456	437,456	100%	No change	N/A	Unknown	Unknown	437,456
OCC	1,036,865	733,333	71%	100%	303,532	$10,765	$8,347	1,036,865
Polypropylene Battery Components**	103,342	103,342	100%	No change	N/A	Unknown	$31,003	103,342
Administrative-Generated Waste (Paper)	137,338	37,560	27%	50%	30,859	$1,095	$2,973	68,419
Oxide Inserts	84,514	84,514	100%	No change	N/A	Unknown	Unknown	84,514
Wood	79,474	20,000	25%	90%	53,527	$1,203	N/A	73,527
Label Backing	34,562	0	0%	90%	31,106	$1,103	N/A	31,106
Metal Banding	31,535	31,535	100%	No change	N/A	N/A	N/A	31,535
Film Plastic	18,845	0	0%	100%	18,845	$3,342	Unknown	18,845
Aluminum Cans (UBC)+	4,645	Unknown	Unknown	100%	4,645	$329	$1,858	4,645
Total	1,968,576	1,447,740			442,514	$17,837	$44,181	1,890,254
Total Current Recycling Activity	73.50%				96.00%			
Total Proposed Recycling Activity				Total Proposed Recycling Activity				
Total Proposed Cost Savings and Revenue				$62,018				

*The numbers calculated in this analysis do not include in-process recycling, which adds significantly to the given quantities.

**Currently, polypropylene is sent to the smelter. This analysis presents the option of grinding and recycling the polypropylene material, and the costs given reflect only the potential revenue to be earned from the sale of the material.

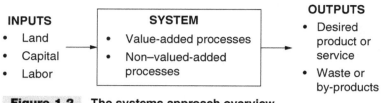

Figure 1.2   The systems approach overview.

The starting point of any system is to clearly define its goal or purpose. Once the goal is defined, the required inputs, outputs, and process designs can be determined concurrently, focusing on the system goals. For example, suppose an organization sets a goal to manufacture 30,000 television sets per year at a rate of return of 14 percent. Now the system can be defined in terms of processes, inputs, and outputs. The inputs, also known as the factors of production, are land, capital, and labor:

- Land or natural resources—naturally occurring goods such as water, air, soil, minerals used in the creation of products.
- Labor—human effort used in production, which also includes technical and marketing expertise.
- Capital—human-made goods (or means of production) that are used in the production of other goods. These include machinery, tools, and buildings.

From the television-manufacturing example, the inputs would include the raw materials, such as the electronics, plastics, and metals. Considering these raw materials, the processes would be designed concurrently. Attention would be given to value added and non–value added activities. The process design will answer the following questions:

- What is to be produced?
- How are the products to be produced?
- When are the products to be produced?
- How much of each product will be produced?
- For how long will the products be produced?
- Where are the products to be produced?

The answers to these questions are obtained from product, process, and schedule design and are not independently and sequentially determined, but developed concurrently. A clear vision is critical and the success of a firm is dependent on having an efficient production system. The product design phase defines which products will be produced and provides detailed drawings of the part. A quality deployment function (QDF) is generally applied to accomplish this. A QDF is an organized approach to identify customer needs and translate the needs to product characteristics, process design, and the tolerances required. Following is an exploded part drawing that identifies the raw material and components that make up an assembly or final product (Fig. 1.3). The process design phase examines the following issues:

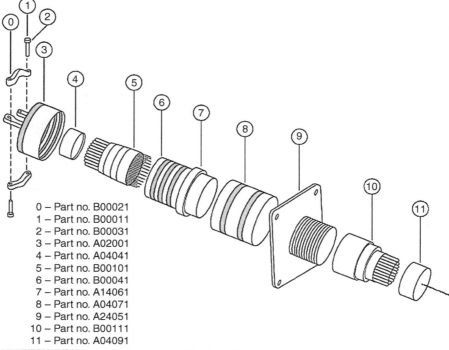

0 – Part no. B00021
1 – Part no. B00011
2 – Part no. B00031
3 – Part no. A02001
4 – Part no. A04041
5 – Part no. B00101
6 – Part no. B00041
7 – Part no. A14061
8 – Part no. A04071
9 – Part no. A24051
10 – Part no. B00111
11 – Part no. A04091

Figure 1.3   **Exploded assembly drawing.**

■ How is the product to be produced?
■ Make or buy decision
■ Process selection
■ Equipment selection
■ Process time determination

The desired outputs of the design phase are

■ Processes
■ Equipment
■ Raw materials required
■ Formal documents
    ■ Route sheet
    ■ Assembly chart
    ■ Operation process chart
    ■ Precedence diagram

Table 1.2 gives an example of a route sheet. Figures 1.4, 1.5 and 1.6 give examples of the assembly chart, operation process chart, and precedence diagram respectively.

TABLE 1.2 ROUTING SHEET DATA REQUIREMENT	
DATA	PRODUCTION EXAMPLE
Component name and number	Plunger housing
Operation description and number	Shape, drill, and cut off
Equipment description	Automatic screw machining
Unit times	Setup time: 5 h operation time and 0.005 h per component
Raw material requirement	1 in diameter × 12 ft of aluminum bar per 80 components

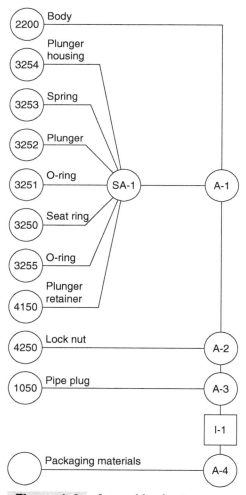

Figure 1.4   Assembly chart.

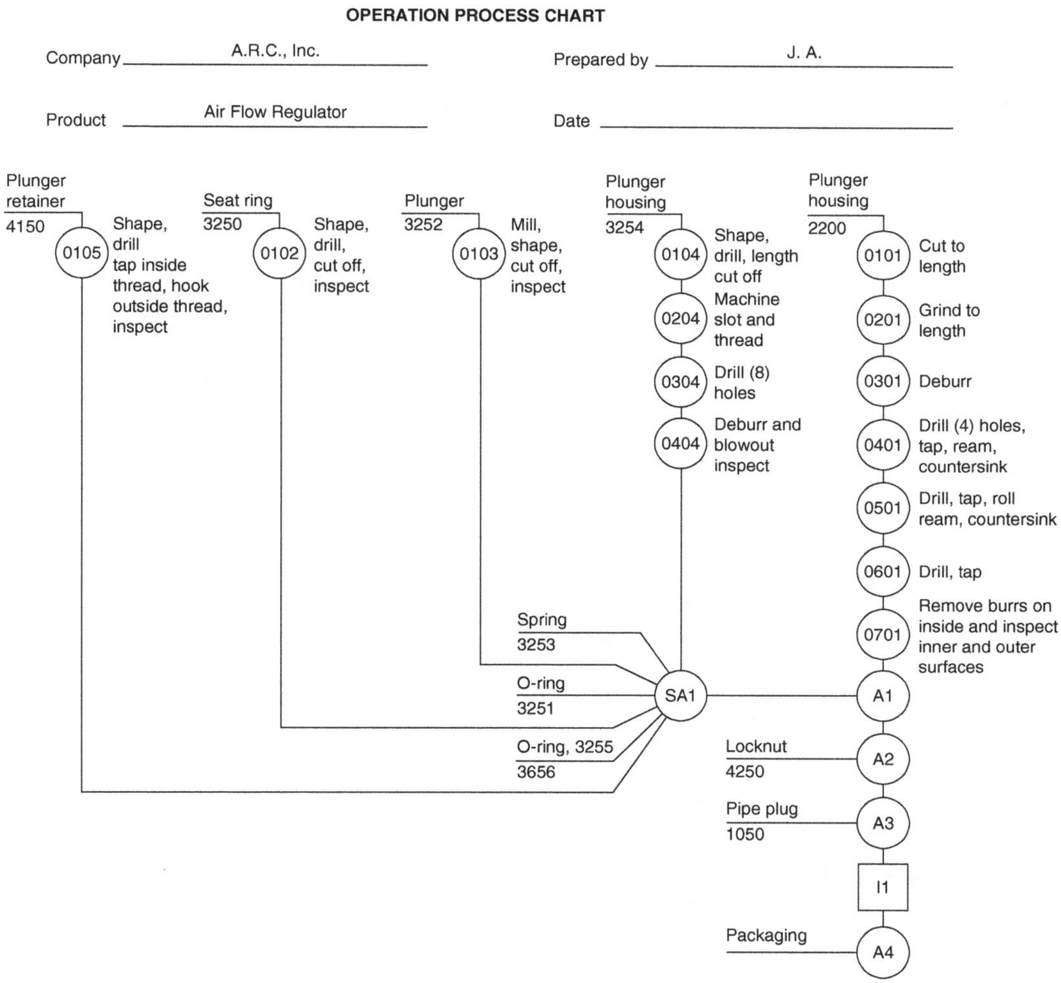

Figure 1.5   Operation process chart.

The schedule design phase examines the following issues:

- How much to produce? (lot sizing)
- When to produce? (production scheduling)
- How long to produce? (market forecasts)

The determination of the outputs of a system seems very straightforward at first examination, as it should be defined in the goal. In the television example, it is 30,000 televisions per year. This is only a partial answer as there will also be unintended outputs, which includes waste and by-products, that must also be considered. Waste

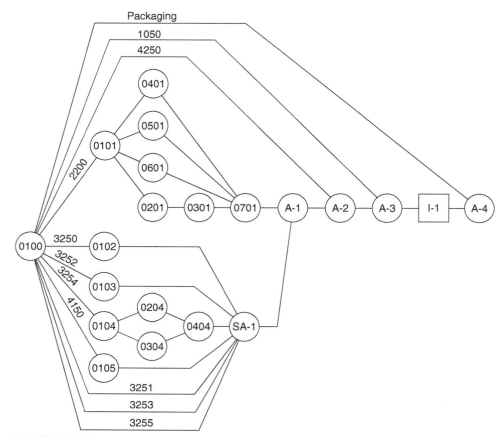

Figure 1.6   Precedence diagram.

management becomes a new unintended problem created by the system. In the previously discusssed 30,000 unit television set manufactoring example, this includes wire fragments to food waste from employees' lunches. In terms of solid waste, managers and decisions makers must take a reactive role to resolve these problems, ranging from recycling to reuse. A better approach is to handle these issues proactively when determining processes and inputs. From a system approach, this book will provide waste analysis and minimization techniques and strategies for all phases of the supply chain.

1.6   Summary

The systems approach in an integral concept to the successful implementation of solid waste minimization and serves as the basis for the solution methodologies. This holistic approach considers the entire supply chain and addresses waste generation issues at the earliest phases possible, often preventing the generation of the waste and eliminating waste management issues all together. The solid waste management hierarchy ensures that solution alternatives are applied in the preferred order and serve as guidelines thought the analysis process. In a manner of thinking, the systems approach examines the solid waste problem from a macro level and the hierarchy examines it from a micro level.

An understanding and agreement to the meaning of the terms used in solid waste management are essential in communication and problem resolution. These terms, which include source reduction, recycling, pollution prevention, and solid waste analysis and minimization, are often used as synonyms, but as discussed in this chapter can carry different meanings. These terms are not intended to be comprehensive, but provide a general framework to aid in a meaningful discussion of these issues. When promoting solid waste minimization to decision makers, it is critical to relate the benefits to both the environment and the economics of the company. Creating this win-win situation in a manager's eyes will significantly enhance the likelihood that solid waste minimization recommendation will be implemented.

2

BACKGROUND AND FUNDAMENTALS OF SOLID WASTE ANALYSIS AND MINIMIZATION

2.1 Introduction

Solid waste generation continues to increase in the United States, and as discussed in this chapter, solid waste measurement and reduction tools have not kept pace. Each year, U.S. industrial facilities generate and dispose of 7.6 billion tons of nonhazardous industrial waste (www.epa.gov, retrieved August 12, 2002). This solid waste is measured and monitored by various U.S. government agencies, such as the U.S. Environmental Protection Agency and similar state-level organizations. These agencies also develop methods to reduce the solid waste generation levels by promoting waste reduction and recycling.

As mentioned in Chap. 1, a common problem the government agencies face from U.S. corporations is perceived poor economics of solid waste reduction programs. Many corporate leaders believe waste reduction and recycling are not profitable for their company. U.S. corporations feel they are in the business of manufacturing products and providing services, not recycling and waste disposal. This common attitude is problematic and hinders waste reduction efforts across the United States. The paradigm must be shifted from this attitude to one that stresses concern for the environment makes good business sense. Corporate environmental concern makes sense because it can be economically beneficial, positively raises pubic opinion, and assists corporations to comply with environmental regulations. The U.S. Department of Labor estimates over 4.8 million corporations operated in the United States in the year 2001. The fact that such a large number of corporations are operating in the United States, each generating various amounts and compositions of solid waste, stresses the importance of widespread waste reduction. Economical justification of waste reduction will increase top corporations' managements to improve the environment and minimize the waste each disposes. The evaluation and prediction system developed in this research may be applied to estimate the economic cost-benefits from increased recycling. This application could assist U.S. businesses in reducing solid waste and increasing recycling to bolster profits. A solid waste prediction and evaluation system would be mutually beneficial to both government and businesses. This will assist in creating a win-win situation considering the large amount of industrial waste generated in the United States, the government's need to monitor and reduce the waste, and their goal to decrease solid waste levels and increase recycling.

This chapter provides relevant background information and discusses the fundamentals of solid waste analysis and minimization. These concepts are critical to have an overall understanding of the field and aid in developing optimal solutions. Included in this chapter are a brief history of the field; an overview of regulations and laws; and a discussion of trends, data, and statistics in the global field. This information gives a macro view of the problem and the need for solid waste minimization. In addition, overviews of the tools that have been developed to analyze and improve the situation are discussed, including risk assessment, industrial ecology, life cycle assessment, and fundamentals of recycling processes. These items are not the focus of this book, but highlight some of the tools and methods available in the field. Finally a discussion of common problems and an overview of the government support available to combat the issues are provided.

2.2   A Brief History of Solid Waste Management

From a global standpoint, the evolution of solid waste management has followed humankind's development and progress as it transitioned from an agricultural base, through the industrial revolution, and now to the information age. As societies addressed the critical waste management needs that were associated with these transitions, methods, policies, and regulations were created to keep pace. The earliest documented waste management regulation dates back to 3000 B.C., in Cretan capital, Knossos. This and other policies and events associated with the evolution of solid waste management are provided in this section. The transition of the solid waste management movement revolved around two initiatives; first, the protection of public health and second a shift to protection of the environment. Following is a brief general timeline of these movements and the remainder of this section highlights the major events during each movement:

- Agricultural Age: 500 B.C. to the mid-1700s
- The Industrial Revolution and the Machine Age: mid-1700s to the early 1910s
- The War and Interwar Period: early 1910s to the 1950s
- The Information Revolution and the Space Age: 1950s through the 2000s

2.2.1 AGRICULTURAL AGE: 500 B.C. TO THE MID-1700s

The neolithic revolution was the first agricultural revolution and was the starting point for the transition from nomadic hunting and gathering communities and bands, to agriculture and settlement. It occurred in various independent prehistoric human societies between 10 to 12 thousand years ago. During this time frame, the concept and need for solid waste management was not a key concern. Trash removal was strictly convenience related, where there were no guidelines to disposing of solid waste; individuals could dump their waste wherever they saw fit, with little to no intervention. Very few documented records exist that relate to solid waste management prior to 3000 B.C. What is known is that societies operated under a convenience mentality, lacking guidelines or regulations related to solid waste management. During this time frame most households and small communities deposited waste within or just outside their villages based on group consensus. The first recorded landfill sites were discovered in the Cretan capital, Knossos, in 3000 B.C. Waste was placed in large pits and covered with earth at various levels. The first documented solid waste management regulation occurred in the city of Athens, Greece, in 500 B.C. At that time, the city of Athens organized the first municipal dump in the Western world where citizens were required to dispose of their waste at least 1 mile from the city walls.

2.2.2 THE INDUSTRIAL REVOLUTION AND THE MACHINE AGE: MID-1700s TO THE EARLY 1910s

In the mid-1700s, the world economies underwent a major shift from an agricultural base to an industrial base, as societies became larger and more organized. The Industrial Revolution was a period in the late 18th and early 19th centuries when major changes in agriculture, manufacturing, and transportation had a profound effect on socioeconomic and cultural conditions around the world. It started with the mechanization of the textile industries, the development of iron-making techniques, and the increased use of refined coal. Trade expansion was enabled by the introduction of canals, improved roads, and railways. The introduction of steam power (fuelled primarily by coal) and powered machinery (mainly in textile manufacturing) underpinned the dramatic increases in production capacity. The development of all-metal machine tools in the first two decades of the 19th century facilitated the manufacture of more production machines for manufacturing in other industries. The effects spread throughout Western Europe and North America during the 19th century, eventually affecting most of the world. The shift to an industrial base had major impacts on the types and amounts of waste being generated, and introduced a new set of concerns. Specifically, solid waste management shifted toward a cleanliness and public health movement to eradicate diseases. Following is a brief timeline of the major events related to solid waste management during this time frame:

1690—The first paper recycling mill in the United States using recycled fibers (including waste paper and old rags) is established at the Rittenhouse Mill near Philadelphia.

1757—Benjamin Franklin institutes the first municipal street-cleaning service in the United States, in Philadelphia; at the same time, American households begin digging refuse pits instead of throwing garbage out of windows and doors.

1776—The first metal recycling in the United States occurs when patriots in New York City melt down a statue of King George III and make it into bullets.

1842—A report in England links disease to unsanitary environmental conditions, helping to launch the "age of sanitation."

1860s—In Washington, D.C., people still dump garbage and slop on the street, while pigs, rats, and cockroaches flourish.

1874—In Nottingham, England, a new technology called "the destructor" provides the first systematic incineration of municipal solid waste.

1885—The nation's first garbage incinerator is built on Governor's Island, New York.

1895—The New York City Street Cleaning Commissioner sets up the first comprehensive system for public sector garbage management in the country.

1896—Waste reduction plants, which compress organic wastes to extract grease, oils, and other by-products, are introduced to the United States from Vienna, Austria. The plants are later closed, since they emit noxious odors.

The first recycling center in the United States is established in New York City.

1899—New York City Street Cleaning Commissioner organizes the first rubbish-sorting plant for recycling in the United States.

1900s—"Piggeries" are developed in small- to medium-sized towns in the United States. At these facilities, swine eat fresh or cooked food waste. It is estimated that 75 pigs consume 1 ton of refuse per day. Food waste is recycled as pig feed until the late 1960s.

1902—Seventy-nine percent of 161 cities in the United States surveyed in a Massachusetts Institute of Technology study provide regular collection of waste materials from people's homes.

2.2.3 THE WAR AND INTERWAR PERIOD: EARLY 1910s TO THE 1950s

On June 14, 1914, Archduke Franz Ferdinand, heir to the Austro-Hungarian throne, was assassinated in Sarajevo by a member of Young Bosnia, a group whose aims included the unification of the South Slavs and independence from Austria-Hungary. The assassination in Sarajevo set into motion a series of fast-moving events that eventually escalated into full-scale war. Major European powers were at war within weeks because of overlapping agreements for collective defense and the complex nature of international alliances. World War I had officially began. This point marked the beginning of wartime economies around the globe and lasted until the 1950s, upon the end of World War II. Wartime economies are very different from peacetime economies. All resources must be mobilized and conserved to support the war efforts. The war effort led to some major material exchanges to reuse items. The British government established the earliest documented industrial waste exchange, called the National Industrial

Materials Recovery Association, in 1942. This waste exchange was created to conserve materials for the war effort during World War II. During this time frame, the first landfills were established to dispose of wastes and the initial recycling efforts began to aid with the war effort. Following is a timeline of the major events:

1904—The nation's first major aluminum-recycling plants open in Cleveland and Chicago.

1909—More than 100 incinerators close in the United States due to noxious smoke.

1914—After a shaky start, incinerators increase in popularity in North American cities. About 300 incinerators operate in the United States and Canada.

1916—Cities begin switching from horse-drawn to motorized refuse-collection equipment.

1920s—Using wetlands located near cities as a garbage-disposal facility becomes popular. Garbage is placed in the wetlands in layers, with ash and dirt layers on top as cover.

1934—The Supreme Court bans the dumping of municipal waste into the ocean, a common practice until this time.

1940s—The Fresno, California, Director of Public Works leads the effort in developing sanitary methods for disposing of trash in large urban areas.

1942—Great Britain establishes the first materials exchange to converse resources for the war effort.

1942 to 1945—Americans collect and industry recycles rubber, paper, scrap metal, fats, and tin cans—about 25 percent of the waste stream—to help the war effort.
    During the war, Army troops bury trash in the ground, providing the initial idea for the sanitary landfill.

1945—Almost 100 cities in the United States are using sanitary landfills.

2.2.4   THE INFORMATION REVOLUTION AND THE SPACE AGE: 1950s THROUGH THE 2000s

Upon the end of World War II, world economies began to shift back toward normal production. Also, during this time, more information became available on the dangers of pollution and contamination. The information revolution served as a catalyst to quickly disseminate environmental data around the world and pinpoint issues for further analysis. The solid waste movement shifted from a public health and wartime conservation focus to an environmental protection and natural resource conservation focus. During this time, initiatives and focus on the environment exploded, with many governments taking a very active role in establishing polices and regulations. The creation of environmental regulations during this time frame greatly surpassed all previous time periods immensely. Following is a timeline of the major events associated with the information revolution as it relates to the evolution of solid waste management:

1950s—Many urban areas use close-in, open-burning dumps because they reduce the volume of refuse and extend the usability of the site. But by the end of the decade, open burning of refuse is prohibited in many areas.

1954—Olympia, Washington, enacts one of the first pay-per-can programs.

1955—With consumer prosperity at an all-time high in the United States, *Life* magazine heralds the advent of the throwaway society.

1958 to 1976—The amount of packaging produced and disposed of in the United States increases by 67 percent, due to the increase in consumerism after World War II.

1959—The American Society of Civil Engineers publishes the standard guide to sanitary landfilling. To guard against rodents and odors, the guide suggests compacting the refuse and covering it with a new layer of soil each day.

1965—The first federal solid waste management law, the Solid Waste Disposal Act, authorizes research and provides for state solid waste grants. These include site inventory programs, resource recovery systems, and constructing new or improved solid waste disposal facilities.

1968—More than 33 percent of U.S. cities collect waste that is separated in some manner.

1970—The U.S. Environmental Protection Agency (EPA) is created by President Nixon. Its first administrator is William Ruckelshaus.

1971—Oregon passes the nation's first bottle bill, paving the way for nine other states to offer refunds of 5 or 10 cents for returned containers.

1972—The first buy-back centers for recyclables open in Washington State. These centers accept beer bottles, aluminum cans, and newspapers.

1974—The first city-wide use of curbside bins occurs in University City, Missouri, for collecting newspapers.

1975—All 50 states have some form of solid waste regulations in place, although the requirements vary widely.

1976—The Resource Conservation and Recovery Act (RCRA) creates the first significant role for the federal government in waste management. The law emphasizes recycling, resource conservation, and proper waste management.

1979—EPA prohibits open dumping and sets first standards for landfills.

1980—The first community-wide household hazardous waste collection day is held.

1987—Mobro 4000, the garbage barge, sails from New York up and down the U.S. East Coast, looking for a place to dispose of its waste. Rejected by facilities in six states and three countries, the barge draws public attention to the perceived landfill capacity shortage in the Northeast. The garbage is finally incinerated in Brooklyn and the ash is disposed of in a landfill near Islip, Long Island.

1989—EPA sets a 25 percent national waste reduction and recycling goal.

Twenty-six states have comprehensive laws making recycling an integral part of solid waste management.

1991—EPA sets improved solid waste landfill standards that include requirements for location, groundwater protection, monitoring, and postclosure care. EPA also issues new performance and emissions standards for municipal solid waste (MSW) combustors.

More than 3000 household hazardous waste community collection programs have been documented in all 50 states.

1992—President Bush issues Executive Order 12780, to stimulate waste reduction, recycling, and procurement of recycled goods in all federal agencies.

1994—EPA launches the WasteWise program to help businesses, educational institutions, and other large facilities reduce waste and recycle more materials.

1994—EPA launches its Jobs Through Recycling initiative to bring together the economic development and recycling communities through grants, networking, and information sharing.

President Clinton issues Executive Order 12873, which requires federal agencies to establish waste prevention and recycling programs and to buy and use recycled and environmentally preferable products and services. Clinton creates the Office of the Federal Environmental Executive to enforce this executive order.

1995—EPA issues the first Comprehensive Procurement Guideline, designating 19 recycled-content products for which the federal government should give procurement preference.

1996—The nation reaches a 25 percent recycling rate. EPA sets a new recycling goal of 35 percent.

1996—The first voluntary recycling and composting initiatives are held at the Olympics, at the 1996 Olympic Games in Atlanta. Organizers aim to divert 12 million aluminum cans, 20 million PET bottles, and 3000 tons of paper for recycling.

1999—EPA's "Municipal Solid Waste in the United States: Facts and Figures," updated ever year, provides data on U.S. waste generation, recovery, and disposal rates.

2000—EPA establishes a link between global climate change and solid waste management, showing that waste reduction and recycling help stop global climate change.

2000—More than 5000 U.S. cities are using EPA's pay-as-you-throw programs, in which residents pay for MSW collection based on the amount of waste they throw away—encouraging recycling and waste reduction.

2001—EPA policy requires its offices to use paper with 100 percent recycled content and 50 percent postconsumer content.

2002—EPA kicks off Resource Conservation Challenge urging Americans to meet or beat two goals by 2005: boosting the national recycling rate from 30 percent to at least 35 percent and curbing by 50 percent the generation of 30 harmful chemicals normally found in hazardous waste.

2.3   Environmental Laws and Regulations

2.3.1   INTRODUCTION

Governments play a critical role in managing the environment including the atmosphere, land, water bodies, and all natural resources. Governments are valuable institutions for resolving problems involving natural resources at both the local and global scales. Although in recent decades the economic market has been identified as a suitable mechanism for managing environmental quality, markets have serious failures and governmental intervention, regulation, and the rule of law is still required for the proper, just, and sustainable management of the environment. This section discusses several of the key environmental laws and regulations that pertain to solid waste management.

2.3.2   SOLID WASTE DISPOSAL ACT OF 1965

The Solid Waste Disposal Act (SWDA) (P.L. 89-272, 79 Stat. 992) became law on October 20, 1965. In its original form, it was a broad attempt to address the solid waste problems confronting the nation through a series of research projects, investigations, experiments, training, demonstrations, surveys, and studies. The key points of the SWDA were

- Promote better management of solid wastes.
- Support resource recovery.
- Directed that the U.S. Public Health Service (PHS) promulgate and enforce regulations for solid waste collection, transportation, recycling, and disposal. (The EPA was not formed until 1970.)
- Provided financial assistance for states to study and develop solid waste management plans.
- Provided support for research and development of improved methods of solid waste management.

The decade following its passage revealed that the SWDA was not sufficiently structured to resolve the growing mountain of waste-disposal issues facing the country.

As a result, significant amendments were made to the act with the passage of the Resource Conservation and Recovery Act of 1976 (RCRA).

2.3.3 THE NATIONAL ENVIRONMENTAL POLICY ACT OF 1969

The National Environmental Policy Act (NEPA), established in 1969 by the U.S. government (42 U.S.C. 4321), was one of the first laws written related to environmental protection. NEPA, which is pronounced NEE-pa, established a basic policy to assure that all branches of government give proper consideration to the environment prior to undertaking any major federal action that significantly affects the environment. One of the key phrases from this document that relates to recycling is "to enhance the quality of renewable resources and approach the maximum attainable recycling of depletable resources."

NEPA requires federal agencies to integrate environmental values into their decision-making processes by considering the environmental impacts of their proposed actions and reasonable alternatives to those actions. To meet this requirement, federal agencies prepare a detailed statement known as an environmental impact statement (EIS). EPA reviews and comments on EIS prepared by other federal agencies, maintains a national filing system for all EISs, and assures that its own actions comply with NEPA.

2.3.4 RESOURCE RECOVERY ACT OF 1970

Increasing concerns over protection for human health and the environment lead to amendments of the 1965 SWDA, and the 1970 Resource Recovery Act (RRA) was passed. The RRA increased federal involvement with management of solid waste; it encouraged waste reduction and resource recovery and created national disposal criteria for hazardous wastes. This act was the forerunner of the Resource Conservation and Recovery Act (RCRA) of 1976. Following are the key points of RRA:

- Directed that the nation would change its emphasis from solid waste disposal to recycling and energy recovery.
- Required the U.S. Public Health Systems to investigate and report on the disposal of hazardous waste in the nation.

This was an important guidance document for the early stages of solid and hazardous waste management. The RRA also marked the birth of the EPA, as it was formed in the interim.

2.3.5 RESOURCE CONSERVATION AND RECOVERY ACT OF 1976

In the United States RCRA is the primary law governing the disposal of solid and hazardous waste. Congress passed RCRA on October 21, 1976 to address the increasing problems the nation faced from the growing volume of municipal and industrial waste. RCRA, which amended the Solid Waste Disposal Act of 1965, set national goals for

- Protecting human health and the environment from the potential hazards of waste disposal.
- Conserving energy and natural resources.
- Reducing the amount of waste generated.
- Ensuring that wastes are managed in an environmentally sound manner.

RCRA also included directives that the EPA establishes regulations to control solid waste disposal. To achieve these goals, RCRA established three distinct, yet interrelated, programs:

1 The *solid waste program* under RCRA Subtitle D encourages states to develop comprehensive plans to manage nonhazardous industrial solid waste and municipal solid waste, sets criteria for municipal solid waste landfills and other solid waste disposal facilities, and prohibits the open dumping of solid waste.

2 The *hazardous waste program* under RCRA Subtitle C establishes a system for controlling hazardous waste from the time it is generated until its ultimate disposal—in effect, from cradle to grave.

3 The *underground storage tank (UST) program* under RCRA Subtitle I regulates underground storage tanks containing hazardous substances and petroleum products.

RCRA banned all open dumping of waste, encouraged source reduction and recycling, and promoted the safe disposal of municipal waste. RCRA also mandated strict controls over the treatment, storage, and disposal of hazardous waste. The first RCRA regulations, "Hazardous Waste and Consolidated Permit Regulations," published in the Federal Register on May 19, 1980 (45 FR 33066; May 19, 1980), established the basic cradle-to-grave approach to hazardous waste management that exists today. RCRA focuses only on active and future facilities and does not address abandoned or historical sites which are managed under the Comprehensive Environmental Response, Compensation, and Liability Act (CERCLA)—commonly known as Superfund.

2.3.6 THE HAZARDOUS AND SOLID WASTE AMENDMENT OF 1984

RCRA was amended and strengthened by Congress in November 1984 with the passing of the federal Hazardous and Solid Waste Amendments (HSWA). These amendments to RCRA required phasing out land disposal of hazardous waste. Some of the other mandates of this strict law include increased enforcement authority for EPA, more stringent hazardous waste management standards, and a comprehensive underground storage tank program. Key points from the HSWA are

- Direct the EPA to revise criteria for landfills that receive hazardous household waste or small quantities of industrial hazardous waste.

- Require treatment of all contaminated surface water running off of landfills.
- Methods of disposing of wastewater sewage sludge at landfills are included in the Clean Water Act as amended.
- Increased enforcement authority for EPA.
- Provided more stringent hazardous waste management standards.
- Created a comprehensive underground storage tank program.

The HSWA marked the most significant set of amendments to RCRA with a complex law with many detailed technical requirements. Additional restrictions on land disposal and the inclusion of small-quantity hazardous waste generators (those producing between 100 and 1000 kg of waste per month) in the hazardous waste regulatory scheme were added. The EPA was directed to issue regulations governing those who produce, distribute, and use fuels produced from hazardous waste, including used oil. Under HSWA, hazardous waste facilities owned or operated by federal, state, or local government agencies must be inspected annually, and privately owned facilities must be inspected at least every 2 years. Each federal agency was required to submit to EPA an inventory of hazardous waste facilities it has possessed in its history. The HSWA also imposed on EPA a timetable for issuing or denying permits for treatment, storage, and disposal facilities; required permits to be for fixed terms not exceeding 10 years; terminated in 1985 the interim status of land disposal facilities that existed prior to RCRA's enactment, unless they met certain requirements; required permit applications to be accompanied by information regarding the potential for public exposure to hazardous substances in connection with the facility; and authorized EPA to issue experimental permits for facilities demonstrating new technologies. EPA's enforcement powers were increased, the list of prohibited actions constituting crimes was expanded, penalties were increased, and the citizen suit provisions were expanded. Other provisions prohibited the export of hazardous waste unless the government of the receiving country formally consented to accept it; created an ombudsman's office in EPA to deal with RCRA-associated complaints, grievances, and requests for information; and reauthorized RCRA through FY88 at a level of about $250 million per year. Finally, HSWA called for a National Groundwater Commission to assess and report to the Congress in 2 years on groundwater issues and contamination from hazardous wastes. The commission was never funded and never established, however.

2.3.7 POLLUTION PREVENTION ACT OF 1990

The Pollution Prevention Act of 1990 (PPA) was passed as part of the Omnibus Budget Reconciliation Act of 1991. The measure declared pollution prevention to be the national policy, and directed EPA to undertake a series of activities aimed at preventing the generation of pollutants, rather than controlling pollutants after they are created. Matching grants were authorized for states to establish technical assistance programs for businesses, and EPA was directed to establish a Source Reduction Clearinghouse to disseminate information. The Act also imposed new reporting requirements on industry. Firms that were required to file an annual toxic chemical release form under the Emergency Planning and Community Right-to-Know Act of

1986 must also file a report detailing their source reduction and recycling efforts over the previous year. A more complete description of the act, which addresses air and water pollution as well as waste, is provided in the first section of this report.

2.3.8 FEDERAL FACILITY COMPLIANCE ACT OF 1992

The Federal Facility Compliance Act, 1992 (FFCA) amended RCRA with the primary purpose of ensuring a complete and unambiguous waiver of sovereign immunity with regard to imposition of administrative and civil fines and penalties on federal facilities. It requires compliance with all federal, state, interstate, and local requirements in the same manner and extent as any person. The objectives of the FFCA bring all federal facilities into compliance with applicable federal and state hazardous waste laws, of waiving federal sovereign immunity under those laws, and of allowing the imposition of fines and penalties. The law also requires the U.S. Department of Energy (DOE) to submit an inventory of all its mixed waste and to develop a treatment plan for mixed waste. The FFCA has ten sections, concerning extent of waiver of sovereign immunity, application of RCRA to radioactive mixed wastes, application to public vessels, waste munitions and to federally owned treatment works. The language of RCRA waiver of sovereign immunity prior to FFCA was unclear on whether federal facilities had to pay civil and administrative penalties for violations of hazardous waste provisions. Cases have involved the determination of whether federal facilities have to pay penalties, even though there was no waiver of sovereign immunity. Legislatively, there were many failed attempts at passing the act. But due to increasing support for environmental movements and frustration at state regulatory agencies that could not hammer federal facilities with fines and penalties, the act was passed in 1992. Key points of the FFCA are

- All federal agencies are subject to all substantive and procedural requirements of federal, state, and local solid and hazardous waste laws in the same manner as any private party.
- The sovereign immunity of the United States is expressly waived in all such cases.
- Substantive and procedural requirements of such law include all administrative orders, civil and administrative fines and penalties, and reasonable service charges imposed for issuing and reviewing permits, plans, and studies, and inspecting facilities.
- Employees, officers, and agents of the United States may not be liable for civil penalties under any such law for actions committed within the scope of that person's official duties, but such persons may be liable for criminal penalties.
- The administrator of EPA is authorized to commence an administrative enforcement action against any federal agency or department in the same manner as against a private party.
- Agencies must reimburse EPA for the required annual inspections of agency hazardous waste facilities, and for EPA to conduct a comprehensive groundwater monitoring at the first inspection of each such site conducted after October 6, 1992 (unless such an evaluation has been conducted within the preceding year).

- In consultation with the Secretary of Defense, EPA is required to propose regulations identifying when military munitions become hazardous waste, and providing for the safe transportation and storage of such waste.
- Federally owned wastewater-treatment works are not to be considered hazardous waste facilities if most of the water treated consists of domestic sewage, and certain other specified requirements are met. Introduction of a hazardous waste into a federally owned wastewater-treatment works is prohibited.

2.3.9 LAND DISPOSAL PROGRAM FLEXIBILITY ACT OF 1996

Land Disposal Program Flexibility Act of 1996 provided regulatory flexibility for land disposal of certain wastes as amendments to RCRA. This act exempts hazardous waste from RCRA regulation if it is treated to a point where it no longer exhibits the characteristic that made it hazardous, and is subsequently disposed in a facility regulated under the Clean Water Act or in a Class I deep injection well regulated under the Safe Drinking Water Act. A second provision of the bill exempted small landfills located in arid or remote areas from groundwater-monitoring requirements, provided there is no evidence of groundwater contamination.

2.4   Solid Waste Generation around the World

2.4.1 INTRODUCTION

This section provides a bird's-eye view of the solid waste generation trends around the world. This analysis is useful for several key reasons

- Data on solid waste generation provides the public with a viewpoint on the solid waste problem.
- For a solid waste manager or engineer confronted with a new waste stream, national figures and trends may be useful in determining how other facilities or industries have managed similar waste streams.
- The data provides a gauge to judge the environmental performance of public, regulatory, and advocacy groups around the world.

Governments in various nations have established and implemented numerous programs to reduce solid waste levels, but do not have a clear understanding of the quantities and compositions that individual companies contribute to the overall levels. Compounding this problem, there is no single source for data regarding solid waste generation and developing a comprehensive picture can be somewhat challenging. An improved understanding of the solid waste generation rates and characteristics of individual companies and industries provides insights for communities and government to understand, control, and reduce solid waste generation. This section compiles data from various sources to create a comprehensive picture and provides a discussion on commonly used solid waste metrics.

2.4.2 WASTE GENERATION RATES IN THE UNITED STATES

In the year 2000, U.S. residents, businesses, and institutions generated more than 230 million tons of municipal solid waste (MSW), which is approximately 4.6 lb of waste per person per day, up from 2.7 lb per person per day in 1960 (www.epa.gov, retrieved December 12, 2007). Figure 2.1 displays the increased solid waste generation trends for the United States in terms of total waste generated (millions of tons). This graph indicates the increased solid waste generation in the United States and stresses the need for waste reduction activities to combat this growth. The data used for this graph was provided from "Municipal Solid Waste Generation, Recycling, and Disposal in the United States: 2000 Facts and Figures," Franklin and Associates for the EPA, June 2002.

Currently, no solid waste generation standards exist for U.S. companies. Such standards would be useful for communities, government agencies, or businesses to monitor generation levels, compare performance, and subsequently reduce solid waste generation. To successfully implement a solid waste reduction strategy, government and businesses must be able to quantify the amounts and types of solid waste a company generates and compare these amounts with similar industries. A common business adage summarizes this well, stating, "If you can't measure it, you can't manage it." Currently no comprehensive measures of solid waste generation quantities or waste stream compositions exist for individual companies. Various U.S. government agencies at the state and local levels often use nonscientific and nonstandardized approaches to estimate solid waste generation rates for industrial and business sectors under their jurisdiction. The methods currently used involve convenient sampling and

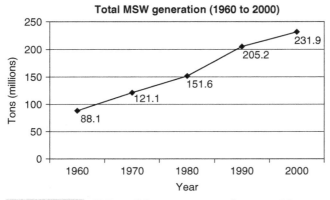

Figure 2.1 **U.S. solid waste generation trend from 1960 to 2000.**

applying limited statistical analyses (usually only the mean are calculated) and rarely calculating variances. These current methods significantly increase the variability and inaccuracy of the estimates.

This type of research had not been conducted earlier for several reasons. First, most solid waste studies are conducted by the government, which is concerned with aggregate data, not individual company data. Federal, state, and local level governments collect data from individual companies and use it for overall waste generation and recycling statistics. This provides an overview of the total generation, but offers few insights on specific waste generation quantities and recycling levels on the individual companies or groups that comprise the total.

Another reason why this type of research has not been conducted earlier is because the appropriate data had not been consolidated or analyzed. A portion of the data required for this research existed, but was never processed into useful information. The data could have been collected, but the effort was never taken until this research. Contributing to data collection issues, some companies view solid waste data as confidential and are hesitant to release it to the government. Finally, most solid waste generation studies involved minimal statistical modeling.

MSW, more commonly known as trash or garbage, consists of everyday items such as product packaging, grass clippings, furniture, clothing, bottles, food scraps, newspapers, appliances, paint, and batteries. The pie chart in Fig. 2.2 was created with data provided from the EPA (www.epa.gov, retrieved December 12, 2007).

In 2006, U.S. residents, businesses, and institutions produced more than 251 million tons of MSW, which is approximately 4.6 lb of waste per person per day. The annual U.S. trends for MSW can be found in Figs. 2.3 and 2.4. Several MSW management practices, such as source reduction, recycling, and composting, prevent or divert materials from the waste stream. Source reduction involves altering the design,

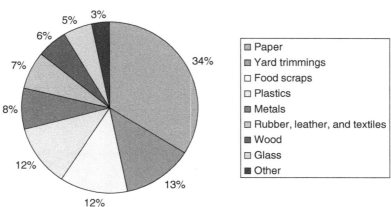

2006 total U.S. waste generation (before recycling)

- Paper
- Yard trimmings
- Food scraps
- Plastics
- Metals
- Rubber, leather, and textiles
- Wood
- Glass
- Other

34%
3%
5%
6%
7%
8%
12%
12%
13%

Figure 2.2 **2006 solid waste generation in the United States.**

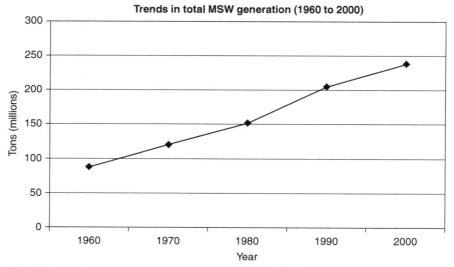

Figure 2.3   Annual U.S. trends in total MSW generation.

manufacture, or use of products and materials to reduce the amount and toxicity of what gets thrown away. Recycling diverts items, such as paper, glass, plastic, and metals, from the waste stream. These materials are sorted, collected, and processed and then manufactured, sold, and bought as new products. Composting decomposes

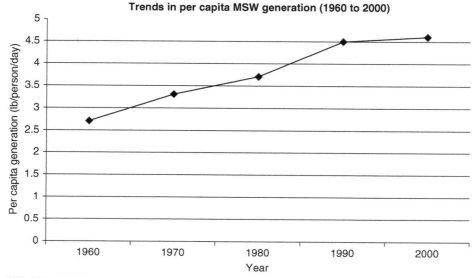

Figure 2.4   Annual U.S. trends in per capita MSW generation.

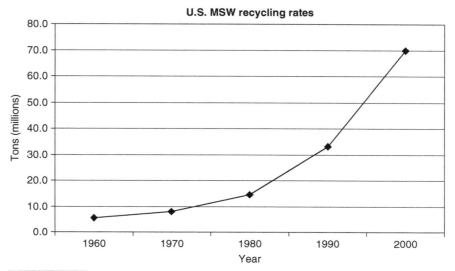

Figure 2.5   U.S. MSW recycling tonnage trends.

organic waste, such as food scraps and yard trimmings, with microorganisms (mainly bacteria and fungi), producing a humus-like substance.

Other practices address those materials that require disposal. Landfills are engineered areas where waste is placed into the land. Landfills usually have liner systems and other safeguards to prevent groundwater contamination. Combustion is another MSW practice that has helped reduce the amount of landfill space needed. Combustion facilities burn MSW at a high temperature, reducing waste volume and generating electricity.

Currently, in the United States, 32.5 percent is recovered and recycled or composted, 12.5 percent is burned at combustion facilities, and the remaining 55 percent is disposed of in landfills. The annual U.S. trends for recycling can be found in Figs. 2.5 and 2.6. Source reduction can be a successful method of reducing waste generation. Practices such as grass recycling, backyard composting, two-sided copying of paper, and transport packaging reduction by industry have yielded substantial benefits through source reduction. Source reduction has many environmental benefits. It prevents emissions of many greenhouse gases, reduces pollutants, saves energy, conserves resources, and reduces the need for new landfills and combustors.

Recycling, including composting, diverted 82 million tons of material away from disposal in 2006, up from 15 million tons in 1980, when the recycle rate was just 10 percent and 90 percent of MSW was being combusted with energy recovery or disposed of by landfilling.

Typical materials that are recycled include batteries, recycled at a rate of 99 percent, paper and paperboard at 52 percent, and yard trimmings at 62 percent. These materials and others may be recycled through curbside programs, drop-off centers, buy-back programs, and deposit systems. Figure 2.7 shows the recycling rates for many common materials.

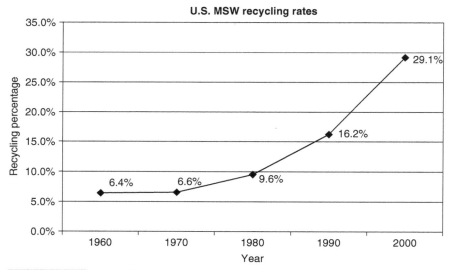

Figure 2.6   U.S. MSW recycling rate trends.

Recycling prevents the emission of many greenhouse gases and water pollutants, saves energy, supplies valuable raw materials to industry, creates jobs, stimulates the development of greener technologies, conserves resources for our children's future, and reduces the need for new landfills and combustors.

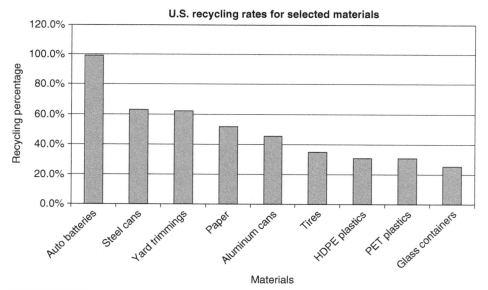

Figure 2.7   Recycling rates of selected materials.

Recycling also helps reduce greenhouse gas emissions that affect global climate. In 1996, recycling of solid waste in the United States prevented the release of 33 million tons of carbon into the air—roughly the amount emitted annually by 25 million cars. Burning MSW can generate energy while reducing the amount of waste by up to 90 percent in volume and 75 percent in weight. EPA's Office of Air and Radiation is primarily responsible for regulating combustors because air emissions from combustion pose the greatest environmental concern and provided this data.

In 2006, the national recycling rate of 32.5 percent (82 million tons recycled) prevented the release of approximately 49.7 million metric tons of carbon into the air— roughly the amount emitted annually by 39 million cars, or 1300 trillion Btu, saving energy equivalent to 10 billion gallons of gasoline.

The number of landfills in the United States is steadily decreasing—from 8000 in 1988 to 1754 in 2006. The capacity, however, has remained relatively constant. New landfills are much larger than in the past.

2.4.3 INTERNATIONAL WASTE GENERATION COMPARISON RATES

As discussed in the book *Germany, Garbage, and the Green Dot: Challenging the Throwaway Society* by Bette Fishbein, it is very difficult to make international comparisons regarding waste generation. For example, Ms. Fishbein points out that "According to the data published by the Organization of Economic Cooperation and Development, waste generation in Germany is 318 kilograms per person per year as compared to 864 kilograms per person per year in the United States. This might suggest that the average person in the United States generates two or three times as much garbage as the average person in Germany. However, data from the two countries are not comparable: the German data neither include materials collected from recycling, nor do they include some commercial waste, both of which are included in U.S. data. International comparisons of waste generation are usually unreliable because countries use different data collection mythologies and different definitions of waste." Table 2.1 shows the annual waste generation rates for various countries.

2.5   An Overview of Environmental Concerns

The world has changed significantly over the past century. Societies are shifting to a convenience-oriented mind-set, world populations are increasing, and subsequently waste generation is shifting, which is creating new environmental impacts. For example, from a convenience standpoint, solid waste generation rates are increasing due to the proliferation of individually packaged food servings, fast-food containers, and disposable diapers. Figures 2.8 and 2.9 display the total and per capita waste-generation rates in the United States.

TABLE 2.1 2006 INTERNATIONAL WASTE GENERATION RATES

| | AIR | | | | WASTE | | | ENVIRONMENT R&D BUDGET |
| | SULFUR OXIDES | | NITROGEN OXIDES | | QUANTITY GENERATED | | | |
	kg / CAPITA	% CHANGE SINCE 1990	kg / CAPITA	% CHANGE SINCE 1990	INDUSTRIAL WASTE PER UNIT OF GDP (T/MILLION USD)	MUNICIPAL WASTE (kg / CAPITA)	NUCLEAR WASTE PER UNIT OF ENERGY (T/MTOE)	AS % OF TOTAL GOVERNMENT R&D BUDGET
Australia	127	59	84	29	20	690	N/A	4.2
Austria	4	−55	25	−3	N/A	560	N/A	1.9
Belgium	22	−88	32	−40	50	460	2.2	2.3
Canada	76	−27	78	−6	N/A	420	6.2	4.4
Czech Republic	15	−58	26	−24	30	290	1.7	2.9
Denmark	4	−87	32	−36	10	740	N/A	1.7
Finland	16	−64	41	−32	110	470	1.9	1.8
France	9	−60	23	−29	50	540	4.2	2.7
Germany	7	−89	17	−48	20	600	1.2	3.4
Greece	46	4	29	11	N/A	440	N/A	4
Hungary	24	−76	18	−24	30	460	1.7	9.7
Iceland	35	22	90	−2	10	520	N/A	0.4
Ireland	25	−48	31	5	40	740	N/A	0.9
Italy	11	−63	22	−34	20	540	N/A	2.7
Japan	7	−14	16	−2	40	400	1.5	0.8
Korea	10	−46	24	47	40	380	3.2	4.5

Luxembourg	7	−80	38	−27	30	710	N/A	N/A
Mexico	12	N/A	12	18	N/A	340	0.1	1
Netherlands	5	−58	27	−28	40	620	0.1	1.2
New Zealand	19	39	39	16	10	400	N/A	N/A
Norway	5	−58	47	−5	20	760	N/A	2.1
Poland	38	−55	21	−38	120	250	N/A	2.4
Portugal	28	−9	28	13	50	470	N/A	3.5
Slovak Republic	19	−81	19	−53	130	270	3	1
Spain	37	−29	35	14	30	650	1.2	3
Sweden	6	−45	27	−25	110	480	4.1	2.2
Switzerland	2	−60	11	−46	N/A	650	1.9	0.1
Turkey	25	18	13	35	30	440	N/A	N/A
United Kingdom	17	−73	26	−43	30	580	1	1.8
United States	49	−31	64	−19	N/A	750	1	0.4
G7	29	−45	41	−22	50	630	1.6	N/A
EU	15	−65	24	−30	40	570	1.6	N/A
OECD total	27	−41	34	−17	50	560	1.5	N/A

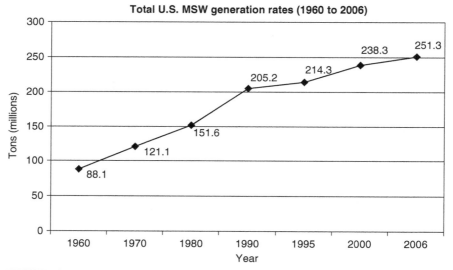

Figure 2.8   Total U.S. MSW generation rates.

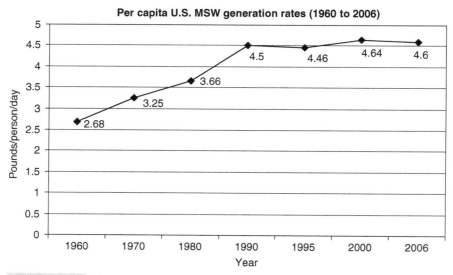

Figure 2.9   Per capita U.S. MSW generation rates.

The average American discards 4.60 lb of garbage every day. Most of this waste is compacted and buried in landfills, and as the waste continues to grow, so will the pressure on the landfills, resources, and environment. The impacts are intensified by an ever increasing population. The U.S. Census Bureau estimates the current world population at 6.6 billion people, with a projected annual growth rate of

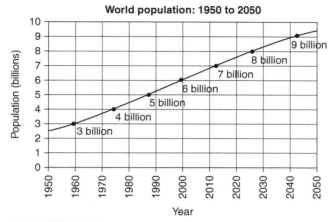

Figure 2.10   **World population trends.**

approximately 1.2 percent. By 2050, the world population is projected to be over 9 billion. Globalization and the development of third world countries are compounding these issues as well. More people generating more waste is not a good combination for the environment. Figure 2.10 provides a visual representation of the projected population increases.

One of the central purposes of solid waste analysis and minimization is to reduce or eliminate the environmental impacts of individual companies and industries. An understanding of these impacts is critical, when addressing solid waste issues, to provide direction for reduction efforts. Some of these impacts are more important than others, thus a comprehensive understanding will allow managers and engineers to focus on more serious problems. This section provides an overview of these impacts with a focus on the effects, not the sources. These impacts are

■ *Space availability*—As the world population increases and cities grow, the available space to dispose of solid waste decreases. By minimizing solid waste levels, disposal space will also decrease making land available for other uses.

■ *Landfill leachate*—Leachate is the liquid generated from a landfill that is created from decomposing waste. It's created after rainwater mixes with the chemical wastes or liquids present in the landfill. Once it enters the environment, the leachate is at risk of mixing with groundwater near the site, which can have very negative effects. Groundwater is the source of drinking water for over 40 percent of the urban population, and up to 90 percent of the rural population. It was formerly assumed that this source of water was not subject to contamination, but recent studies have shown that groundwater can in fact become contaminated. Landfill leachate may be virtually harmless or dangerously toxic, depending upon the characteristics of the material in the landfill. Typically, landfill leachate has high concentrations of nitrogen, iron, organic carbon, manganese, chloride, and phenols. Other chemicals including pesticides, solvents, and heavy metals may also be present. Modern landfill sites require that the landfill leachate be collected and treated. Since there is no method to ensure that rainwater cannot enter the landfill site, landfill sites must now have an impermeable layer at the bottom. The landfill leachate that collects at the bottom must be monitored and treated if required. This liquid can be treated in a similar manner to sewage, and the treated water can then be safely released into the environment.

■ *Global warming*—One study reports that U.S. landfills are responsible for 3.8 percent of the global warming damage from human sources in the United States. Municipal solid waste landfills are the largest source of human-related methane emissions in the United States, accounting for about 25 percent of these emissions in 2004 (www.epa.gov, 2005). This gas consists of about 50 percent methane (CH_4), the primary component of natural gas, about 50 percent carbon dioxide (CO_2), and a small amount of nonmethane organic compounds. In 2003, U.S. landfills generated 131.2 teragrams methane in terms of carbon dioxide equivalents (where 1 teragram represents 1 million metric tons). Reducing the amounts of solid waste disposed in landfills would reduce methane generation and subsequently reduce global warming.

■ *Consumption of natural resources*—A large component of solid waste minimization is recycling. Recycling reduces the consumption of virgin natural resources by utilizing perceived waste materials. For example, production of recycled paper uses 80 percent less water and 65 percent less energy, and produces 95 percent less air pollution than virgin paper production. If every American recycled their newspaper just 1 day a week, the United States would save approximately 36 million trees a year. For every 4-ft stack of paper recycled, one tree is saved and deforestation is minimized. Recycling also reduced environmental impacts due to mining. For example, by recycling aluminum, the need for the raw mineral bauxite is eliminated, which in turn eliminates the need for bauxite mining and smelting.

■ *Loss of habitat*—Although it is difficult to accurately quantify habitat loss, many animal species are displaced by the creation of landfills and the effects of deforestation. By minimizing solid waste levels and increasing recycling, available habitats for animals will not be disrupted due to development or expansion of landfills and the effects of deforestation to acquire virgin raw materials.

2.6 Industrial Ecology and Solid Waste Exchanges

2.6.1 INTRODUCTION

Industrial ecology is the field of research that studies waste generation from a macro level for all industries. From a solid waste standpoint, industrial ecology is concerned with the conversion or reuse of undesirable materials into something useful for another company or industry, in other words: waste exchanges and material efficiency. Material efficiency is defined as the percentage of process by-products that are recycled or reused divided by the total by-product generation for a company or industry.

Industrial ecology and waste exchanges examine the material efficiency and methods to improve that efficiency. Whereas waste audits examine an individual company's ability to reduce, reuse, or recycle, waste exchanges examine an entire industry's or region's ability to reduce, reuse, or recycle. In essence, waste exchanges examine methods for one company to use another company's by-products as a raw material, diverting this material from entering a landfill. Waste exchanges are a great tool that can enhance a company's recycling levels and generate economic benefits as part of the solid waste auditing process. Waste streams identified during a solid audit that the company cannot reduce or reuse could be sent to another company using one of many solid waste exchanges operating around the world.

With increased pressures on companies to improve profitability and reduce environmental impacts, waste exchanges are more popular then ever. Many companies and nonprofit organizations are turning to these exchanges to bolster corporate images and reduce costs. The Internet has simplified, streamlined, and reduced the costs associated with the administration of waste exchanges, as well. Information is available in real time, 24 hours per day, which makes such systems more accurate and user friendly, while allowing the exchanges to reach a larger client base.

2.6.2 HISTORY AND BACKGROUND

Waste commodity exchange is defined as the ability of a company or organization to use another company's waste as its raw material. As the old adage goes, "One person's trash is another person's treasure." Instead of sending seemingly worthless items or process by-products away to a landfill, the goal of the waste commodity exchange is to find a company that may get more use out of these products.

A good household example of this is garage sales, which are an excellent way to reuse products. Another alternative is to find different ways to reuse items. Baby food jars, for example, can be reused to store miscellaneous nuts, bolts, and washers in a workshop.

Waste exchanges have been around for over 60 years. The British government established the earliest documented industrial waste exchange, called the National Industrial Materials Recovery Association, in 1942. This waste exchange was created to conserve materials for the war effort during World War II. The first North American waste exchange was started in Canada in 1974 for hazardous waste. The National Industrial Materials Recovery Association is no longer active as it disbanded after the war. The Canadian waste exchange is still active as the Canadian Waste Materials Exchange (CWME).

Waste commodity exchanges are reuse and recycling services that help these types of material exchanges to occur on a much larger scale for businesses. These services help businesses save money, as well as helping the environment by diverting waste into usable raw materials.

2.6.3 WASTE COMMODITY EXCHANGES IN NORTH AMERICA

Over 200 waste commodity exchanges are currently operating in North America. These exchanges differ in terms of the service area, materials exchanged, exchange processes, and fee structures. Many of these exchanges are coordinated by state and local governments, while others are for-profit businesses. The U.S. Environmental Protection Agency (Washington) provides an excellent reference list of waste exchanges and contact information at www.epa.gov/epaoswer/non-hw/recycle/jtr/comm/ exchange.htm.

More than 35 national and 150 state-specific waste exchanges exist in the United States, and Canada has more than seven national waste exchanges. The majority of the waste exchanges are specific to certain regions or states. The drawback to regional or state-specific exchanges is that they expose the available materials to fewer potential companies. The benefits of regional exchanges, though, are that they significantly reduce transportation costs, especially for heavy or bulky items and large quantities.

Regional exchanges are appealing to companies that may continually exchange waste items over an extended period of time due to longstanding process by-products. An example of this is plastic scrap from a manufacturing process. Another company may be able to grind the scrap, use it as a raw material, and establish dedicated routes to transport the material. On the other hand, national exchanges expose materials to a much larger numbers of companies, but transportation fees may make some options infeasible.

The material and waste focus of the various exchanges differs significantly. Some are very broad and deal with a wide variety of materials. For example, in terms of the national exchanges, Recycler's World (www.recycle.net) and the Reuse Development Organization (www.redo.org) handle any waste that users post on the Web site.

On the other hand, some exchanges are very narrowly focused. Good national examples of this are the American Plastics Exchange (www.apexq.com), which deals solely with plastics, and Planet Salvage (www.planetsalvage.com), which deals only with used automobile parts. Overall, any material that is available from one business and wanted by another can become an exchange item, and a waste exchange most likely exists for it.

Materials that are available for exchange are generated from a variety of sources, which include

- By-products
- Damaged materials
- Expired products
- Obsolete and off-specification goods
- Overstock virgin products
- Surplus

Common materials that are available and wanted for exchange include categories such as

- Acids
- Agricultural by-products

- Alkalis
- Ash and combustion by-products
- Chemicals
- Computers and electronics
- Construction and demolition debris
- Durables and furniture
- Glass
- Metals
- Miscellaneous
- Oils and waxes
- Paints and coatings
- Paper
- Plastics
- Refractory material
- Rubber
- Sand
- Services
- Shipping materials
- Solvents
- Textiles and leather
- Wood

Waste exchanges are used by a variety of organizations, including private sector waste generators, government agencies, solid waste district staff, recycling organizations, and material brokers. Materials exchange users can be anyone who handles surplus or unwanted materials, such as architects, administrative assistants, buyers, engineers, residents, consultants, custodians, environmental managers, government employees, procurement specialists, purchasing representatives, recycling brokers, shipping clerks, and storeroom managers.

Differences in the business models and processes for the waste commodity exchanges are also evident. Many of these exchanges serve as a meeting place for companies that would like to list materials and potential respondents, who then work out the details of payment, transportation, and storage themselves to facilitate exchanges. Some exchanges have an eBay-type Web posting system, whereas others produce printed periodicals. Some handle requests via the phone or fax, however, most utilize the Internet.

According to the EPA materials exchange Web site, "Typically, the exchanges allow subscribers to post materials available or wanted on a Web page listing. Organizations interested in trading posted commodities then contact each other directly. As more and more individuals recognize the power of this unique tool, the number of internet-accessible materials exchanges continues to grow, particularly in the area of national commodity-specific exchanges."

Finally, the major difference among the exchanges dealt with the fee structures. Most exchanges are no cost, but some charge periodic membership fees or fees per transaction. Overall, waste exchanges have very minimal fees—just enough to cover the administrative costs. The American Plastics Exchange, for example, is the most expensive waste exchange, with a $360 per year membership fee and $0 per exchange—still, a very cost-effective exchange. The typical per exchange fee, for exchanges that did charge, was $5 to $10.

2.6.4 SUCCESS STORIES

Waste exchanges have played an important role in assisting companies identify and implement recycling and reuse opportunities. These efforts result in lower operating costs, reduced purchasing costs, reduced storage costs, enhanced corporate images, diminished demand for landfill space and incinerator capacity, and, ultimately, a cleaner environment.

"It is estimated that, by promoting the reuse and recycling of industrial materials through waste exchanges, the industry currently saves $27 million in raw material and disposal costs and the energy equivalent of more than 100,000 barrels of oil annually," as determined by the National Materials Exchange Network operating in Silver Spring, Maryland. These savings often translate directly to the companies' bottom lines with stronger financial performance.

In 1998, its first year of operation, the Ohio Materials Exchange (Columbus, Ohio) exceeded initial expectations by exchanging over 2600 tons of waste and saving Ohio businesses $103,000 in disposable costs. According to Dale Gallion, manager of quality assurance at Diamond Products (Elyria, Ohio), the company used to pay to have leftover metal powder scrapped. After joining the waste exchange program, Diamond Products sold 8000 lb of metal powder for $14. Additionally, the company accumulates about 1000 lb of metal powder each month, and plans to continue using the exchange.

Another good example is the Massachusetts Materials Exchange (Pittsfield, Massachusetts). In the past 4 years, the Massachusetts Materials Exchange has moved over 2000 tons of materials, saving participants more than $100,000 in avoided disposal and purchasing costs.

Waste exchanges are a cost-effective means of helping businesses save money, as well as helping them divert waste into usable raw materials. Advances in information technology over the past decade have served as a catalyst to promote the exchanges, and further allow exchanges to provide current information on both the materials available for use and the materials wanted, which helps business make better environmental and financial decisions.

These exchanges also play an important role in assisting waste generators in identifying and implementing recycling and reuse opportunities. For example, since the inception of the Ohio Materials Exchange, businesses using the service reported savings of over $13.5 million in disposal costs and diverted over 340,000 tons from landfills. These efforts result in lower operating costs, diminished demand for landfill space and burning capacity, which all lead to a cleaner environment. Natural resources are limited, and resources need to be conserved as much as possible. The fewer raw materials used, the greater supply for future generations, and waste exchanges are helping achieve this goal.

2.7   Life Cycle Assessment

2.7.1   OVERVIEW

Life cycle assessment (LCA) is a proactive product-focused approach to minimizing waste or by-products. LCA analyzes the wastes generated that are associated with a specific product over its entire life from cradle to grave. In other words, LCA analyzes the product's life from procuring raw materials to final disposal (or reclamation). This is in contrast to waste exchanges that take a reactive approach, which aid in diverting wastes from landfills after they are generated with little thought to reducing the waste from the start of the process. The term "closing the loop" is often used with the LCA process. The concept involves examining the entire cradle-to-grave process as a continuous loop. Figures 2.11 and 2.12 display the components and overview of the life cycle analysis respectively. The idea is to prevent waste materials or by-products from leaving the loop, hence closing the loop. By closing the loop, solid waste can be minimized by preventing its generation in the first place. For example, bottle deposits in

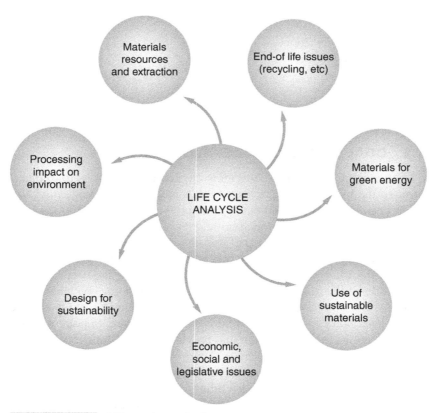

Figure 2.11   **Life cycle analysis components.**

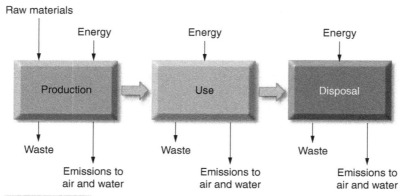

Figure 2.12   **Life cycle process overview.**

Michigan are an attempt to close the loop by reclaiming used beverage bottles by collecting a deposit at the time of purchase and refunding it at the time of recycling. The LCA process consists of four phases

1 Goal and scope development
2 Life cycle inventory
3 Impact audits
4 Analysis and action

2.7.2   GOAL AND SCOPE DEVELOPMENT

The LCA process stems from developing environmental indices to evaluate the seriousness of the by-products generated from a process and provide direction to managers in deciding which by-products generate the largest threat to the environment. The first step is to determine the specific goals and scope of study in relation to the intended application. One key step of this phase is to specify the functional unit. The functional unit is the measurement that will be used as the central reference point throughout the LCA process. For the example, if a company were comparing glass versus plastic bottles, the functional unit could be "1-L bottle container for carbonated beverages." The company would then rate the environmental and economic impacts for using glass bottles versus plastic bottles. It is important to point out that the containers may not be the same weight, but provide the same functional use, in this case containing 1 L of carbonated beverage. Apart from describing the functional unit, the goal and scope should address the overall approach used to establish the system boundaries. The system boundary determines which unit processes are included in the LCA and reflects the goal of the study. Finally the goal and scope phase includes a description of the method applied for assessing potential environmental impacts and which impact categories included. A common approach to this has been developed by the Swedish Environmental Institute and Volvo by creating six environmental indices. These indices are

1 *Scope*—The general impression of the environmental impact
2 *Distribution*—The extent of the affected area
3 *Frequency or intensity*—The regularity and intensity of the problem in the affected area
4 *Durability*—The permanence of the effect
5 *Contribution*—The significance of 1 kg of the emission of the substance in relation to the total effect
6 *Remediability*—The relative cost to reduce the emission by 1 kg

2.7.3 LIFE CYCLE INVENTORY

The second phase of LCA involves performing an inventory analysis. Usually the starting point is to develop a process flowchart that sequentially describes the entire process from cradle to grave and includes

■ The process step
■ Inputs into the process (raw materials, chemicals, and energy)
■ Outputs and emissions from the process (air emissions, waster emissions, and solid waste)

The data must be related to the functional unit defined in the goal and scope definition. Data may be presented in tables and some preliminary interpretations may be completed at this stage. The result of the life cycle inventory is a list which provides information about all inputs and outputs in the form of elementary flow to and from the environment from all the unit processes involved in the study.

2.7.4 IMPACT AUDITS

The third phase LCA is aimed at evaluating the contribution to the impact categories specified in the scope and goal. In this phase, impact potentials are calculated based on the results from the inventory phase and are normalized and weighed based on the functional unit. Normalization provides a basis for comparing different types of environmental impact categories (all impacts get the same unit). Weighing implies assigning a weighing factor to each impact category depending on the relative importance.

2.7.5 ANALYSIS AND ACTION

The fourth phase of LCA focuses on using the results of the inventory and audit portions to make decisions. An analysis of the major contributions, a sensitivity analysis and an uncertainty analysis are used to draw the conclusion and recommendations to meet the goals of the project.

2.7.6 APPLICATIONS

Many value-added applications of LCA have been developed to reduce the environmental impact of many companies. In 1994, the Swedish Waste Research Council conducted a survey of product manufacturers to determine areas that have successfully applied LCA. Key findings from the survey included

- Compare different options within a particular process with the object of minimizing environmental impacts.
- Identify processes, ingredients, and systems that are major contributors to environmental impacts.
- Provide guidance in long-term strategic planning concerning trends in product design and materials.
- Help to train product designers in the use of environmentally sound materials.

Also many countries have developed environmental regulations related to products based on findings from LCA. Some notable examples in the beverage container industry include

- Germany—Mandatory deposit refund on plastic beverage containers (expect milk)
- Norway—Tax on nonreturnable beverage containers
- Denmark—Ban on domestically produced nonrefillable bottles and aluminum cans

2.7.7 SUMMARY

LCA takes a holistic viewpoint on process analysis and design from cradle to grave for all aspects of a product from procurement to final disposal. Some key benefits of LCA include

- Reduces waste generation proactively for all stages of the product and process from beginning to end.
- Aids in the selection of waste reduction activities by ranking the emissions bases on severity and the impact to the environment.
- Minimizes environmental impact based on organizational goals and severity of emissions.
- Allows for the direct comparison of process, product, or material options based on the function unit and relative impacts.
- Improves corporate images as more environmentally friendly or green.
- Reduces costs.

On the other hand the LCA process also has some inherent drawbacks. The majority of these drawbacks stem from uncertainties and incorrect estimates during the life cycle inventories. Many organizations have found a great deal of difficulty in developing accurate estimates in terms of emission for all stages of the product. In addition, determining relevant importance of factors during the goal and scope phases potentially introduces a subjective bias.

2.8   Fundamentals of Recycling Processes

2.8.1   INTRODUCTION

An understanding of the recycling industry and the related processes can be very beneficial when evaluating the life cycle of a product. In addition, a basic understanding can aid managers and engineers in making better decisions in regards to recycling and disposal options. Following is a brief summary of some of these benefits:

- Gain a better understanding of recycling options and the different roles that material recovery facilities, processors, and material brokers play in the process.
- Gain an understanding of the recycling process and the process flow for materials as they leave a facility.
- Gain an understanding of the material separation needs based on the recycling process for each material type.
- Gain an understanding of the recycling process to design better processes and products to reduce the environmental impact.
- Learn more about the LCA process and develop more accurate inventory audits.

This section provides an overview of the recycling industry which includes brief discussions of the business entities operating in the field. In addition, overviews of the recycling processes for major waste items are also provided. The intent of this section is to expose the reader to basic terms and processes in the field, not to provide a detailed or comprehensive analysis.

2.8.2   RECYCLING INDUSTRY OVERVIEW

The recycling industry is comprised of five primary entities that work together to get recyclable materials from the point of generation as a by-product or waste to the stage where they can be used again as raw materials. Figure 2.13 shows the recycling process overview. These entities are

1 *Haulers*—These companies transport materials between entities, including the generation facility, consolidation points, and processing facilities. Oftentimes, these companies will lease a semitruck trailer to the generating facilities to consolidate and store recyclable materials before transportation to a consolidator or depot.
2 *Material recovery facilities (MRFs)*—An MRF is a specialized plant that receives, separates, and prepares recyclable materials for marketing to end-user manufacturers. There are two types of MRFs: clean and dirty. A clean MRF (Fig. 2.14) accepts recyclable materials that have been collected in comingled wastes from curbside collection separated at source from municipal solid waste generated by either residential or commercial sources. There are a variety of clean MRFs. The most common currently are two-stream MRFs, where source-separated recyclables are

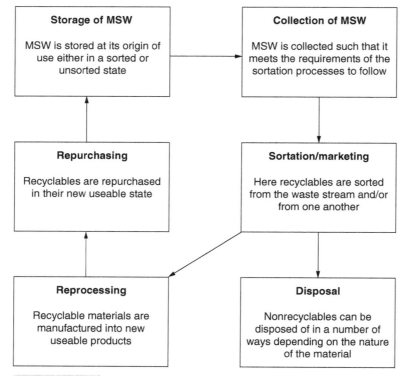

Figure 2.13 Recycling process overview.

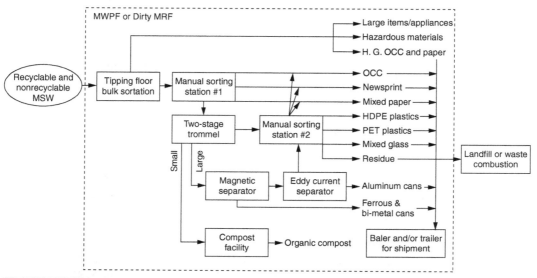

Figure 2.14 Dirty MRF overview.

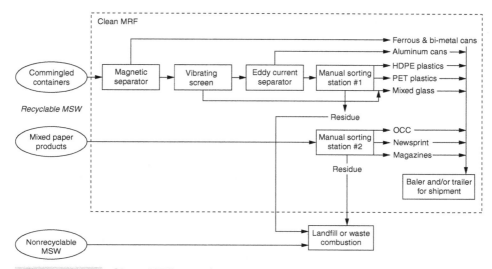

Figure 2.15 Clean MRF overview.

delivered in the form of a mixed food and beverage container stream (typically glass, ferrous metal, aluminum and other nonferrous metals, PET [No.1] and HDPE [No.2] plastics) and a mixed paper stream. A dirty MRF (Fig. 2.15) accepts a mixed solid waste stream and then proceeds to separate out designated recyclable materials through a combination of manual and mechanical sorting. The sorted recyclable materials may undergo further processing required to meet technical specifications established by end markets while the balance of the mixed waste stream is sent to a disposal facility such as a landfill.

3 *Consolidators and depots*—A consolidator or depot is similar to a MRF, but it does not perform any sorting operations. These entities hold or store materials until a specified batch size is reached or when a recycling processing facility is ready to process the material.

4 *Material brokers*—Material brokers buy recyclable materials from cities, businesses, depots, or MRFs and sell the materials to a processing facility.

5 *Processing facilities*—These facilities perform the actual processes to recycle materials. Many different processing facilities exist for different materials, such as metals, glass, and papers.

2.8.3 ALUMINUM RECYCLING

Aluminum recycling can be broken down into three steps:

1 Sorting
2 Baling
3 Compressing

This section provides an overview of each step. Before being transported to an aluminum recycling facility, the material is passed under a magnet to remove any steel. Aluminum is a nonferrous metal, so it is not magnetic. Steel is a ferrous metal, so it is magnet. The magnet picks up the steel cans and separates them from aluminum. The remaining cans are then crushed and compressed to form bales and transported to the aluminum recycling facility.

Once at the facility, the cans are shredded into small pieces about 1 inch in diameter. These pieces then pass through a magnet which removes any remaining ferrous metals such as steel. The shredded aluminum then travels to a de-coater, where hot air (at 930°F) removes any coating or decoration.

The hot cans go straight from the de-coater to the furnace, where they are melted at a temperature of 1300°F. Once melted, the liquid aluminum is transferred to a holding furnace which clears any remaining contaminants, and a degasser which removes any gas.

The liquid aluminum is poured into cooled rectangular shaped moulds. The cooling transforms the aluminum back into a solid metal. This solid metal is taken to a saw where the ends are made square and transported to the rolling facility. At the rolling facility the aluminum is rolled into large sheets to be used as raw material.

2.8.4   GLASS RECYCLING

Different types of glass go through different recycling processes. For example, cookware melts at a much higher temperature than container glass and must be processed separately. This section follows the typical recycling process of container glass (such as beverage bottles). There are four types of glass related to recycling processes:

1 Container glass (wine and beer bottles)
2 Float glass (windows)
3 Cookware (plates and dishes)
4 Automotive glass (windshields)

Glass for recycling is mostly collected from businesses or community drop-off sites. Trucks collect the bottles and transport them to be stored in a depot. When a processing batch of glass has been collected and delivered to the depot, it is all transported to a glass-recycling facility.

Once at the recycling facility, the glass is crushed. Crushed glass is called cullet. Cullet goes through many processes to remove nonglass items. To remove ferrous metal, the cullet is passed through a strong magnet which removes the ferrous metals such as steel and iron. The removed nonferrous metals, the cullet, passes by powerful air jets which separate the metal pieces from the cullet. To remove lightweight items, such as paper, the cullet goes through a vacuum. To remove any remaining items that are not glass, such as ceramics, the cullet passes under a laser which rejects them. The cullet is now ready to be made into new glass. To make new glass, the cullet goes into a furnace where it is melted at a temperature of 2700°F. The high temperature turns the cullet into a liquid called molten glass. The molten glass is shaped into molds to make bottles or jars. Recycled glass is melted at a lower temperature than virgin glass, which saves 30 percent of the energy used.

2.8.5 PAPER RECYCLING

Collected paper must be sorted and graded before being recycled. The recycling process itself can be separated into eight steps

1 Sorting
2 Baling
3 Pulping
4 Screening
5 De-inking
6 Pouring
7 Rolling
8 Packing

This section provides a brief overview of each of these steps. Paper recycling can be challenging because there are over 50 grades of waste paper. The main four groups are

1 Low grade (mixed paper, corrugated board)
2 De-inking grade (newspapers, magazines, office paper)
3 Kraft grade (unbleached brown backing)
4 High grade (printer cut-offs and unprinted paper)

Large amounts of paper, including shredded paper, are baled before being transported to a paper mill. Once at the paper mill, the paper is placed into a large vat and mixed with water. The process breaks down the paper into tiny strands of cellulose fibers. Eventually, this turns into a mushy mixture called pulp.

The pulp is then filtered and screened. The screens are made of a series of holes and slots of different shapes and sizes, and remove any remaining contaminants such as bits of plastic of glue. For certain uses, pulp must also be de-inked. There are two main methods of de-inking

1 *Washing*—Chemicals can be used to separate the ink from the paper and then washed away with water. Although this process requires the use of chemicals and waster the quantities used are much less than the manufacture of new paper and the water can often be cleaned and reused.
2 *Floatation*—Air can be passed through the pulp to produce foam. The foam holds at least half of the ink and can be skimmed off.

Pulp is poured into a huge flat wire screen. On the screen, water starts to drain from the pulp and the recycled fibers quickly begin to bond together to form a watery sheet. The sheet, which now resembles paper, passes though a series of heavy rollers, which squeeze out more water, some heated cylinders, which dry the paper, and an iron roller, which irons the paper. Next, the paper is wound into a giant roll. One roll can be as wide as 30 ft and weigh as much as 20 tons. The roll of paper is cut into smaller rolls, or sometimes sheets, before being dispatched for use.

2.8.6 PLASTIC RECYCLING

Plastic recycling can be separated into six steps

1 Sorting
2 Shredding
3 Cleaning
4 Melting
5 Extrusion
6 Pelletizing

Plastics are synthetic polymers made from oil and natural gas and are one of the world's most used raw materials. Plastics are blended in different formulas and modified with additives to create the 40 categories of plastic and the several specific grades within these. All plastics are labeled with an identification code, a number from 1 to 7. Before plastics are processed, they are sorted into seven different polymer types. The polymer type indicates both the properties and characteristics of the material, such as the melt temperature and its suitability for recycling. The symbols used to classify the different polymer types can be found on the plastic bottles. The seven different polymer types have been listed in detail in Chap. 1.

Once sorted, the plastics are baled before being transported to a plastics reprocessing plant. Once at the reprocessing plant, the plastic is shredded into small pieces which are then washed. After washing, the plastic pieces are passed under a metal detector to remove any metal, and a de-dusting unit, which removes any lighter particles.

The clean plastic pieces are dried and melted so they can be made into new shapes. The melted plastic is then filtered to remove any remaining contaminants and extruded to form fine spaghetti-like strands. The plastic strands are then cut into pellets, cooled in water, then dried and stored ready to be processed and molded as new plastic items.

2.9 Common Problems and the Human Factors of Recycling Systems

There are many common problems involved in creating a successful recycling program. The two biggest problems found in the duration of this research were material misplacement and hindrances inherent in the company itself.

2.9.1 MATERIAL MISPLACEMENT

One problem often encountered when companies attempt to separate recyclables is employee involvement. At most sites surveyed, containers dedicated to certain materials were found to contain other materials as well. This often results in higher costs for the companies. For instance, at one company, recyclable material was found in a dumpster dedicated for hazardous wastes and therefore costs more to dispose. However, wood, old corrugated containers (OCC), and other recyclables were finding their way into this dumpster.

At another company, recyclables were found in the landfilled trash stream even though there were recycling containers nearby. One such example was in a warehouse. The warehouse had recycling containers for shrink wrap and clean cullet (broken glass) positioned throughout the aisles, and yet the shrink wrap, glass, and wood were all found in the garbage cans, which were also throughout the aisles. In addition, there were aluminum collections barrels containing glass, glass collection barrels containing aluminum, and miscellaneous other trash mixed into the bins labeled for recyclables.

A third company, a heavy manufacturer and assembly plant, also encountered this problem of material misplacement. Upon a preliminary tour of the plant, it was noted that in clearly marked recycling containers, other materials such as Styrofoam cups, paper, and mixed trash were found. Similarly, in a large office building with over 600 employees the paper collection boxes also contained film plastic, OCC, and mixed trash.

These observations raised several serious questions. First, why do people place the wrong materials into collection containers or place recyclables into a landfill stream even when the appropriate container is located within close proximity? What is the motivation that causes some people to obey the recycling policies and what is the motivation that causes others not to? What can be done to improve involvement and deter carelessness in regards to recycling? What human factors come into play in determining the signs, which should be used to mark collection containers and in the placement of containers? Finally, what motivational ideas work best in regards to recycling, trash separation, and waste minimization?

It has been noted that when recycling containers are distinctive according to acceptable material, the employees are much more likely to place the proper material in the container. This may involve a difference in the type of container used, the color of the recycling container, or the size and shape of the opening or of the container itself. For instance, some companies use large wire bins which are always for recyclables only. Other used colored containers to designate the end point of the materials. Still others use containers of different shapes and sizes for each different material. Often the same types of similar containers are used with only a difference in size and shape of the opening for each material.

It is also obvious that when there are clear signs designating the acceptable materials for each container that accidental wrong placement of materials is much less likely. These signs usually work better if they are in color, have pictures, and are placed in easy sight. The signs must be easily recognizable and distinguishable in order to facilitate employee involvement. People do not like to be bothered by extra time and energy needed in order to recycle. If it requires extra effort, people will often not participate.

In a similar venue, the collection containers must be properly placed or people will not bother to find them. Individuals who have a personal interest in recycling will hold onto recyclables until they can be properly disposed of or placed in the proper container. Such people have often been found to do such things as take aluminum cans home from work if the company does not recycle them. These people will take the added time and energy necessary to get the materials in their proper containers. However, if one does not have such personal convictions, then in fast-paced society taking the extra time and effort to find the proper recycling container is often too time consuming and problematic to bother with.

If there is a motivation to recycle, the outcome is much better. There is not one cure-all for motivating employees, but there are some highly successful motivation factors. These include money, fun, and free time. The three motivation factors can be utilized in a variety of ways. One option is to keep track of recycling by department and then award the best department(s) with cash, a party, or a paid afternoon off. There are countless other creative options available. The feasibility of these options depends on the particular company and its schedule, policies, and recycling revenue.

According to human nature, there will always be some people who are too careless, apathetic, or lazy to obey recycling mandates. However, the proper use of human factors and motivation can minimize the number of such people and in turn minimize the incident of contamination of recyclables and recycling collection containers.

2.9.2 OBSERVED COMMON HINDRANCES

There are many common hindrances to recycling that occur on a higher level than employee involvement. These include management perceptions, company policies, union rules and regulations, poor past performance in recycling attempts, and many other reasons. These hindrances must be overcome or successful recycling will be impossible. It is often very difficult to overcome these hindrances. Very often, they are due to a misunderstanding or wrong perceptions. This makes it vitally important for the assessment team to understand the hindrances and how to combat them.

Though it may seem impossible to overcome polices or company rules, recommendations may still be made that do not conform to the problematic rules. The recommendations can be provided as win-win situations, and may go a long way to adjusting the policies inhibiting recycling practices. The most important issue is that the economically and ecologically best scenario is chosen. This can always be accomplished with win-win situations if those participating are willing to be creative. In short, the assessment team cannot direct the company to change its policies, but can present alternatives, which allow the company, management, or union to see the downfall in the policy and the benefits of alterations.

2.10   Related Research

2.10.1   INTRODUCTION

Research has been conducted in the area of solid waste generation prediction and evaluation of individual companies. The research that has been conducted was done primarily by the U.S. government and at the state or local level. Four major research studies have been conducted from 1995 until 2006 regarding solid waste estimation or prediction. This section discusses each study with emphasis on contributions and drawbacks.

2.10.2   1999 CALIFORNIA WASTE CHARACTERIZATION STUDY

The California Integrated Waste Management Board (CIWMB) (www.ciwmb.ca.gov/WasteChar/) has conducted research in characterizing solid waste for individual companies based on Standard Industrial Classification (SIC) codes in the State of California. The CIWMB has developed a solid waste characterization database that contains waste stream data for different types of businesses (Statewide Waste Characterization Study Results and Final Report, 1999). The database segmented businesses into 38 different groups based on products or services provided. The study collected and analyzed three types of information, basic business data (number of employees, location, and daily disposal amounts), business waste compositions (the types and amounts of materials typically disposed by an entire business grouping), and waste disposal rates for each business type (how much waste was typically disposed by each of the 38 business groupings, based on cubic yards disposed per employee per year). The data was collected from 1207 businesses using a 1-day sample at each facility. The study only included materials disposed at landfills, not recycled materials. Additionally, the CIBWM combined some business groups. This so-called lumped group consisted of 14 business groups whose combined contribution to the statewide waste stream was less than 5 percent.

A major contribution of the CIWMB study was that it analyzed which businesses generated similar percentages of components in their waste streams (such as paper, metal, and plastics). The CIWMB based these groupings on the SIC codes established by the U.S. government; Department of Commerce. The SIC code system has been in use since the 1930s and was established to promote uniformity and comparability of data collected and published by U.S. agencies (U.S. Department of Labor, www.dol.gov, retrieved May 4, 2002). A major drawback of this study was the waste grouping procedure. The procedure used in this study was subjective and not based on a statistical method.

Benefits and contributions of this study were

- The research team physically collected the data, increasing the reliability.
- Businesses were categorized into 38 groups that generate similar waste streams. These groups were based on SIC codes. This was the most significant and most heavily researched contribution of the study.
- Annual per employee waste volumes (in terms of cubic yards) were estimated for each of the 38 groups.

Drawbacks and limitations of the study were

■ This study applied limited mathematics and no modeling to evaluate and predict solid waste generation. Only means and variances were calculated, not expected ranges.
■ Few insights were derived on solid waste generation and limited comparisons between business groups.
■ The study was limited in that it only calculated per employee annual waste volumes for each business group.
■ Did not include recycling waste generation or any recycling analysis, skewing total waste generation depending on recycling rates of companies studied.
■ Only studied California businesses.
■ Used a 1-day sampling method to annualize generation rates, significantly increasing the variance.
■ Measured generation in terms of volume not weight, increasing error due to level of compaction (density).
■ Did not analyze financial issues of waste generation or recycling.

This study provided a good baseline and a simple estimation tool for solid waste generation rates of individual Californian companies.

2.10.3 1995 COMMERCIAL GENERATION STUDY, PALM BEACH COUNTY, FLORIDA

From 1993 though 1995 the Solid Waste Authority (SWA) of Palm Beach County conducted a research study to determine the waste generation characteristics of commercial property in Palm Beach County, Florida. The purpose of the study was to determine the quantity of commercial solid waste generated and the relative generation rates of various types of commercial property (Solid Waste Authority of Palm Beach County, 1995).

This study classified commercial properties into 54 different business groups. Approximately 10 percent of all businesses in each type were sampled. The study analyzed the total tons of solid waste disposed of by each company surveyed. Waste levels were normalized using floor space square footages of the facilities that generated the waste. A total of 15,371 observations were made on 1501 outside containers (dumpsters), an average of 10.24 observations per container. The final result of this study was a listing of the 54 business types identified and the annual tonnage of waste each business type generates per square foot of facility space. No useful raw data from this study was available for this research.

Benefits and contributions of this study were

■ The research team physically collected the data, increasing the reliability.
■ Businesses were categorized into 54 groups that generate similar waste streams.
■ Annual solid waste generated per facility square foot for each group was calculated.

Drawbacks and limitations of the study were

- This study applied limited mathematics and no modeling to evaluate and predict solid waste generation. Only means and variances were calculated.
- Few insights were given on solid waste generation and limited comparisons between business groups. The study calculated waste generation based on floor spaces (in terms annual waste tonnages for each business group per square foot).
- Waste compositions or recycling levels were not gathered or analyzed; only total solid waste generation per year was determined. This significantly narrows the scope of the study and limits its benefit to waste measurement and management.
- Normalizing the data based on square footage of facilities may be misleading because unutilized space is included in the figure; some companies may be more efficient with space utilization than others. Also some companies may warehouse material whereas others may not; this will also significantly skew the numbers.
- Waste reduction or recycling was not included as part of this study. A financial or cost dimension was not included in this study.
- Limited sample size consisting only of Palm Beach County businesses was analyzed.

As with the CIWMB study, this study provided a useful baseline and a simple estimation tool for solid waste generation rates of individual Palm Beach County companies. This study and the CIWMB study are the only solid waste estimation or prediction analyses conducted on individual company generation. Both studies, although offering some contributions to the field contains major deficiencies.

2.10.4 MEASURING ENVIRONMENTAL PERFORMANCE OF INDUSTRY PROJECT

Completed in July 2000, research was conducted to correlate individual company environmental performance to several indicator variables for six industries in the European Union (Science and Technology Policy Research Center, University of Sussex, United Kingdom, 2001). The name of the project was Measuring Environmental Performance of Industry (MEPI). Science and Technology Policy Research at the University of Sussex, United Kingdom, coordinated the project.

The key objectives of the research project were to

- Develop quantitative indices for the environmental performance of six industrial sectors by collecting environmental and financial data for a large number of European firms. The sectors studied were the electricity, pulp and paper, fertilizer, printing, textile finishing, and computer manufacturing industries.
- Analyze and determine indices to deepen understanding of the causes of changes in industrial environmental performance.

The research objectives were accomplished by

■ Collecting environmental and financial performance data for 280 European companies and 430 production sites in six industrial sectors.
■ Conducting a statistical analysis of the environmental performance of companies and production sites.
■ Developing a benchmarking tool allowing companies to compare their environmental performance to the performance of other European companies that operate in the same sector.
■ Creating a ranking of European companies according to their performance on key environmental indicators.

This research involved a regression analysis, which was conducted to identify a set of core indices that could give a good representation of the overall environmental performance of a firm. The core indices were the amounts of air emissions, water emissions, waste generation, energy and resource input. Principal component analysis (PCA), similar to factor analysis, was used to identify which variables statistically predicted the indices and therefore explain the most variability within data sets. The variables examined to establish relationships with the indices were management data such as sales, profits, employee turnover, and environmental certifications. These variables represented a simplified account of the environmental performance of a company.

Contributions of the MEPI project were

■ Developed quantitative and standardized indices comparing and evaluating individual company environmental performance with similar companies.
■ Utilized mathematical modeling to statistically draw conclusions and develop relationships.
■ Designed as a comprehensive evaluation tool including many factors to rate environmental impact, initially including solid waste generation rates.

Drawbacks and limitations of the MEPI project were

■ Focused mostly on management data such as sales, profits, employee turnover, and environmental certifications to predict rating outputs, not solid waste generation or other waste outputs.
■ Examined only six, very specific industrial sectors.
■ Conducted only in the United Kingdom.
■ This study placed little emphasis on solid waste generation of companies. More emphasis was placed on water and air emission of the companies to evaluate environmental performance.
■ This study focused on identifying factors to evaluate environmental performance by establishing correlations. Limited mathematical modeling was applied, only the variables that were significant were identified, not modeled or analyzed in depth.

This research was very innovative and was the pioneer effort in rating environmental performance of individual companies using mathematical analyses. A major drawback for this study was that no significant variables were found to rate or predict solid waste generation performance of companies in each sector. The study found air and wastewater emissions could be statistically evaluated using indicators, but not solid waste generation. This research placed more emphasis on these variables over solid waste generation. As mentioned, a serious drawback of this study was that no model was developed; the study just identified the variables that are significant in evaluating environmental performance (the core indices). This study solely analyzed data; it did not model the data.

2.10.5 PREDICTION METHODOLOGY FOR WASTE GENERATION AND COMPOSITION

A study conducted in Europe developed a mathematical model to predict solid waste generation for entire countries. The research is published in an article titled "Municipal Solid Waste: A Prediction Methodology for Generation Rate and Composition in the European Union and the United States of America" (Daskalopoulos, 1998). The research conducted examines historical data from 1980 through 1993 to determine correlations between United States and European Union MSW and indicator values. In this research the significant parameters considered are population and the mean standard of living for each country [as measured by gross domestic product (GDP)].

For this research, models were developed to express the relationship between GDP, population, and MSW, in tons per year for the European Union and the United States. The following equations were developed to describe the relationship for the European and American cases:

$$\text{Europe: MSW} = 0.1292 \times \text{GDP}^{0.4414} \times \text{Population}^{0.4855}$$

$$\text{America: MSW} = 4.08413 \times 10^{-3} \times \text{GDP}^{0.458} \times \text{Population}^{1.24075}$$

The degree of accuracy of these models is determined by the reliability of the published information, which has been provided by international organizations. The models developed can be used to predict the future amount of the MSW generated in a country, provided that gross domestic product and population forecasts are available. Although this research studied waste generation for entire counties, the indictors may also be significant for individual companies in terms of sales and number of employees (instead of GDP and population). The model outputs for this study were evaluated with the model outputs for this research. In particular the equations were linearized by taking the logs of both sides (Walpole and Myers, 1993). The results of the linearized equation for America were compared to the results from the model developed for this research. This method is discussed in Chap. 7; it involves validating the system.

Contributions of this study were

■ Developed quantitative prediction models for solid waste generation.
■ Identified the variables that aid in the prediction of solid waste.

Drawbacks and limitations of this study were

■ Solid waste generation rates were only determined for entire countries, not individual companies.
■ The study did not examine waste stream compositions or recycling.

This study was useful in that it mathematically predicted solid waste generation quantities. The study used historical data to aid in prediction. The major drawback is that only entire county generation data was studied and no compositions were analyzed. This research identified the variables that were significant in predicting the solid waste amounts for entire countries. The variables were population and GDP. Applying this to individual company generation quantities, the number of employees, and sales were tested to examine if they were significant in predicting generation.

2.10.6   SOLID WASTE ESTIMATION METHODS

Various waste-assessment methods have been developed to estimate solid waste generation for businesses. These methods assess waste, but do not predict or evaluate it. Most of these methods lack the versatility and scalability to apply nationally. Most of these methods require substantial data collection, before the waste estimation can begin. This significantly adds to research costs, hence reducing the usefulness of the methods. This research overcomes these problems by developing a standardized statistical system that is scalable and versatile.

The U.S. government developed and researched some of these waste-estimation methods (U.S. Army Corps of Engineers, 1990). For example, the U.S. Army Corps of Engineers developed four forecasting techniques for solid waste service plans at military facilities. The four techniques outlined by the Corps provide varying degrees of accuracy. The research noted that the more precise an estimate must be, the more it will cost to obtain. The solid waste forecasting techniques that the Corps developed and researched include moving average forecasting, per capita forecasting, and two-sampling forecasting methods that vary in the amount of samples taken. The Corps rated each forecasting method based on cost and accuracy using a low, medium, and high scale.

One common problem when measuring solid waste is the unit of measurement used. Two primary solid waste measurements exist, volume and weight. To avoid confusion, solid waste quantities should be expressed in terms of weight. Weight is the more accurate measure because weight can be measured directly, regardless of the degree of compaction. Weight records are also necessary in the transportation of solid waste because the quantity that may be hauled on highways is usually restricted by weight limits rather than volume limits.

Because some recycling and solid waste data are obtained by volume (for example, cubic yards), the use of standard volume-to-weight conversion factors is an essential element of the recycling measurement method. EPA developed numerous conversion factors for volume to weight from past research (www.epa.gov). The conversion factors are given in terms of density (mass/volume). The use of conversions factors is often important when conducting waste assessments and recycling surveys. Many

TABLE 2.2   EPA DENSITY CONVERSION FACTORS

MATERIAL	CONDITION (LEVEL OF COMPACTION)	DENSITY CONVERSIONS FACTOR
Food scraps	Uncompacted	7.49 lb/gallon
Glass bottles	Whole	500–700 lb/yd^3
Glass bottles	Crushed	1,800–2,700 lb/yd^3
Aluminum cans	Whole	50–75 lb/yd^3
Aluminum cans	Compacted	250–430 lb/yd^3
Ferrous metals	Whole	150 lb/yd^3
Ferrous metals	Flattened	850 lb/yd^3
Newspaper	Uncompacted	360–505 lb/yd^3
Newspaper	Compacted/baled	720–1,000 lb/yd^3
Old corrugated containers (cardboard)	Uncompacted	50–150 lb/yd^3
Old corrugated containers (cardboard)	Baled or compacted	700–1,100 lb/yd^3
White ledger	Loose	110–205 lb/yd^3
White ledger	Compacted	325 lb/yd^3
PET plastic	Uncompacted	30–40 lb/yd^3
PET plastic	Compacted	120 lb/yd^3
HDPE plastic	Uncompacted	24 lb/yd^3
HDPE plastic	Compacted	85 lb/yd^3
Mixed textiles	Loose	175 lb/yd^3
Car tires	Whole tire	21 lb each
Wood	Chipped	625 lb/yd^3
Wood	Whole pallet	40 lb each

times it is easier or less costly to gather data in terms of volume than weight; the conversions factors allow for efficient low cost conversions. Table 2.2 displays EPA density conversion factors. The conversions factors were obtained from EPA (http://www.epa.gov/epaoswer/non-hw/recycle/recmeas/, retrieved August 10, 2002).

2.10.7 UNIVERSITY-SPONSORED PROGRAMS

In 1996, an innovative partnership between the Lucas County Solid Waste Management District and The University of Toledo College of Engineering was formed to help improve environmental and economic conditions in Lucas County, Ohio. The partnership created the Waste Analysis and Minimization Research Project and later, in 2003, the Business Waste Reduction Assistance Program. The primary purpose of the Project is to provide no-cost solid waste assessments to Lucas County manufacturers and businesses. Since the inception of the project, over 70 waste assessments have been completed, over 109,000 tons of solid waste has been identified for reuse, reduction, or recycling and over $3.1 million have been identified as potential cost savings for Lucas County business.

The first and foremost goal of the project is to provide a valuable service to the Lucas County community. The project uses the knowledge and expertise of The University of Toledo's faculty and students to identify cost savings for local businesses through waste minimization and process efficiency solutions. The project's objectives are

- Increase manufacturing competitiveness through reduced solid waste disposal costs and optimize use of raw materials, packaging and floor space.
- Improve corporate image.
- Decrease reliance on landfills for disposal.

The University of Toledo graduate and undergraduate students majoring in industrial engineering perform all assessments. The district oversees all assessments. Typically, assessments consist of

- An analysis of the company's process and overall waste generation.
- Recommendations designed to maximize process efficiency and reduce solid waste disposal costs.
- Detailed reference list of vendors that complement recommendations.

All waste assessments are provided on a confidential basis and at no cost to businesses residing within Lucas County. Funding for the Project's assessments is provided by a grant from the Lucas County Solid Waste Management District with support from The University of Toledo. Much of the data collected for the cases studies presented in this book were gathered from this research project.

Comparison to similar university projects   Throughout the United States there have been several programs and research studies linking colleges, government agencies, and businesses involved with environmental improvement. This section provides an overview concerning some of these research studies and programs and compares them to the WAMRP. Table 2.3 provides a snapshot comparison between the university-sponsored programs. This includes a study conducted by the National Wildlife Federation titled "State of the Campus Environment" which critiqued U.S. colleges' environmental performance (1), the U.S. Department of Energy's Industrial Assessment Centers (2), Youngtown State University's Industrial Waste Minimization Project (3), Cornell University's Waste Management Institute (4), Indiana University's Institute on Recycling (5), and Iowa State University's Total Assessment Audits (6).

State of the campus environment (National Wildlife Federation)   In 2000, the National Wildlife Federation (NWF) conducted a survey of 891 institutions of

TABLE 2.3   UNIVERSITY WASTE REDUCTION PROGRAM COMPARISON

PROJECT	SERVICE PROVIDED	COMPANY SIZE	COST	UNIVERSITY DEPARTMENT AFFILIATION	COMMENTS
The University of Toledo Waste Minimization Project	Solid waste	all sizes	No fee	Industrial engineering	
The U.S. Dept. of Energy Assessment Centers	Energy, solid waste, hazardous waste, productivity	Small- to medium-sized (<500 employees)	Nominal fee	Mechanical engineering	26 participating universities
Youngstown State University Waste Minimization Project	Solid waste	Small- to medium-sized (<500 employees)	No fee	Environmental science	
Cornell University Waste Management Institute	Agricultural waste	Farms	No fee	Agricultural	Only performs farm audits
Indiana State University Institute on Recycling	College waste audits	Colleges	No fee	Environmental science	College communities

higher education in the United States. The survey measured environmental performance and sustainability of U.S. college campuses in many categories. The survey rated engineering programs poorly in regards to integrating environmental consciousness into the curriculums (the NWF rated engineering programs as a D on a scale of A, B, C, D, and F with A being the highest rating). The results of this survey indicate the need for increased efforts of engineering college directors to improve environmental awareness in their programs. The WAMRP at The University of Toledo and other programs discussed in this section describe the pioneer efforts to improve integration of environmental consciousness into college curriculums and which will improve the NWF rating.

Industrial assessment centers   The federal government has been funding industrial assessments for small- and medium-sized manufacturing firms under the Industrial Assessment Center (IAC) program (formerly called the Energy Analysis and

Diagnostic Center [EADC] program) since 1976. The program is funded through the U.S. Department of Energy's Office of Industrial Technologies (U.S. Department of Energy, 2001).

The industrial assessments provide an in-depth assessment of a plant site; its facilities, services, and manufacturing operations. This term is used to refer to a process which involves a thorough examination of potential savings from energy efficiency improvements, waste minimization and pollution prevention, and productivity improvements. Assessments are performed by local teams of engineering faculty and students from 26 participating universities across the country. The 26 participating universities are

1 Arizona State University, Tempe, AZ
2 Bradley University, Peoria, IL
3 Colorado State University, Fort Collins, CO
4 Loyola Marymount University, Los Angeles, CA
5 Iowa State University, Ames, IA
6 Lehigh University, Bethlehem, PA
7 Mississippi State University, Mississippi State, MS
8 North Carolina State University, Raleigh, NC
9 Oklahoma State University, Stillwater, OK
10 Oregon State University, Corvallis, OR
11 Rutgers University, New Brunswick, NJ
12 San Diego State University, San Diego, CA
13 San Francisco State University, San Francisco, CA
14 Syracuse University, Syracuse, NY
15 Texas A&M University, College Station, TX
16 University of Dayton, Dayton, OH
17 University of Florida, Gainesville, FL
18 University of Illinois at Chicago, Chicago, IL
19 University of Louisiana at Lafayette, Lafayette, LA
20 University of Massachusetts, Amherst, MA
21 University of Miami, Coral Gables, FL
22 University of Michigan, Ann Arbor, MI
23 University of Texas at Arlington, Arlington, TX
24 University of Utah, Salt Lake City, UT
25 University of Wisconsin-Milwaukee, Milwaukee, WI
26 West Virginia University, Morgantown, WV

The assessment begins with a university-based IAC team conducting a survey of the eligible plant, followed by a 1- or 2-day site visit, taking engineering measurements as a basis for assessment recommendations. The team then performs a detailed analysis for specific recommendations with related estimates of costs, performance and payback times. Within 60 days, a confidential report, detailing the analysis, findings, and recommendations of the team is sent to the plant. In 2 to 6 months, follow-up phone calls are placed to the plant manager to verify recommendations that will be implemented.

Manufacturers are not the only benefactors of the IAC program. Students involved in the program have a unique opportunity to see a range of manufacturing operations first hand. This results in more motivated students who more often than not opt for the energy management field as a career.

The IAC is similar to the WAMRP in that college students conduct waste assessments to aid companies in reducing waste. The primary differences are that the WAMRP focuses on solid waste assessments and IAC performs energy audits, including solid waste. The IAC only audits small- to medium-sized companies (less than 500 employees) and the WAMRP audits all-sized companies. The WAMRP focus also allow the team to analyze solid waste in more depth.

Industrial waste minimization research project (Youngtown State University, Ohio)   In 1995, the Mahoning County Industrial Waste Minimization Project, in a cooperative effort between the Mahoning County Solid Waste District and Youngstown State University began using interns to provide free waste minimization audits (Covey, 2000). The project team attended a training seminar titled "Integrated Manufacturing Assessments" to develop their waste audit process. Between 1995 and 1999, the project conducted 22 waste audits. In 1997, the project received the Governor's Award for Outstanding Achievement in Pollution Prevention and received additional grants in 1998 (Ohio EPA, 2004). In 1999, the project was discontinued due to lack of resources. This project was very similar to the WAMRP at The University of Toledo. Both projects involved a joint effort between government and academia and conducted no-cost waste audits. The major difference between the programs was the college departments involved. The WAMRP used industrial engineering students and faculty, while this project used environmental science majors. For this type of waste audits, industrial engineering majors have an advantage over environmental science majors because the industrial engineering curriculum is based on process analysis, engineering economics, and systems improvement.

Waste Management Institute (Cornell University)   The Cornell Waste Management Institute (CWMI) was established in 1987 to address environmental and social issues associated with waste management (Harris, 2000). Researchers and educators work to develop technical solutions to waste management problems and to address broader issues of waste generation and composition, waste reduction, risk management, environmental equity, and public decision-making. Current areas of research for the CWMI are composting as a component of integrated waste management; assessing the benefits and impacts of agricultural application of sewage sludges, manures and fertilizers; and source reduction.

Major goals of the CWMI outreach program are to improve the ability of local officials, businesses, and the public to make informed waste management decisions and to enhance the competency of solid waste professionals through increased training opportunities. The Cornell Cooperative Extension network, with offices in every New York State county, provides useful means of reaching these audiences. Outreach activities are based on research, with a goal of extending up-to-date, objective, research-based knowledge to a wide range of audiences from state agencies to America's youth.

The institute develops and sponsors technical and management workshops and conferences. The CWMI staff also produces reports, audio-visual, and computer-based training materials. The CWMI houses a publication library with over 5700 reference materials and is open to the public. A database of the library materials can be accessed and searched through the CWMI Web site.

This project is similar to the WAMRP in that both integrate colleges and government agencies to increase environmental awareness. The primary difference rests in the focus of each project. The WAMRP focuses on no-cost waste assessments and the CWMI focuses on waste characterization.

Institute on Recycling (Indiana University) The Indiana Institute on Recycling, a state-run agency, was created in 1989 by the General Assembly of Indiana. It is located at Indiana State University, in Terre Haute, Indiana. The institute developed concepts, methods, and procedures for assisting Indiana residents in reducing and recycling solid waste. The institute has conducted over 90 solid waste reduction studies with emphasis on cost-benefits. The studies have developed economical and environmental solutions for companies to address specific solid waste issues at their facilities; for example, cost analyses on the use of reusable shipping containers.

The institute is similar to the WAMRP in that both are supported by state funding, utilize college students to conduct the studies, and apply economic benefit analyses as an incentive for companies to reduce waste. The primary difference is that the institute addresses very focused, specific waste reduction needs of companies covering a broad range of areas and the WAMRP specifically conducts solid waste audits to characterize the companies' waste streams and target areas for reduction.

Total Assessment Audits (Iowa State University) In 1993, Iowa State University and the Iowa Energy Center developed a Total Assessment Audit (TAA) project that conducted energy, waste reduction and productivity audits simultaneously for manufacturing facilities (Haman, 2000). The TAA concept originated from the belief that a manufacturing facility is better served using a holistic approach to problem solving rather than the more conventional isolated approach. The TAA approach assumes energy, waste reduction, and productivity objectives are interrelated and that simultaneous evaluation will produce synergistic results.

The TAA methodology utilized in Iowa has evolved over a 6-year time frame and utilized on 25 audits. The TAA procedure emphasizes top management support and the use of cross-functional teams. Benchmarking exercises and energy usage analyses have been conducted to begin the audit process identify improvement opportunities.

An initial 2- to 3-day on-site audit is conducted with the aid of key facility operating and management personnel. The TAA team members are paired with the appropriate facility personnel to collect specific data, gather a working knowledge of operations, and identify opportunities for improvement. Workshops are conducted at the end of the day to exchange information and are structured to encourage open discussion and debate regarding the merit of each identified opportunity.

The TAA method is a very powerful tool for small- and medium-sized manufacturers and identifies and evaluates improvement opportunities, while accounting for the interconnected levels of energy, waste, and process analyses. This method allows for an efficient, cost-effective analysis, in essence combining three audits into one. This TAA concept is similar to the WAMRP in that both involve universities and aid companies in reducing waste. The major differences are that TAAs analyze more functions for the business, are only conducted for manufacturers, have less involvement from students, and are not provided free of charge.

2.11   Summary

This chapter provided a broad overview of the solid waste industry and solid waste minimization. An understanding of these concepts is critical for an overall understanding of the field and aid in developing optimal solutions. The worldwide generation rates provide an overview of the solid waste problem and the need for engineering solutions. The government's response to these problems was also introduced with a discussion of laws and regulations. Finally, from the practitioner's standpoint several tools were discussed to minimize solid waste, including solid waste exchanges, life cycle assessment, and the human factors of recycling.

3

BENEFITS OF SOLID WASTE
MANAGEMENT AND MINIMIZATION

3.1 Introduction

The purpose of a project, plan, or initiative (solid waste minimization or otherwise) is to achieve measurable results that can be tied into the original goal. These results or benefits are often critical in determining the fesasibility or acceptance of a project proposal. These benefits are also the key selling points used when promoting solid waste minimization to stakeholders and decisions makers. The benefits of solid waste minimization and recycling can be separated into four areas:

1 Environmental
2 Economic
3 Corporate image
4 Personal and social

Ideally, an organization would like to create a situation where multiple benefits can be realized from a single project. This synergistic approach allows for the creation of win-win situations when applied appropriately using the system approach discussed in this book. Specifically, the company will realize cost-benefits and enhanced public image, the environment will be protected, and the stakeholders of the organization (including employees) often gain a sense of well-being and harmony with the environment as the organization is protecting the greater good for society. This chapter discusses in more detail these benefits and includes examples that may be used to promote solid waste minimization to decision makers.

3.2   Environmental Benefits

Waste minimization efforts are a big step forward in moving toward a sustainable environment. The results are clear: cleaner air and water, less pollution, more forested land and open space, and reduced greenhouse gases. It is obvious that recycling translates into less trash entering landfills. But the greatest environmental benefits of recycling are not related to landfills, but to the conservation of energy and natural resources and prevention of pollution when a recycled material, rather than a raw material, is used to make a new product. Since recycled materials have been refined and processed once, manufacturing the second time around is much cleaner and less energy intensive than the first. The following list summarizes the key benefits to the environment that can be derived from solid waste minimization:

- Conservation of natural resources (water, trees, energy, and land)
- Healthier environment via landfill emissions reduction (carbon dioxide, methane, and leachate)
- Global warming reduction
- Conservation of habitats

The primary environmental benefit of solid waste minimization is resource conservation. The Medical University of South Carolina (MUSC) reports that the college recycled 1269 tons of paper, metals, organics, and other materials in 2003. Based on the school's calculations, this saved a total of about 13,756 Btu of energy, enough energy to power nearly 137 homes for 1 year. In addition, products made using recovered rather than virgin or raw materials use significantly less energy. Less energy used means less burning of fossil fuels such as coal, oil, and natural gas. When burned, these fuels release pollutants, such as sulfur dioxide, nitrogen oxide, and carbon monoxide, into the air. By using recycled materials instead of trees, metal ores, minerals, oil and other raw materials harvested from the earth, recycling-based manufacturing conserves the world's scarce natural resources. This conservation reduces pressure to expand forests cutting and mining operations.

Recycling and composting diverted nearly 70 million tons of material away from landfills and incinerators in 2000 as reported by the National Recycling Coalition. This total is up from 34 million tons in 1990, doubling in just 10 years. Below are some interesting facts about the relationship between recycling and resource conservation. These facts can have great emotional appeal when promoting waste minimization and can serve as part of a comprehensive strategy to promote recycling:

- Every ton of paper that is recycled saves 17 trees.
- The energy we save when we recycle one glass bottle is enough to light a light bulb for 4 hours.
- Recycling benefits the air and water by creating a net reduction in 10 major categories of air pollutants and 8 major categories of water pollutants.

- In the United States, processing minerals contributes almost half of all reported toxic emissions from industry, sending 1.5 million tons of pollution into the air and water each year. Recycling can significantly reduce these emissions.
- It is important to reduce our reliance on foreign oil. Recycling helps the nation accomplish this by saving energy.
- Manufacturing with recycled materials, with very few exceptions, saves energy and water and produces less air and water pollution than manufacturing with virgin materials.
- It takes 95 percent less energy to recycle aluminum than it does to make it from raw materials. Making recycled steel saves 60 percent, recycled newspaper 40 percent, recycled plastics 70 percent, and recycled glass 40 percent. These savings far outweigh the energy created as by-products of incineration and landfilling.
- In 2000, recycling resulted in an annual energy savings equal to the amount of energy used in 6 million homes (over 660 trillion Btu). In 2005, recycling is conservatively projected to save the amount of energy used in 9 million homes (900 trillion Btu).
- A national recycling rate of 30 percent reduces greenhouse gas emissions as much as removing nearly 25 million cars from the road.
- Recycling conserves natural resources, such as timber, water, and minerals.
- Every bit of recycling makes a difference. For example, 1 year of recycling in just one college campus, Stanford University, saved the equivalent of 33,913 trees and the need for 636 tons of iron ore, coal, and limestone.
- Recycled paper supplies more than 37 percent of the raw materials used to make new paper products in the United States. Without recycling, this material would come from trees. Every ton of newsprint or mixed paper recycled is the equivalent of 12 trees. Every ton of office paper recycled is the equivalent of 24 trees.
- When 1 ton of steel is recycled, 2500 lb of iron ore, 1400 lb of coal and 120 lb of limestone are conserved.
- Brutal wars over natural resources, including timber and minerals, have killed or displaced more than 20 million people and are raising at least $12 billion a year for rebels, warlords, and repressive governments. Recycling eases the demand for the resources.
- Mining is the world's most deadly occupation. On an average, 40 mine workers are killed on the job each day, and many more are injured. Recycling reduces the need for mining.
- Tree farms and reclaimed mines are not ecologically equivalent to natural forests and ecosystems.
- Recycling prevents habitat destruction, loss of biodiversity, and soil erosion associated with logging and mining.

Solid waste minimization also aids in creating a healthier environment by reducing landfill emissions. As discussed in Sec. 2.5 of the previous chapter, under environmental concerns, landfills emit a liquid called leachate. Leachate is a liquid that is generated from a landfill, which is created from decomposing waste, created after rainwater mixes with the chemical waste in a landfill, or liquids present in the landfill. Once it enters the environment, the leachate is at risk of mixing with groundwater near the site, which can have very negative effects. This liquid can be treated in a similar manner to sewage, and the treated water can then be safely released into the environment.

The Medical University of South Carolina reported that in 2003, recycling reduced overall air emissions by 24.9 tons (excluding carbon dioxide and methane) and reduced waterborne waste by 4.2 tons. By reducing air and water pollution and saving energy, recycling offers an important environmental benefit: it reduces emissions of greenhouse gases, such as carbon dioxide, methane, nitrous oxide, and chlorofluorocarbons, that contribute to global climate change. Section 2.5 in the previous chapter gives more detail on global warming. Recycling and composting reduce greenhouse gas by

- Decreasing the energy needed to make products from raw materials.
- Reducing emissions from incinerators and landfills, which are the largest source of methane gas emissions in the United States.
- Slowing the harvest of trees, thereby maintaining the carbon dioxide storage benefit provided by forests.

3.3   Economic Benefits

The economic benefits of waste minimization is often one of the key selling points when promoting environmentally conscious initiatives to businesses. Other than regulatory compliance, the cost benefits from waste minimization can turn an environmental decision into a wise business decision that will increase an organization's financial statements. Often times, when promoting a solid waste minimization program to the decision makers of an organization, the most influence benefits are the cost savings and potential revenue generated from the program. Often, when the creation of a recycling program is first discussed with management the first response is "we do not have a budget for recycling." This is far from the truth; the budget does exist and the starting point is the funds that the company is currently paying for waste hauling and removal. The systems approach to solid waste minimization explores cost-effective methods to better utilize these funds and protect the environment. The three areas cost benefits are usually derived from

1 Cost avoidance in solid waste hauling and disposal.
2 Cost savings in material purchases due to reuse and reduction.
3 Revenue generation from the sale of recyclable material.

Many organizations are surprised to learn that recycling and waste minimization can make strong business sense. A common environmental adage is "become green to make green." The Business Waste Reduction Assistance Program at the University of Toledo has identified over $3.1 million in annual savings for Northwest Ohio businesses in the 70 waste assessments that the program has completed. For example, at a plastic manufacturer with 100 employees, approximately $16,000 in annual cost-benefits were identified via increased plastic, paper, and cardboard recycling. The waste stream amounts and revenues also cost-justified the purchase of a baling system.

In the state of Pennsylvania, recycling adds significant value to the state's economy. In the state, collection and processing, the first step in the recycling process, involves sorting and aggregating recyclable materials. It includes municipal and private collectors, material recovery and composting facilities, and recyclable material wholesalers. These activities employ nearly 10,000 people in Pennsylvania, with a payroll of $284 million and annual sales of $2.3 billion. Recycling manufacturing involves the actual conversion of recyclables into products. The primary recycling manufacturers in Pennsylvania in order of magnitude are steel mills, plastic converters, paper and paperboard mills, and nonferrous metal manufacturers. Recycling manufacturing employs over 64,000 people with a payroll of almost $2.5 billion and annual sales of over $15.5 billion. Reuse and remanufacturing focus on the refurbishing and repair of products to be reused in their original form. The largest activities are retail sales of used merchandise and reuse of used motor vehicle parts. The amount of value that can be added via this process is limited because of competition from new products. Nevertheless, reuse and manufacturing contributes over 7000 jobs, a payroll of $115 million, and sales of over a half billion dollars.

On a national scale, the recycling industry continues to grow at a rate greater than that of the economy as a whole. According to the Institute for Local Self-Reliance, total employment in the recycling industry from 1967 to 2000 grew by 8.3 percent annually while total U.S. employment during the same period grew by only 2.1 percent annually. The recycling industry also outperformed several major industrial sectors in regard to gross annual sales as its sales rose by 12.7 percent annually during this period. Furthermore, the number of recycling industries in the United States increased from 8000 in 1967 to 56,000 in 2000. These facilities employ 1.1 million people across the country.

For many items, recycling can be more cost effective versus disposal. Following is a summary of select construction materials based on survey results of 63 companies:

- Average cost to recycle
  - Asphalt debris: $5.70 per ton
  - Concrete rubble: $4.85 per ton
  - Used bricks and blocks: $5.49 per ton
  - Trees and stumps: $37.69 per ton
  - Wood scrap: $46.43 per ton
- Average cost of disposal
  - Over $75.00 per ton and can be as high as $98.00 per ton

Recycling saves money for manufacturers by reducing energy costs. In 2001, New Jersey's recycling efforts saved a total of 128 trillion Btu of energy, equal to nearly 17.2 percent of all energy used by industry in the state, with a value of $570 million.

The sale of recycled products is an increasingly important component of the retail sector, and commerce in general. There are over 1000 different types of recycled products on the market and due to changes in technology and increased demand, today's recycled products meet the highest quality standards. Recycled products are also more readily available than ever before and are affordable. By purchasing recycled products, consumers are helping to create long-term stable markets for the recyclable materials that are collected from New Jersey homes, businesses, and institutions.

The economic value of clean air, water, and land is significant, but difficult to quantify. Since recycling plays an important role in protecting these natural resources, it must be an attributed economic value in this context as well.

3.4   Corporate Image Benefits

Corporate imaging and product branding play a critical role in the profitability of any organization. Successfully maintaining and strengthening these concepts are one of the chief duties of any marketing department and environmental initiatives can go a long way to bolster them. Specifically, by focusing on solid waste minimization and publicizing these efforts, an organization can

- Increase sales by attracting environmentally conscious consumers.
- Improve the recruitment of employees who share similar values.
- Attract environmentally conscious partners.
- Attain free corporate publicity.
- Increase employee involvement gateway to other programs (heart and mind).
- Maintain cleaner facilities.

3.5   Personal and Social Benefits

Solid waste minimization also offers personal and social benefits. Although many of these benefits are somewhat intangible and difficult to measure they are worth mentioning. They are worth mentioning because they can be selling points when promoting an environmental program. Below is a list of some of these benefits

- Personal satisfaction for helping the environment.
- Sustainable environment for future generations.
- Cleaner facilities.
- Buy-in at work programs (EI).
- Healthier environments and a higher standard of living.
- Generate money to assist local programs such as the sale of aluminum cans to benefit a children's burn unit at a hospital.

3.6   Summary

This chapter discussed the benefits of solid waste analysis and minimization. Specifically, the benefits were grouped into environmental, economic, corporate image, and personal. These benefits are also the key selling points used when promoting solid waste minimization to stakeholders and decisions makers.

Part 2

SOLID WASTE ASSESSMENT

STRATEGIES

4

DEPLOYMENT ALTERNATIVES

4.1 The Deployment of Solid Waste Minimization

The environmental and economic benefits of solid waste analysis and minimization are clear. Many organizations can significantly improve their bottom line performance and reduce their environmental impact by focusing efforts on solid waste reduction as demonstrated in Part 1 of this book. This chapter covers the next phase of that process, after the need has been identified that solid waste reduction is an important organization goal. The next phase is implementation and execution. There are a variety of alternatives to deploy solid waste minimization. These alternatives range from a massive organization-wide launch that covers multiple facilities to a smaller scope short-term project. The following is a list of deployment alternatives ranging from large-scale organizational initiatives to smaller project-based launches:

- Corporate-wide launch
- Single facility launch
- Department-based launch
- Specific waste stream analysis
- Product-based or life cycle analysis (LCA)
- Project-based launch

Each method has advantages and disadvantages and the deployment mechanism should be selected based on the available resources (financial, employee, and technology), the project timeline, the project goals, and the corporate culture of the organization. Table 4.1 summaries key benefits and drawbacks of each method.

For an organization just beginning the solid waste analysis and minimization process, the project-based launch is most often recommended. The primary reasons for this are that it requires fewer resources and has a shorter timeline. The key idea is that early, quick results can lead to bigger projects in the future based on the success

TABLE 4.1   DEPLOYMENT METHOD BENEFITS AND DRAWBACKS

	BENEFITS	DRAWBACKS
Corporate-wide launch	1. Biggest results 2. Consistent processes across the entire organization 3. Optimize the entire business system 4. Economy of scale	1. Highest cost 2. Longest implementation time 3. Requires intensive planning 4. Requires intensive data collection
Single facility launch	1. Can be used as a pilot project for other facilities 2. Analyzes entire facility for complete optimization	1. High cost 2. Required intensive data collection
Department-based launch	1. Simplified implementation analysis 2. Promotes departmental team work	1. Can miss opportunities in other departments
Specific waste stream analysis	1. Simplified implementation analysis	1. Can miss opportunities with other waste streams
Product-based or life cycle analysis (LCA)	1. Analyzes more than just waste generated in facility	1. Intensive data collection
Project-based launch	1. Short timeline 2. Low cost	1. Smallest impact

of smaller projects. A project is also easier to manage, for example examining the feasibility of installing a cardboard baler in a facility. Other advantages include

- The utilization of tools in a more focused and productive way.
- Increases communication between management and practitioners.
- Facilitates a detailed understanding of critical business processes.
- Gives employees and management views of how solid waste analysis tools can be of significant value to organizations.

The central concept, regardless of which deployment method is chosen, is to tie project results into bottom line and environmental benefits. The same concept is applied when prioritizing organizational needs or selecting among potential projects. These benefits should be expressed in terms of key process output variables such as

- Tons of solid waste per year
- Cubic yards of solid waste per year
- Annual disposal costs
- Process cycle time

■ Customer satisfaction
■ Scrap or defect rate

Several key concepts to keep in mind during the deployment planning phase include

■ Communicating the benefits of Six Sigma as a business strategy across the organization.
■ Aligning with management in the deployment of solid waste minimization.
■ Building a successful infrastructure for solid waste minimization deployment.
■ Integration of solid waste minimization with other lean manufacturing, theory of constraints (TOC), and other improvement methods.
■ Selection and orchestration of successful solid waste minimization projects and project teams.
■ Utilization of the right metric to drive the right activity.
■ Planning and execution of projects.
■ Selection of the right statistical tools.

Finally, the recommended overall approach to the deployment of solid waste analysis and minimization is based on proven Six Sigma methodologies. The Six Sigma process involves the DMAIC methodology described in Chap. 6. Chapter 8 discusses about the solid waste audit process.

4.2   Choosing a Waste Minimization Provider or Partner

When an organization chooses to implement solid waste minimization, an outside organization can be very useful and cost-effective to help with the implementation. This may appear to be an expensive proposition when compared to an in-house implementation. However, the organization has to consider the real cost and time required for effective program development, trainer development, and so forth. In addition, it is important for organizations to consider the value of time lost when instructional material is not satisfactory, not effective, and/or is inefficiently presented.

A good solid waste minimization provider can help with determining the deployment strategy, conducting initial training, and providing project coaching. The decision of which group is chosen can dramatically affect the success of the program. However, choosing the best group to help an organization implement solid waste minimization can be a challenge. Often the sales pitch given by a consultant sounds good; however, the strategy and/or training do not match the needs of the organization.

The following is a suggested list of questions to present to the solid waste minimization providers being considered:

■ What is your basic solid waste minimization strategy and flow?
■ What do you suggest doing next if we would like to compare the solid waste minimization program offered by your organization with that offered by other organizations?

■ What reference material do you utilize that follows the information used in your program (so that people can get further information or review on concept at a later date)?

■ During solid waste minimization training do you use multimedia presentations such as PowerPoint?

■ What is the basic format of your waste minimization course for executive training?

■ How do you address business and service processes?

■ What topics do you cover in your workshops?

■ How do you address the application of the techniques to real-world situations?

■ What software do you use?

■ What have others said about your training/consulting in the application of solid waste minimization? Can they be contacted?

■ What is the experience level of you staff?

■ What companies have you helped successfully implement solid waste minimization?

Once the list of providers is narrowed down, consider requesting that they describe their basic implementation strategy to prioritize projects within organizations. It is also recommended to visit each prospect to see firsthand a 1-day training session. Send a decision maker to view this session. Some providers or consultants might initially look good, and then appear less desirable after their training approach and material are reviewed first-hand.

4.3   Essential Elements of the Deployment Plan

Applying solid waste minimization so that bottom line benefits are significant and lasting change results requires a well thought out and detailed plan. Once a waste minimization partner or provider is selected (if used), it should assist the organization with the development of an effective implementation plan. Table 4.2 displays a sample implementation schedule for the kick off of a solid waste minimization plan. Although, every plan is unique, it should consider the essential elements (Table 4.2), which are described in detail in subsequent chapters.

Several of the key elements for the solid waste minimization implementation plan are

■ *Create support infrastructure for the first wave*—The support infrastructure for solid waste minimization is the first and critical step to ensure a successful program launch. For solid waste minimization to be successful there must be a commitment from upper level management and an infrastructure that supports that commitment. The key factors for creating a successful infrastructure are discussed in the next chapter.

■ *Select key players*—Deployment of solid waste minimization is most effective with the assistance of trained practitioners in the field. This can be accomplished by hiring

TABLE 4.2 KICK-OFF SCHEDULE FOR SOLID WASTE MINIMIZATION	
TASK	TIMELINE (DAYS)
Choose a provider	30
Executive training	3
Project champion training	5
Create support infrastructure for first wave	21
Select solid waste minimization project(s)	21
Solid waste minimization training: measurement	5
Executive/champion meeting: preanalysis phase	3
Solid waste minimization training: analysis phase	5
Executive/champion meeting: preimprovement phase	3
Solid waste minimization training: improvement phase	5
Executive/champion meeting: pre-control phase	3
Solid waste minimization training: control phase	5
Solid waste minimization training/infrastructure postmortem	14
Start second solid waste minimization training wave	14
Total	137 days

a consultant. Internally, the organization should select a project champion to serve as a team leader of a diversified cross-functional team that has knowledge in the various business areas of the organization, including operations, engineering, human resources, maintenance, finance, and accounting.

■ *Select key projects*—Initial projects should focus on tangible waste streams that have high visibility within the facility. This will help to build confidence within the team and strengthen support within the organization. Some good examples would be studying the use of returnable containers or reducing cardboard usage. Early small wins will lead to future big wins.

■ *Training and coaching*—Prior to training registration, team members should have been assigned to a defined project on which they will be working. This will help them to understand how to specifically utilize tools as they apply to processes that they are analyzing. It will also give them an opportunity to ask detailed questions and receive coaching on their projects.

■ *Project report outs*—Although difficult to schedule, ongoing top-down and bottom-up communications are also essential elements of a deployment plan. Communication plans are often the first thing overlooked on a busy manager's schedule, but are essential to break down the barriers between managers and practitioners.

■ *Postmortem*—Successful deployment of solid waste management and minimization is best achieved in a series of waves focusing on strategic change areas. Between waves there is time for evaluating effectiveness, compiling lessons learned, and integrating improvements into the infrastructure. Some examples of improvement opportunities are

■ Were projects aligned with the strategic focus of the organization?
■ Was communication adequate between practitioners and management?
■ Were the forecasted timelines met?
■ Did the phase achieve the desired results?

4.4   Summary

The deployment phase planning process and implementation is a critical step to solid waste minimization. This phase lays the foundation for all other steps, provides the needed training and establishes communication channels. Poor planning in this phase can lead to entire program failure.

CREATING A SUCCESSFUL SOLID
WASTE MINIMIZATION LAUNCH

5.1   Introduction

Any project or organizational program can be a failure or a success, depending on how it is implemented. For every organization, the implementation of solid waste minimization will be unique, but there are several common factors to successful programs. The key drivers for a successful program are

- Executive leadership
- Strategic goals
- Project selection
- Training and execution
- Resources
- Metrics
- Culture
- Communications
- Planning
- Results

The factors listed above have been identified as a general list of the most critical items to ensure a successful solid waste minimization program. This chapter discusses these factors in greater detail and provides a lessons learned section that contains tips and strategies from organizations that have been successful in launching solid waste minimization.

5.2   Executive Leadership

Support from top management is the key ingredient to successful and lasting solid waste minimization implementation. Commitment and leadership from executive management is the most influential factor for a strong program and significant results. Solid waste minimization can not be treated as a fad or flavor of the month within an organization. Top management must take a strong and vocal stand on supporting it and devote the necessary resources to ensure success.

Enthusiasm must be generated at all levels of the organization, but must start from the top. For a solid waste minimization or environmental program to be successful, top management must be the biggest cheerleaders and show authentic enthusiasm and support. It is critical that the executive leadership team understands the benefits and methodologies of the program and how they complement current business strategies. An understanding of these items is important to allocate resources and match team members to projects based on skill sets. The executive team also is responsible for creating the implantation plan. The success of a solid waste minimization plan depends largely on how well the executive team understands the value of the program.

5.3   Strategic Goals

The solid waste minimization targets and plans should be aligned with the strategic goals for the organization. At this stage in the implementation process, quality function deployment (QFD), or House of Quality, is a very useful tool to drill projects down to a strategic focus that will have significant impacts on the organization's environmental and economic performance. Figure 5.1 summarizes the House of Quality process. The overall approach is an iterative process in which a cross-functional team completes a series of houses using the following guidelines:

1 *Primary what*—Insert a comprehensive list of the customer and organizational expectations from the solid waste minimization program in the first column of House 1 with the importance ratings from customer and organizational surveys inserted into the adjacent column. Government regulations and employee needs should also be considered.

2 *How*—A brainstorming session is held with a cross-functional team in the second sub-step to determine the important "hows" relating to the initial list of "whats". These "hows" form the primary row of House 1. They are an organization's high level business processes and address how requirements can be met.

3 *Relationship matrix*—Relationships of high, medium, low, and none are assigned to how business processes affect customer requirements. Numerical weights such as 9, 3, 1, or 0 can be used to describe these relationships. Calculations are then performed: cross multiplication with the importance rankings. The values in these columns are then totaled.

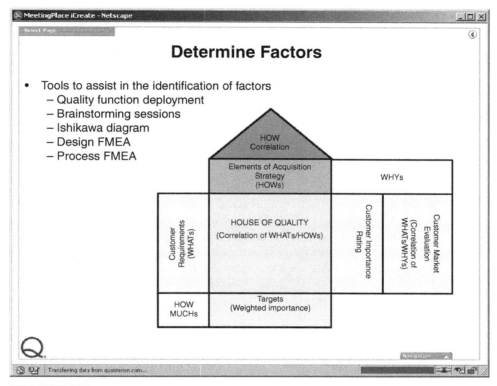

Figure 5.1 House of quality.

4 *How much*—The output is a ranked list of initial "hows" that can be used for strategic focus areas. These items are transferred to House 2, becoming the next "whats" if more detail is needed.
5 The process continues until the ranked "hows" are specific enough to assign strategic goals and measurements.

Upon successful completion of this process, there will be a list of key focus areas, which can be assigned strategic goals and objective measures. This process provides confidence that project resources are focused on meeting critical needs of the business. The following are tips when creating the House of Quality:

- The more specific the processes the less time the process will require.
- Combine customer input with internal business strategies, employee needs, and government regulations to obtain a comprehensive list of customer needs.
- Consider first organizing all the "whats" into categories with similar detail, and then perform the iterative process, inserting new "whats" at the appropriate House level.

- The process may require a significant amount of time and resources to complete, but will lead to projects that aligned with strategic goals.
- Sometimes it takes many repetitions of the process to achieve meaningful results.

The House of Quality process is a valuable tool to ensure meaningful lasting solid waste minimization projects. It also aids in resource planning and helping to prioritize projects considering resource constraints.

5.4   Resources

To ensure the most effective solid waste minimization program, the right people must be assigned to the right tasks and duties to ensure positive results that meet the desired goals. The project manager serves as the leader, but cannot do the work alone.

The primary goal is to align resources with the strategic priorities of the organization. This includes assigning team members and their associated tasks that align with these goals. Management should create a supportive environment and realign resources as priorities change. Below is a sample list of the roles and responsibilities of a well-staffed solid waste minimization team.

Champion (executive level manager)
- Remove barriers to success
- Develop inventive programs
- Question methodology
- Approve or reject project-improvement recommendations
- Implement change
- Communicate vision
- Determine project-selection criteria
- Allocate resources to ensure project success

Technical expert (solid waste management expert)
- Complete training
- Communicate vision
- Leverage project resources
- Share waste-minimization expertise
- Function as a change agent to leverage new ideas and best practices
- Improve overall project-execution efficiency
- Conduct and oversee training
- Formulate business strategies with senior management
- Aid in selecting projects that fit strategic business needs
- Motivate others toward a common vision
- Remove barriers to success
- Participate in multiple projects
- Aid in formulating effective presentations to upper management

Project manager (facilitator and coordinator)
- Lead projects
- Complete extensive waste minimization training
- Communicate vision
- Lead the team in the effective utilization of methodologies
- Select, teach, and use the most effective tools
- Utilize interpersonal and meeting facilitation skills
- Develop and manage a detailed project plan
- Schedule and lead team meetings
- Oversee data collection and analysis
- Establish reliable measurement systems
- Sustain team motivation and stability
- Communicate the benefit of the project to all associated with the process
- Track and report milestones and tasks
- Calculate environmental benefits and economic savings
- Prepare and present executive level presentations

Team member
- Contribute process expertise
- Communicate change with other coworkers not on the team
- Collect data
- Accept and complete all assigned action items
- Implement improvements
- Attend and participate in all meetings

Sponsor/process owner (manager of process)
- Ensure that process improvements are implemented and sustained
- Communicate process knowledge
- Obtain necessary approval for any process changes
- Select team members
- Maintain team motivation and accountability

An organization may choose more roles than those listed above, but these represent a typical list. One key point is that an executive level manager champions the project within the organization. This will ensure that goals are met and the proper resources are allocated. This high-level attention will also keep the entire organization focused on the goals. The process owners should be selected from the managers within the department responsible for the business activities. The technical expert may handle multiple projects and serves as the content expert on the new tools and applications for the solid waste minimization efforts.

Champions, sponsors, and the technical experts are the leaders of the waste minimization and environmental change and care should be taken to select the best suited team. Having dedicated team members is just as important as having well-trained leaders. In regards to selecting the project manager, the chosen individual should posses the following characteristics:

- Value collaboration
- Value helping others build on positive relationships
- Value being a supporter
- Consistently monitor what is going on
- Think before they speak
- View themselves as instruments to help the team
- Clearly define what is expected of the team
- Help groups relate to conflict as normal and productive
- Listen actively and effectively to the team

Individuals selected as project managers need to be respected for their ability to lead change, effectively analyze a problem, facilitate a team, and successfully manage a project.

5.5   Metrics

This section covers metrics as they relate to the overall implementation of a solid waste minimization business strategy. Metrics that track the effectiveness of the waste minimization infrastructure are a very important component in the process to ensure that goals are being met and that data-driven decisions are being made. The metrics that are created for the solid waste minimization program should be developed to provide valuable information on the success, variability, and efficiency of the process. The metrics should gauge whether the solid waste minimization infrastructure is accurately measuring the success of the programs. Several key items worth tracking to gauge the infrastructure effectiveness are

- Team member names and contact information
- Key process output variables
- Key process input variables
- Estimated completion date
- Actual completion date
- Tools utilized for the project
- Open action items
- Solid waste items analyzed
- Baseline measurements
- Financial calculations
- Actual project environment benefits and financial savings
- Lessons learned concerning project effectiveness

From this information, managers can gain insights into the effectiveness of each solid waste minimization project and the overall success. Successful projects can also be achieved for use in later projects. In general, metrics need to capture the right activity and be designed to give immediate feedback. Successful metrics focus on the process rather than the product or individuals.

5.6   Culture

The culture of an organization can play a significant role in the determining the success of an environmental program or any change effort. Some key organization culture assessment questions to consider when launching a solid waste minimization program are

- How do the stakeholders value environmental protection?
- How has the company historically dealt with change initiatives?
- Does the company make consistent changes that don't last?
- How effective are the project teams?
- Is the company frequently focusing on the same problem?
- What is required within the company culture to make continual process improvement a lasting change?
- What will prevent your organization from achieving success with a solid waste minimization program?

Ultimately, the issue of organizational culture boils down to the organization's ability to change in order to meet changes in the business environment. Specifically, how quickly the organization adapts to change and how well it embraces the continual improvement process. The key factors that support a culture for change, including the modifications for a solid waste minimization program are

- *Creativity*—Teams should be allowed to experiment and explore various process options. A creative environment can lead to higher quality programs and to stronger employee buy-in.
- *Motivation*—Employee support is critical to the success of an environmental program. There are various methods to motivate employees to support such programs, including financial and career advancement incentives. Employee motivation should be taken into account when planning for programs.
- *Empowerment*—Stronger programs can also result when team members are given more authority over the programs.

5.7   Communications

Effective and continuous communication of the solid waste minimization program is needed to create a shared vision within the organization. This is required by all levels of management to emphasize the fit between the program and business needs. A well thought out vision has the power to motivate employees and create shared values. Communicating this vision effectively is an ongoing process and is integral to a well thought out solid waste minimization implementation plan. The four Cs of effective communication plans are

- *Concise*—Common and clear language should be used to communicate the vision and goals of the program.
- *Consistent*—The same message should be continually repeated; frequent changes in goals and programs will reduce employee buy-in.
- *Complete*—All members of the organization should receive the same message; this may be challenging for shop floor workers.
- *Creative*—A novel communication approach will increase employee retention of the message. This could include an employee involvement program to create and spread program details.

5.8   Lessons Learned

The success of a solid waste minimization program depends on the existence of a solid infrastructure and top management leadership. Leaders need to take personal responsibility for driving waste minimization efforts, including the participation in projects. Some topics that many organizations have mentioned that drive the waste minimization change efforts are

- To be successful, the environmental effort must be launched organization-wide.
- Ensure the necessary resources are invested in the new initiative and monitor the payback.
- Spread the word about the new initiative as quickly as possible.
- Senior management needs to lead the effort and take an active and visible role.
- Provide monetary incentives for success.
- Reward aggressive team members.
- Monitor all aspects of success.

In terms of establishing waste minimization goals, it may be beneficial to keep the acronym SMART in mind, to establish goals that are Simple, Measurable, Agreed to, Reasonable, and Time based. Using this approach tends to lead to greater buy-in from employees and team members. For example, if an organization is exploring ways to reduce the amount of waste they send to the landfill, they could brainstorm as a team to establish this SMART goal:

- *Simple*—Reduce the solid waste that the organization sends to the landfill by 5 percent versus the same period last year.
- *Measurable*—Gather and monitor detailed weekly records of the amounts of waste recorded by the waste hauling company.
- *Agreed to*—The entire team should agree to this goal and at a minimum had opportunity to provide input.
- *Reasonable*—A 5 percent reduction is very achievable by modifying process without heavy investment; on the other hand, a 20 percent would not be reasonable in most cases for a 1 year improvement on a new project.

■ *Time based*—The goal is tracking performance for 1 year based on last year's reported results.

It is also important to examine why projects fail. Oftentimes many projects fail because a team is not properly empowered or supported. Below is a list developed by John Kotter (1994), in his article "Leading Change: Why Transformation Efforts Fail":

■ Not establishing a great enough sense of urgency
■ Not creating a powerful enough guiding coalition
■ Lacking a vision
■ Under-communicating the vision to stakeholders
■ Not removing obstacles to the new vision
■ Not systematically planning for short-term wins
■ Declaring victory too soon
■ Not anchoring changes to the corporate culture

Sharing lessons learned within the organization can lead to stronger future projects and results and at the same time reduce committing the same mistakes twice. When communicating lessons learned, the sessions should not be treated as a blame game session, but a true opportunity to improve processes and increase project speed. Some methods to examine lessons learned include, benchmarking with competitors, holding regularly scheduled best practice review meetings, and collecting information on project performance in a database.

5.9   Summary

Planning, top management support, and proper resource allocation are the three keys to a successful solid waste minimization program launch. Of these three, top management support plays the most important part because they lead the planning and resource allocation process. Establishing SMART goals are also critical to ensure the program is meeting targets, progress is being measured, and that team members buy into and understand the program.

6

THE SIX SIGMA SYSTEMS APPROACH FOR DEPLOYMENT

6.1 Introduction

Six Sigma is a comprehensive and flexible system for achieving, sustaining, and maximizing business success (Pande et al., 2000). It is uniquely driven by a close understanding of customer needs; disciplined use of facts, data, and statistical analysis; and diligent attention to managing, improving, and reinventing business processes (Pande et al., 2000). Six Sigma strives to improve quality, productivity, and bottom-line financial performance. The Six Sigma approach provides a methodology to achieve these successes and provides a system to base any improvement initiative, including a solid waste minimization program. Six Sigma relies heavily on data, facts, and the use of statistical tools to study whether an improvement has been made. These statistical tools are a powerful aid to conduct experiments and compare data and to provide important information about a process to find the causes of problems and draw conclusions. Six Sigma methodologies are broken into six fundamentals as related to solid waste minimization:

1 Define products or services.
2 Know the stakeholders and customers and their critical needs.
3 Identify processes, methods, and systems to meet stakeholders' critical needs.
4 Establish a process of doing work consistently.
5 Error-proof process and eliminate waste.
6 Measure and analyze performance.

To achieve these fundamental goals, Six Sigma uses the five-step DMAIC methodology described below:

Define
- Define and clarify the project goal and timeline with the focus placed on the patients.
- Establish the cross-functional waste analysis and minimization team.

Measure
- Define the current state and current processes, including the development of a process-flow map to baseline the system and to identify any bottlenecks and establish current financial and environmental performance.
- Collect and display data including task time, resources required, and process statistics.

Analyze
- Determine process capability and speed utilizing statistical tools and charts.
- Determine sources of variation and subsequent time bottlenecks.
- Identify and quantify value-added and non-value-added activities.

Improve
- Generate ideas and alternative data based solutions.
- Conduct experiments and validate improved processes.
- Develop action plans and standard operating procedures.

Control
- Develop a control plan.
- Monitor performance.
- Mistake-proof processes.

Six Sigma has three overall targeted solutions: process improvement, process design/redesign, and process management. Process improvement refers to a strategy of fixing a problem while leaving the basic structure of the process intact (Pande et al., 2000). Process design/redesign is a strategy not to fix a process but to replace the process to fix the problem (Pande et al., 2000). Process management refers to integrating Six Sigma methods into everyday business including

- Processes are documented and managed end-to-end and responsibility has been assigned in such a way as to ensure cross-functional management of critical processes.
- Customer requirements are clearly defined and regularly updated.
- Measures of outputs, process activities, and inputs are thorough and meaningful.
- Mangers and associates use the measures and process knowledge to assess performance in real time and take action to address problems and opportunities.
- Process improvement and process design/redesign are used to constantly raise the company's levels of performance, competitiveness, and profitability.

The remainder of this chapter discusses the general Six Sigma approach as it relates to solid waste minimization from the systems perspective. The DMAIC Six Sigma approach will be revisited in Chap. 8 as it specifically relates to solid waste minimization and solid waste audits.

6.2   Define

The purpose of the define stage is to gather sufficient information to clarify the opportunity for improvement, learn about the process, learn about the organizations barriers to solving the problem, and to develop a plan to address the problem. Common tools that are used throughout the define stage are

- Listening to customers
  - Kano model
  - Affinity diagram
  - Pareto analysis
  - Surveys
- Gaining process knowledge
  - Process mapping
  - SIPOC analysis
  - Work studies
- Sizing the problem
  - Force-field analysis
  - Developing project charter

A Kano model (Fig. 6.1) is a tool developed to compare the relationship between customer satisfaction and customer requirements (Gupta et al., 2007). The requirements are broken into three different groups: assumed requirements (unspoken), marketplace requirements (spoken), and love-to-have requirements (unspoken). Most often many of the requirements of the customer are unspoken, so indentifying the love-to-have requirements makes businesses world class.

The affinity diagram (Fig. 6.2) is a method used to obtain input from various stakeholders because often suggestion boxes get overlooked or meetings are taken over by

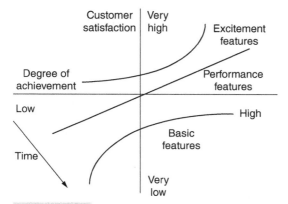

Figure 6.1   **Kano model of customer satisfaction.**

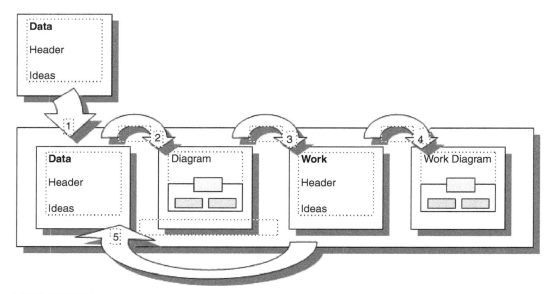

Figure 6.2   Affinity diagram.

a few outspoken individuals; whereas this method allows everyone to contribute. Using this method, a group can generate ideas fast and once they have been generated they are grouped and themes of ideas are developed for identifying action items for creating solutions from the ideas (Gupta et al., 2007).

The Pareto analysis (Fig. 6.3) is a tool used to make decisions based on importance instead of convenience. It is known as the 80:20 rule which means that usually 80 percent of the problems come from 20 percent of the processes. The Pareto chart is a bar chart showing attributes of the problem on the x axis and frequency of occurrence on the y axis (Gupta et al., 2007).

Process mapping (Fig. 6.4) is a way to identify the various activities of the process and show their interrelationships (Gupta et al., 2007). A well-defined map must address the following:

- Process has a purpose.
- Process has beginning and end states.
- Process has needs or inputs.
- Process must have a clear target performance.
- Process output does vary due to uncontrolled sources of variation.
- Process must be evaluated based on its mean or typical performance, as well as range between worse and better performance levels.

The SIPOC (supplier, input, process, output, and customer) diagram (Fig. 6.5) is used to expand the process map to identify players in the operation (Gupta et al., 2007). The

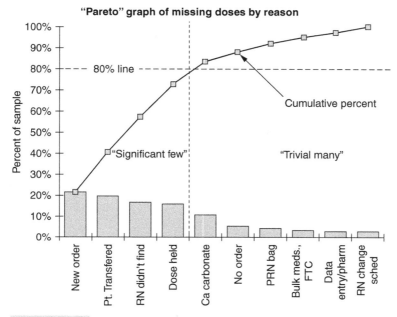

Figure 6.3   Pareto analysis.

benefit of this tool is to identify all variables that affect the process's performance and then prioritize them.

A work study shows how an employee spends his/her time working on different tasks. The analyst records a tally every minute in a specified area and observes the employee for a specified time. After the allotted time the analyst then records each task and can show the percentage of time the employee spends on each task.

The force-field analysis (Fig. 6.6) is a method to identify supportive and resistive resources that could be effectively utilized toward the process goals (Gupta et al., 2007). The objective is to identify factors to accelerate change or resistive resources that would slow process change.

A project charter compiles the problem definition, goals, objectives, and action plans to achieve them. It is a roadmap that

- Justifies the project efforts with financial impact.
- Describes the problem and its scope to be addressed by the project in the specified time frame.
- Declares the goal, objectives, and measures of success.
- Defines the roles of the team members.
- Establishes the timeline, milestones, and key deliverables.
- Identifies required critical resources.

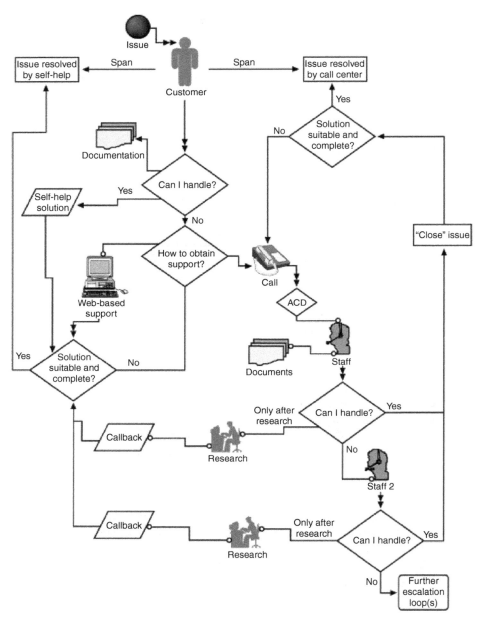

Figure 6.4   Process mapping.

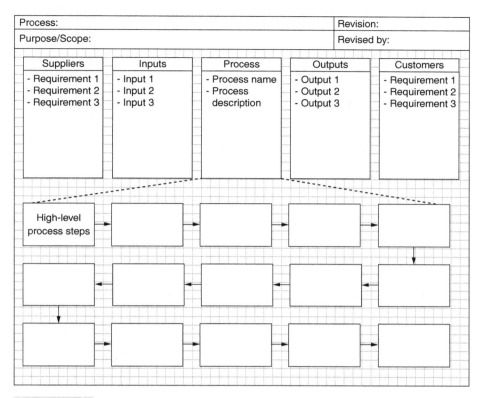

Figure 6.5 SIPOC diagram.

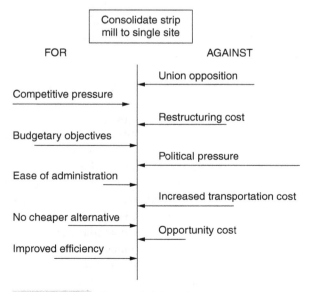

Figure 6.6 Force-field analysis.

6.3   Measure

The measure phase is to identify correct measures, establish a baseline, and eliminate trivial variables (Gupta et al., 2007). The following tools are important to understand:

- Basic statistics
- Statistical thinking
- Cost of quality
- Measurement system analysis
- Critical parameters
- Critical to quality

There are two types of statistics, descriptive and inferential. The descriptive statistics summarize the historical data (Gupta et al., 2007). Basic descriptive statistical analysis consists of the mean, median, mode, range, variance, and standard deviation (Gupta et al., 2007). Also measuring the cost of quality is critical because high variations and inconsistencies can cause high cost and waste valuable resources. Inferential statistics is based on analysis of the sample to infer performance of the process (Gupta et al., 2007). It is generally concerned with the source of the data and seeks to make generalizations beyond the data at hand. Inferential statistics can include regression analysis and hypothesis testing among many others.

6.4   Analyze

The analysis phase begins the convergence of possibilities toward the root cause of the problem (Gupta et al., 2007). Key analysis tools consist of multi-vary analysis, cause and effect diagrams, regression analysis, and failure modes and effects analysis (FMEA). The cause and effect diagram is used to identify the source of the problem of the process. The cause and effect diagram shows each branch of the process and the inputs related. From the diagram one is then able to notice the different contributors to the problem or process.

The regression analysis helps to identify causes and show correlations between variables. Often when solving problems involving many variables most have an inherent relationship. Regression is a way to estimate variable y if there is x amount of variable z. There are independent and dependent variables. Dependent variables are the responses and the independent variables are the causes. Relationships are not always deterministic; that is, x does not always give the same value for y (Walpole and Myers, 1993). As a result relationships cannot be exact, and regression methods are used to predict relationships.

Multi-vary analysis is a way to reduce the number of potential causes by removing the trivial ones (Gupta et al., 2007). Many data sets have hidden or not easily recognized similarities, patterns, or structures (Graham, 1993). Different techniques that can be used to address this are cluster analysis, two-factor analysis of variance, and three-factor experiments.

The failure mode and effect analysis is used to show the potential failures reducing the critical failures of the process (Gupta et al., 2007). Usually when creating a failure mode and effect analysis each potential failure is ranked from 1 to 10 in relation to the damage the failure could cause. Then solutions are thought of to prevent each failure, putting priority on the highest ranked failure.

6.5   Improve

The improvement phase is designed to identify actions remedying the root cause of the waste or inefficiency of a process (Gupta et al., 2007). The following tools can be implemented to solve a problem:

- Systems thinking
- Testing of hypothesis
- Comparative experiments
- Design of experiments

Systems thinking involves making a decision based on data and facts. Often decisions are based on thoughts and studies are not done to find the true causes of problems. Systems thinking is conducting tests and using tool to find these problems, and fix the problems based on these facts.

A statistical hypothesis is an assertion or conjecture concerning one or more populations. Testing of a hypothesis is an inferential technique used to make a statement about an activity or process based on its output (Gupta et al., 2007). One can either fail to reject a hypothesis or reject it. Fail to reject implies that the data does not give sufficient evidence; on the other hand rejection implies the sample evidence refutes the hypothesis. Rejecting the hypothesis means when the hypothesis is true there is a small probability of that sample occurring again. There are always two hypotheses the null and the alternative. The test is against the null; if the test is rejected then the alternative is accepted.

Comparative experiments is measuring a control group then conducting an experiment and seeing if there is a difference in results (Gupta et al., 2007).

The design of experiments relates to comparative experiments. It is a nonbiased way of creating an experiment while getting the wanted answers to the problem. The three goals of an experiment are

1 The experimenter should clearly set forth his/her objectives before proceeding with the experiment.
2 The experiment should be described in detail. The treatments should be clearly defined.
3 An outline of the analysis should be drawn up before the experiment is started (Graham, 1993).

6.6   Control

The control phase is designed to maintain the benefits of the improved process. Tools that are commonly used throughout this process are

- Control charts
- Documentation
- Change management
- Communication
- Reward and recognition
- Check sheets

Control charts (Fig. 6.7) are used to identify variation whether it is random or assignable (Gupta et al., 2007). Control charts record a task over a given amount of time and are plotted against the specification limits to observe if the process is in control. Documentation is the recording of the new process steps to provide standards and guidelines to the employees (Gupta et al., 2007). An example of documentation is ISO 9000. Communication is essential throughout the process but to implement something

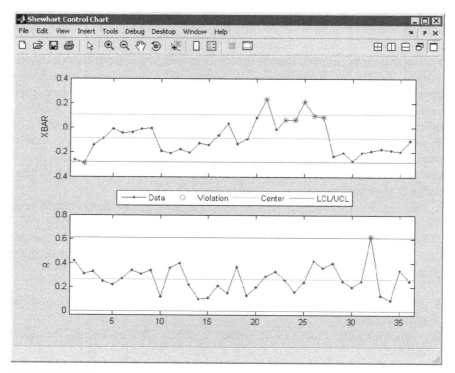

Figure 6.7   Control chart.

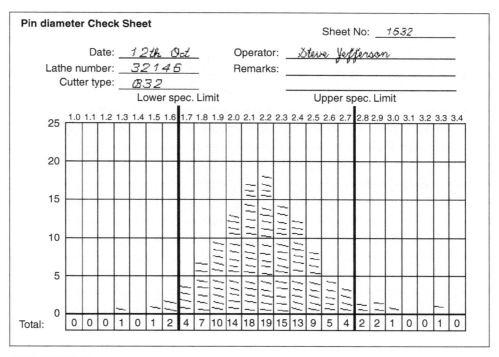

Figure 6.8 Check sheet.

new and to control it is vital to the success of the new process (Gupta et al., 2007). Rewarding and providing recognition to all involved is also important, giving everyone a sense of pride in the hard work put into the project. Check sheets (Fig. 6.8) can be used to show where different problems could arise, and can keep track of different steps throughout the process or times that different products arrive. Check sheets are a good way to monitor a process and can be used for various aspects of the tasks.

6.7 Summary

This chapter gave an overview of the Six Sigma approach to improving system performance. Six Sigma provides an excellent toolbox of approach and methodologies to analyze and improve project deployment or systems. Many of these concepts are incorporated into the systems approach to minimize solid waste. Specifically, the use of a data-driven, goal-focused approach that utilizes a variety statistical and process analysis tools.

7

METRICS AND PERFORMANCE
MEASUREMENT FOR SOLID
WASTE MANAGEMENT

7.1   Introduction

Accurately measuring solid waste minimization performance and tracking trends is a critical step for a successful program. Measurement is needed before a company can begin to manage and improve the solid waste problem. Without knowing a starting point, or baseline measurement, it is very difficult to develop a plan to meet organizational waste minimization goals. This chapter addresses this issue and discuses the various metrics to evaluate solid waste generation and minimization performance. The metrics can be broken down into three categories

1 Environmental impact in terms of solid waste emission rates
2 Business and financial performance
3 Voice of the customers and stakeholders

The remaining of this chapter will address these three categories and discuss their role in developing a strong solid waste minimization program.

7.2   Solid Waste Quantity Measurement

Measuring and tracking solid waste generation trends and the performance of individual companies is critical to its successful management and reduction. For any process improvement, an accurate data-driven baseline must be created and monitored to measure success toward meeting a goal. In this case, measuring a company's waste generation and recycling levels. Traditionally, solid waste generation and recycling levels are reported by total output. For example, the plant generated 150 tons of solid waste this month, of this amount 15 tons, or 10 percent, were recycled. This measurement approach has a very serious shortcoming. The primary flaw is that is does not consider production levels or resource inputs. For example, if two similar manufacturing plants generated the same amounts of solid waste, but the second plant has only half the production volume, the second plant is not doing a good job of managing its waste streams. Several other approaches have been developed to compensate for these flaws and allow for an equal "apples-to-apples" comparison. Five of the most common metrics by which solid waste generation can be measured are

1 Absolute measures
2 Measures indexed to a production output–based quantity
3 Measures indexed to a production input–based quantity
4 Measures indexed to throughput
5 Measures indexed to a production activity

All except the first of these metrics are based on the ratio of generation to some measure of business activity.

7.2.1 ABSOLUTE MEASURES

A comparison of the mass of solid waste generated in 1 year to the mass that is generated in another year is an absolute measure of waste generation. Such measurements are consistent and easy to understand, for example 1 lb of plastic scrap generated last year is equal to 1 lb of plastic scrap generated this year. The major drawback is that they ignore production levels or business activities associated with waste generation. For example, if a manufacturing plant reduces it total solid waste generation by 50 percent, it might indicate a major effort to reduce waste or it might be associated with a large drop in production volume, such as the loss of a large contract that accounts for 50 percent of its total production volume. The solution to this problem is to normalize the data or index solid waste generation to some measure of production.

7.2.2 MEASURES INDEXED TO OUTPUT

Solid waste can be indexed to the mass of products, number of products, or dollar value of products. A good example of this in the automobile industry is waste per vehicle produced. Using this index, the company will now have a more meaningful comparison of waste generation rates at different production levels. A drawback of this approach is that it does not consider significant changes to output. For example, if the automobile plant shifts from producing sport utility vehicles to compact hybrid cars, waste per vehicle would be expected to drop too. This drop may not be linked to a solid waste reduction program, but a major change in business processes.

7.2.3   MEASURES INDEXED TO INPUT

Solid waste can also be indexed to inputs, such as raw materials, dollar value of raw materials, or the number of employees. This is not a commonly used metric, but it can offer several advantages including

- Raw material data and employee numbers are easily available.
- The information is widely applicable to different processes and facilities.

7.2.4   MEASURES INDEXED TO THROUGHPUT

This index is similar to an output index, but links solid waste generation to an intermediate product generated, not a final product. This may be beneficial and more meaningful in some situations, such as a company that would like to measure and track its solid waste generation for an internal unit that supplies another unit within the company. For example, with the final assembly plant for automobiles, the unit that produces the seats for the cars could use this index to measure the waste per car seat assembly produced. This measure would be more meaningful to the seat assembly unit versus a final product index (vehicles produced) since the seat assembly unit may produce a different number of seats for different car models (a sports car may only have two seats whereas a sports utility may have eight seats).

7.2.5   MEASURES INDEXED TO ACTIVITY

Finally, solid waste generation may be indexed to a business activity, such as the number of times a waste generating activity occurs not necessarily related to production levels. In some cases an activity ratio may be a more accurate measure than a production ratio. For example, at a university, how often the school holds orientation sessions or training sessions will impact waste generation in terms of paper and food waste. An index to the number of training sessions conducted may be more meaningful than the number of students or faculty at the school.

7.3   Business and Financial Measurement

From a business standpoint, projects are evaluated based on their impact to the bottom line of the organization. An understanding of the financial benefits of a solid waste minimization project is critical in determining, evaluating, comparing, and selecting projects. In addition, a thorough understanding of the financial impact of the project will aid in promoting the project to upper management and other stakeholders. From a financial standpoint, the three areas on which solid waste minimization projects are evaluated are

1 Initial investment
2 Payback period (and discounted payback period)
3 Internal rate of return

The initial investment is the start-up funds required to begin a given solid waste minimization program. This includes the cost for recycling bins, recycling provider fees, recycling equipment costs (balers, grinders, or electric hand dryers), and training costs. The payback period is the period of time (usually given in years) required for the project's profit or other benefits to equal the project's initial investment. The equation is

$$\text{Payback period} = \frac{\text{cost}}{\text{uniform annual benefits}}$$

The payback period measures how long the project will take to recoup the initial investment, or in other words gauges the rapid return of investment for an organization. Many organizations use the payback period as a litmus test to screen projects based on a predetermined threshold, say 3 to 4 years for many companies. There are some limitations to the payback period, such as

- The payback period is an approximation, not and exact economic analysis.
- All costs and all profits (or savings) of the investment before payback are included without considering differences in timing.
- All economic consequences after the payback period are ignored.
- Being an approximation, the payback period may not select the optimal project.

The internal rate of return (IRR) is the interest rate at which the present worth and equivalent annual worth of a project are equal to zero. Another way to think about the IRR is the annualized interest rate that a project earns over its life. In most cases, organizations have a predetermined minimum attractive rate of return (MARR), which is the minimum interest rate that the organization could accept as the return on a project and still remain profitable. For a given project, if the IRR is greater than or equal to the MARR, it is a profitable decision to accept the project. To solve IRR, the net present worth (NPW) is set equal to zero, as shown in the following equation:

$$\text{NPW} = \sum_{t=1}^{n} \frac{C_t}{(1+r)^t} = 0$$

where t = the time of the cash flow
n = the total time of the project
r = the discount rate (the rate of return that could be earned on an investment in the financial markets with similar risk)
C_t = the net cash flow (the amount of cash) at time t

Solid waste minimization projects should be evaluated based on these three areas, the initial investment, the payback period, and the internal rate of return. The initial investment is important to determine and allocate the start up funds an organization has available to begin the project. This is needed to determine a starting point for the waste minimization efforts. The payback period and the internal rate of return measure the success of the project in financial terms.

7.4   Customer and Stakeholder Satisfaction Measurements

A recent study on employee and customer satisfaction indicated that being green bolsters employees' and customers' opinions of the organization. In February 2008, Brockmann & Company, the customer insight firm released its latest independent report on being green and its relationship to business performance. The report, entitled "The Power of Green" reviews buyer preferences for green brands and considers the state of green in over 100 organizations from around the world. Business people confirm that higher greenness is coincident with higher customer satisfaction, higher employee satisfaction, and higher revenues per employee.

Peter Brockmann, president of Brockmann & Company said, "Business people have worried about the cost of being green. We provide evidence that companies that focus on recycling in the office, reducing energy consumption in the office and use video conferencing or tele-presence technologies intensively, also have higher customer satisfaction, higher employee satisfaction and higher revenues per employee."

This report also showed that top performers scoring high on the green quotient had three times more customer satisfaction than poor performers (organizations scoring low on the green quotient), 4.7 times more employee satisfaction, and 1.7 times more revenue per employee. Other key report findings include

- The reliability, accessibility and quality of video conferencing is the most significant variable influencing the study's result.
- Counterintuitively, top performers don't use disincentives for business travel leaving it to the discretion of employees as to the most effective use of their time and company resources.
- It's not just about green technology adoption—but about green practices and management attitudes that influence corporate culture on questions like encouraging public transit and tele-working.

There are numerous methods to investigate whether customers are satisfied with an organization, products, and the service offered. Common methods are

- Face-to-face interviews; as customers enter or leave your facility or store or office, ask them their opinions.
- Call customers on the phone.

- If you have a customer phone number list, and their permission, consider calling them after their visit and ask how satisfied they are.
- Mail customers a questionnaire.
- E-mail them a customer satisfaction survey.
- E-mail customers an invitation to take a customer satisfaction survey.
- Partner with a market research firm to gather and collect data.

Following are some commonly asked questions related to business and environmental performance perception:

- How satisfied are you with the purchase you made (of a product or service)?
- How satisfied are you with the service you received?
- How satisfied are you with the company overall?
- How do you rate the environmental image of the company?
- And ask the customer loyalty questions, such as
  - How likely are you to buy from us again?
  - How likely are you to recommend our product/service to others?
  - How likely are you to recommend our company to others?
  - Also ask what the customer liked and didn't like about the product, your service, and your company.

Often an essential component of organizational training and development, employee attitude surveys provide a picture of an organization's needs. These surveys can be used to solicit employee opinions on a variety of issues such as the company's success in communicating its mission to employees, or local issues such as quality of the working environment. These surveys often contain a series of multiple-choice items grouped along one or more dimensions of the organization. The types of items included in these surveys may concern areas such as

- Creativity
- Innovation
- Satisfaction
- Senior management
- Interpersonal relations
- Functional expertise
- Compensation
- Ability to listening
- Customer service
- Communication
- Obtaining results
- Analytical thinking
- Mentoring
- Strategic leadership
- Teamwork
- Adaptability

■ Staff development
■ Leadership

The results of this type of feedback process provide an understanding of how the employee perceives the organization along different dimensions. This process helps the organization (human resources department) understand how the employees perceive them. This feedback

■ Is essential to facilitating development and organizational change.
■ Allows the organization to focus on needs and leverage its strengths.
■ Informs the organization on which actions will create problems for the employees.
■ Provides management with employee feedback (both positive and negative) on the internal health of the organization.
■ Measures the impact of current programs, policies and procedures.
■ Can be used to motivate employees and improve job satisfaction.

With both customer and employee surveys, similar information should be collected over a period of time so that trends can be developed and studied. These trends are a tool to gauge the opinions and attitudes of these key stakeholders. Figure 7.1 is an example that demonstrates comparing annual employee satisfaction data, and could be used as a model.

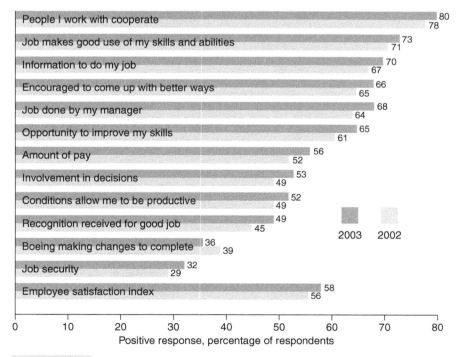

Figure 7.1 Employee satisfaction trends.

7.5   Summary

This chapter discussed the various metrics that are recommended when preparing, conducting, and monitoring a solid waste minimization program. Solid waste emission, financial, and stakeholder feedback should all be incorporated into a strong solid waste measurement system. All three work together to complete the whole picture for these environmental- and business-focused efforts.

THE GENERAL APPROACH FOR
A SOLID WASTE ASSESSMENT

8.1   Introduction to the Systems Approach Framework

After the need has been identified to minimize solid waste generation and top management supports and allocates resources to the effort, a solid waste assessment can be conducted. The solid waste assessment is one of the most important steps of the solid waste minimization process because the data generated provides the solid waste assessment team and management with a much greater understanding of the types and amounts of waste generated by the company. These data can be invaluable in the design and implementation of a waste reduction program. Before rushing in to conduct a solid waste assessment, proper planning should be done to ensure the scope, goal, and timeline of the project fits into the strategic plan of the organization. This chapter provides the systems approach framework to ensure that these goals are met using the five-step Six Sigma DMAIC process discussed earlier in Chap. 6:

- *Define*—Establish the project team, determine the goal, and allocate the necessary resources
- *Measure*—Conduct a solid waste assessment and review existing records
- *Analyze*—Statistically analyze the data collected to identify trends
- *Improve*—Modify the process to meet the organizational goals established in define stage
- *Control*—Ensure that the improvement initiatives stay in place with continuous improvement and feedback

The remainder of this chapter will expand upon the systems approach and break this framework into smaller pieces, and provide an execution model and process flow to accomplish solid waste minimization. As a reminder, the systems approach examines the organization as a whole, or a sum of all business processes to achieve established goals. The concept is to use data to develop comprehensive system-wide changes that will drive environmental and economic performance versus routine incremental improvements. In addition, several examples will be provided to further explore and explain each step of the framework. The solid waste minimization process consists of 11 steps:

1 Establish the solid waste minimization team and charter
2 Review existing solid waste and recycling records
3 Create process flowcharts and conduct throughput analyses
4 Conduct the solid waste sorts at the facility
5 Analyze the data to determine annual generation by work unit or area and establish baseline data
6 Identify major waste minimization opportunities
7 Determine, evaluate, and select waste minimization process, equipment, and method-improvement alternatives
8 Develop the waste minimization deployment and execution plan
9 Execute and implement the waste minimization plan and timeline
10 Validate the program versus goals
11 Monitor and continually improve performance

Figure 8.1 separates this 11-step process as it relates to the Six Sigma approach.

By applying Six Sigma and the systems approach, waste minimization alternatives can be developed and evaluated in a standardized manner to reduce the organization's environmental impact and improve financial performance. With this approach, alternatives are fully described, environmental impacts are quantified, a feasibility analysis is conducted, a financial justification analysis is performed, and feedback is collected from all stakeholders before making a final decision to implement. To aid in describing the 11-step process, a case study is utilized as an example to discuss the real-world application of each step. The case study involves a company in Northwest Ohio that is a leading manufacturer of original equipment batteries, specializing in parts for Chrysler, Ford, Jeep, Nissan, and Mazda. Approximately 18,000 to 20,000 batteries are produced per day in a plant that was opened in 1980. The basic components in battery construction are the case, positive plates, negative plates, separators, plate straps, electrolyte, and terminals. At the company, the separator is made of microporous rubber while the case is made of polypropylene. Other components are lead oxide and acid electrolyte. Waste is generated from each of these components and these waste streams along with the process flow diagrams are discussed in this chapter.

8.2   Step 1: Establish the Team and Define the Project

After identifying the need to minimize solid waste and gaining top management support, the first step involves establishing the team and defining goals. The key outcomes and deliverables of this step are

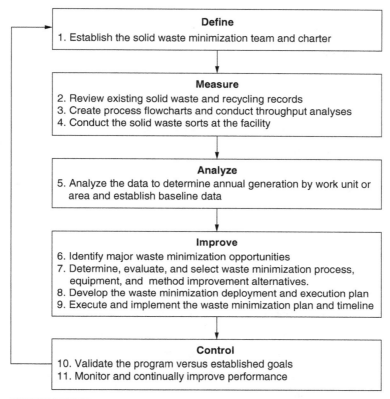

Figure 8.1   Eleven-step Six Sigma approach to solid waste minimization.

- A letter of support from top management to all employees
- Team leader and member identification
- Initial training for the team
- Project goal identification including metrics
- Team charter
- Project timeline
- Project budget

8.2.1 UPPER MANAGEMENT SUPPORT

Upper management support is one of the most important components for any successful project. A letter, memo, or e-mail from the organization president, CEO, or facility manager to all employees is a great way to set the tone of the project and to create awareness among all employees. A clear message must be sent that solid waste minimization is an important focus of the company and that all employees are requested to take part and contribute. Figure 8.2 shows a sample letter that may be used.

January 31, 2010

Memo to all facility employees

Dear coworkers:

As president of the organization, I am pleased to state that we have made a strong, unified commitment to collaboration and cooperation in increasing our recycling levels and reducing waste levels. This commitment has been demonstrated in funded projects, such as our aluminum can–recycling program and the hiring of management and staff to support related environmental areas.

In light of the recent successes and focus on environmental protection, we have placed increased emphasis on unifying our recycling efforts between throughout the organization. As a first step to accomplish the goal, we are forming a solid waste minimization team comprised of your peers to evaluate our current performance and implement process changes to increase our recycling levels and reinforce our focus on protecting the environment. This project will allow us to move forward with these key initiatives and any input that you have is highly welcomed and appreciated. If you have any input, please contact your immediate supervisor to share your thoughts.

In conclusion, we have provided strong support for recycling and waste reduction. We are very pleased with the success of our collaborative efforts at our headquarters office and intend to implement a strong program at our facility. Our continued commitment to the environment will sustain the efforts of this program so that we may become a true environmental leader in our field.

Sincerely,

President and CEO

Figure 8.2   Sample letter.

8.2.2 THE PROJECT TEAM

Selecting the team to lead the solid waste minimization project should not be taken lightly. Thought should be put into selecting a cross-functional team that has process knowledge and good interpersonal skills. The team leader should be in a management position at the facility. This will allow for faster communication with management and a higher degree of authority within the team and throughout the facility. Efforts should also be made not to select an overburdened manager, the team leader must have adequate time that he or she can devote to the project and lead all meetings.

The team itself should be comprised of five to six core members representing different areas in the facility, including both management and hourly employees. For example, in a manufacturing plant, a janitorial worker, an engineer, a maintenance technician, a production supervisor/manager, a line employee, and an accountant would be a good mix. The team members should exhibit concern for the environment, possess strong interpersonal skills, and be well respected among their peers. The team will serve as the internal cheerleaders for the project to help build excitement, participation, and awareness. In addition, a technical expert within the company or contracted as a consultant is a very value-added inclusion in the team and highly recommended. The team will perform the roles discussed in Chap. 5. Temporary team members may be added to the project to assist in data collection during the waste sort discussed later in the chapter. For the example case study, the team was lead by an engineering manager and included a consultant hired on an on-needed basis, a line worker, a custodian, an accountant, a process engineer, a safety captain, and a production supervisor. This cross-functional team collectively possessed a strong understanding of all core and support operations within the facility.

8.2.3 INITIAL TRAINING AND INTRODUCTORY MEETING

The team will require initial training before beginning the project. The goals should be clearly outlined, including the metrics, timeline, and budget. The team should also have the opportunity to give input to the team charter and help in its development. The purposes of the training and introductory meeting are to

- Introduce the team and exchange contact information/work schedules
- Establish clear SMART goals (simple, measurable, agreed to, reasonable, and time-based)
- Define initial team member roles
- Finalize the budget
- Finalize the timeline
- Create the team charter

The technical expert should take the lead role in this process to ensure an adequate training and a clear understanding of the project goal and the path ahead. The training itself should focus details on the 14-step process mentioned in Table 8.1.

8.2.4 PROJECT, GOALS, AND METRICS

The goals of the project should be clearly expressed as soon as possible. This will provide the team with much needed direction and serve as the gauge to evaluate all team activities and accomplishments. The goals should provide a specific direction for the project and not vague generalized improvement slogan such as "to become an environmental leader in the automobile engine manufacturing field" or "to reduce the organization's carbon footprint." A more specific, SMART goal, such as the following should be set:

■ Reduce the amount of solid waste generated per year by 5 percent from the baseline year of 2009.
■ Increase the recycling rate for metals and paper products by 10 percent from the baseline year of 2009.
■ Utilize environmental improvements as a strategic weapon to provide a cost-benefit of 10 percent versus the baseline year of 2009 for expenditures and revenues solid waste removal and recycling efforts.

For the case study project, the goal established was to reduce solid waste generation by 7 percent over a 12-month period versus the same period last year without increasing operating or disposal costs. Such a clear goal provides a specific target for the team and the constraints to achieve it. In this case, the metrics were solid waste tons generated and dollars.

8.2.5 TEAM CHARTER

The team charter is a statement of the scope, objectives, and participants in a project. It provides a preliminary description of roles and responsibilities, outlines the project objectives, identifies the main stakeholders, and defines the authority of the project manager. It serves as a reference of authority for the future of the project.

The purpose of the team and project charter is to document:

■ Reasons for undertaking the project
■ Objectives and constraints of the project
■ Directions concerning the solution
■ Identities of the main stakeholders

The three main uses of the project charter:

1 To authorize the project—Using a comparable format, projects can be ranked and authorized by return on investment.
2 As a primary sales document for the project—It provides stakeholders with a one- to two-page summary to distribute, present, and keep handy when fending off other projects that may attempt to consume the allocated resources.
3 As a focus point throughout the project—The charter may be used as an introductory document for new team members, provide a focal point during team meetings, and use in control or review meetings to ensure tight scope management.

For the case study, a sample team charter is shown in Fig. 8.3.

8.2.6 PROJECT TIMELINE

The project timeline is one of the key deliverables from the define stage. The timeline tracks project performance versus established goals and serves as the strategic

Solid Waste Minimization Team Charter

Purpose Statement and Team Objectives:

The mission of the team is to reduce solid waste emissions by 7 percent versus 2006 generation without added expenses. The team is committed to working effectively together by monitoring process effectiveness, following through on commitments, and helping one another learn. The mission will be accomplished with the following objectives:

- To complete the 14-step assessment process for the company by 31 March.

- To prepare first draft proposals, and present to CEO by 15 April.

- To refine proposals, and present to regional management meeting on 25 April.

- To present the final plan to the CEO by 15 May.

Team Composition and Roles

The team will be made up of representatives from each functional area in the plant including management and craft employees. This range of skills and knowledge will enable the team to understand the issues relating to individual countries, as well as developing solutions to the problems outstanding.

Jack Smith will take the role of team leader. In this role he is responsible for:

- Ensuring this team charter is abided by;

- Managing the day-to-day operations of the team and the team's deliverables;

- Managing the budget;

- Providing support and assistance to individual team members; and

- Providing status reports to the CEO on a weekly basis.

Authority and Empowerment

Jack, as team leader, has the authority to direct and control the team's work, and team members are allocated full time to this project, for its duration.

Resources and Support Available

A budget of $4000 is available to cover travel and initial supplies to launch the program. The CEO will meet with Jack Smith at 4:30 p.m. every Monday afternoon for a progress update and to provide support and coaching appropriately.

Operations

- The first team meeting will be on Monday, 28 February at 2:00 p.m.

- The team will meet every Monday afternoon from 2:00 p.m. to 3:30 p.m. for the duration of the project.

- Each member is expected to present a short status report for the aspect of the project they are working on.

- If a member is unable to attend, a notification must be sent to the team leader and someone else designated to report on the status and communicate further expectations.

- A summary of each meeting will be prepared by Jack and emailed to all members by the morning following the meeting.

Figure 8.3   Sample team charter.

TABLE 8.1   PROJECT TIMELINE

SOLID WASTE MINIMIZATION STEP	TIME (WEEKS)
1. Establish the solid waste minimization team and charter	2
2. Review existing solid waste and recycling records	2
3. Create process flowcharts and conduct throughput analyses	2
4. Conduct the solid waste sorts at the facility	2
5. Analyze the data to determine annual generation by work unit or area	2
6. Establish baseline data	1
7. Identify waste minimization opportunities	6
8. Develop process and method improvement alternatives to minimize solid waste	6
9. Determine vendors and service providers to assist in waste minimization	3
10. Compare and decide among alternatives	1
11. Develop the waste minimization deployment and execution plan	2
12. Execute and implement the waste minimization plan and timeline	4
13. Validate the process versus goals	1
14. Monitor and continually improve performance	1
Total	35

implementation plan. The timeline should be viewed as a control document to evaluate the progress of the team versus preestablished milestones. Proper planning is needed to ensure the timeline is achievable and will meet the goals of the project. Table 8.1 shows some general guidelines for the time required for each step of the solid waste minimization process. Please note that the timeline assumes that the team will be devoting approximately 30 percent of their time to the project and 70 percent of their time to normal job duties. On average a successful project requires approximately 3 to 4 months to complete. Table 8.1 shows the timeline created for the case study.

8.2.7 PROJECT BUDGET

The final deliverable from the define stage of the process is the budget. The budget identifies the financial resources available to the project for process improvements and data collection assistance. The budget could include such items as

- Training and reference materials
- Outside training
- Data collection and trash sort tools (gloves, yard sticks, scales)
- Data collection labor
- Equipment purchases (baler, grinders)

- Container purchases (desk side and workroom floor receptacles)
- Communication funds (newsletters, banners, posters)
- Travel funds to visit top performing sites for benchmarking
- Meeting refreshments (if desired)

Several approaches exist when creating the budget. For example, project start-up funds could be allocated by management and spent at the team's discretion. Management could allocate $1000 to $10,000 for the items listed above and empower the team to best utilize the funds. Many successful companies have taken the approach that no additional dollars will be allocated to the project. The team is given the base budget of the previous year's solid waste removal and recycling costs for a start. The team must then creatively use this same budget to reduce the environmental impact of the organization. All expenditures would be cost justified based on return on investment and available funds in the budget. This approach is generally preferred by management, because no additional funds are needed and the outcome of the project at a minimum will be breakeven for the organization. As a first step in reviewing previous waste removal records, the team determines the budget by calculating last year's expenditures. For the case study example, the team was given $4000 by executive management to cover tools and program start-up costs.

8.3   Step 2: Existing Record Review

Reviewing the existing organizational records for solid waste and recycling usually provides significant insight into the amounts, types, and patterns of waste generation. Very useful data and information can be gained to help focus the efforts of the team and eliminate the need for the collection of existing raw data. Collecting high-quality records will save a great deal of time, money, and effort versus raw data collection via a trash sort or facility walk through. The types of records to collect include

- Purchasing, inventory, maintenance, and operating logs
- Supply, equipment, and raw material invoices
- Equipment service contracts
- Repair invoices
- Waste hauling and disposal records and contracts (including 1 year of amounts and fees collected)
- Contracts with recycling facilities and records of earned revenues from recycling
- Major equipment list
- Production schedule (representative of a year)
- Company brochure or product information
- Material safety data sheets (MSDS)
- Facility layout (hard copy plus CAD copy, if available)
- Process flow diagrams

The purpose of reviewing these records is to determine the total amount of waste generated annually, the total amount of material recycled annually, and the total waste removal and recycling cost structure. A review of purchasing records can also be beneficial to backtrack into an estimate of waste components generated. For example, shipments often arrive in cardboard containers. By researching the number of containers received per year and estimating the weight per carton, the team could estimate the total weight of cardboard boxes disposed each year. Often times, the janitorial staff, maintenance, purchasing, and accounting staff will be most useful when gathering these records. In addition, customer service at the waste hauling and recycling companies may record more detailed records than the information that appears on monthly invoices.

The team should be aware of several limitations that exist from the existing records, such as

■ The records may not provide adequate data if the waste hauling company does not collect this information.
■ The records do not usually provide information about specific waste components (a trash sort is generally needed for this data).
■ The records can be difficult to use if more than one business shares a dumpster.

Figures 8.4 through 8.13 serve as the data collection worksheets for the existing record review. Included in these worksheets are the recommended documents to collect.

8.4   Step 3: Process Mapping and Production Analysis

The goal of this step is to aid the team in fully understanding the business processes and capabilities of the facility. An understanding of these processes is crucial in developing alternatives to reduce solid waste. A process flowchart is a hierarchical method for displaying processes that illustrates how a product or transaction is processed. It is a visual representation of the work flow either within a process or an image of the entire operation. Process mapping comprises a stream of activities that transforms a well-defined input or set of inputs into a predefined set of outputs.

A well-developed process flowchart or map should allow people unfamiliar with the process to understand the interaction of causes during the work flow and contain additional information relating to the solid waste minimization project (such as tons of waste generated per year and annual cost of disposal).

To create a process map, it is important to determine the start and stop points because you will create the process map between those points. The SIPOC (supplier input process output and customer) tool can help. The tool name prompts the team to consider the supplier (the S in SIPOC) of the process, the inputs (the I) to the process, the process (the P) the team is improving, the outputs (the O) of the process, and the customers (the C) that receive the process outputs.

Solid Waste Minimization Records Review

In order for the waste assessment team to begin the waste assessment, please compile the following items and complete the attached questionnaire. If additional space is needed than what is provided on the questionnaire, please attach sheets.

Please provide the following:

1. Major equipment list

2. Production schedule (representative of a year)

3. Company brochure or product information

4. MSDS sheets

5. Facility layout (hard copy plus CAD copy, if available)

6. Process flow diagrams

7. Receipts from each waste hauler (representative of 1 year)

8. Please indicate which safety items the waste assessment team will need to enter your facility:

    ☐ Safety glasses
    ☐ Steeltoe shoes
    ☐ Hard hats
    ☐ Hearing protection
    ☐ Other _____

You will most likely need the following information to complete the attached questionnaire:
1. Terms of disposal for waste streams (e.g., cost, collection schedule, and size).
2. Waste hauler and recycler information, including hazardous waste hauler information.
3. Production and scrap information for your products.
4. Information about end-receptacles, such as capacity, rental cost, pull cost, etc. It would be best to make a chart listing the end receptacles, location, and associated information. Attach chart to this questionnaire.

Figure 8.4   Existing records review worksheet—page 1.

Section 1: Company Information

Date: _____

1. Company name: _____
    Address: _____
    Company contact: _____
    _____
    Phone:_____ Fax: _____
    Purchasing department contact: _____ phone: _____
    Janitorial services contact: _____ phone: _____
    Engineering department contact: _____ phone: _____
    Environmental services contact: _____ phone: _____
2. Primary SIC code: _____ Secondary SIC code (s): _____
3. Number of shifts: _____ Shift times: To:
    Employees per shift: First: _____   First: _____ - _____
                            Second: _____   Second: _____ - _____
                            Third: _____   Third: _____ - _____
    Total employees: _____
    Days per year the plant is in production? _____
4. What is the labor and benefit rate for hourly employees? (if necessary, break out classifications)
    _____

5. Square footage of facility: _____
    What is the cost per square foot of production space or warehousing space? _____
6. What is the number of products produced last year? (breakdown by product) _____
    _____
7. What is the company's minimum attractive rate of return (MARR) and tax rate?_____
    _____
8. What are the approximate annual sales for your company?_____

9. What are your company's primary goals for this waste assessment?

    Free service: ☐ Yes   ☐ No

    Minimization of disposal costs: ☐ Yes   ☐ No

    Minimization of raw material/inventory costs: ☐ Yes   ☐ No

    Other: _____

10. Has a solid waste assessment ever been conducted before?   ☐ Yes   ☐ No

Figure 8.5   Existing records review worksheet—page 2.

Section 2: Process Information

1. Briefly list your company's products and/or services: _____
_____

2. How would you rate current levels of production and the corresponding waste streams?
   ☐ Average
   ☐ Above Average
   ☐ Below Average

3. Please explain any reasons for this variation in production or production trends: _____
_____

4. Please indicate all areas which are present at this location:
   ☐ Storage   ☐ Research and Development
   ☐ Design   ☐ Shipping/Receiving
   ☐ Lab Work   ☐ Manufacturing
   ☐ Office   ☐ Fabrication
   ☐ Retail   ☐ Inventory/Warehousing
                        ☐ Other_____

5. Please indicate all manufacturing processes currently at this location:
   ☐ Anodizing   ☐ Grinding   ☐ Pickling
   ☐ Coating   ☐ Heat Treating   ☐ Polishing
   ☐ Blending   ☐ Kiln Firing   ☐ Printing
   ☐ Brazing   ☐ Machining   ☐ Rolling
   ☐ Cleaning   ☐ Milling   ☐ Shearing
   ☐ Decreasing   ☐ Mining   ☐ Slitting
   ☐ Electroplating ☐ Molding   ☐ Stamping
   ☐ Etching   ☐ Painting   ☐ Welding
   ☐ Extruding   ☐ Paint Strip   ☐ Other _____

6. Which processes are responsible for the majority of your defects? _____
_____

7. What is the scrap rate (per product)? (for instance, for each10 lb of product produced, 1 lb of material is scrapped).
_____

8. Describe how this figure was determined? _____

9. What percent of internal reject and rework is disposed? _____

10. How are rejects processed?_____
_____

Figure 8.6   **Existing records review worksheet—page 3.**

Section 3: Materials Receiving and Handling

1. Do you have a method for quantifying your incoming raw materials?　☐ Yes ☐ No
   Please explain: _____
   _____

2. How is incoming material packaged? _____
   _____

3. Please specify any materials you receive which have a limited shelf life: _____
   _____

4. How is scrap or offal handled? Is it disposed or recycled? _____
   _____

   Are there separate receptacles for scrap or offal: What is the size of the receptacles and how often is it
   pulled or disposed? _____
   _____

   If material is recycled, who is the recycler and what revenue per pound does your company receive for
   the material? (Provide receipts from the recycler equivalent to one year)_____
   _____

   What are the specifications of the material? (type or grade, size of scrap pieces, etc.) _____
   _____

5. Does any of your received material arrive on pallets?　☐ Yes ☐ No
   If yes, what type and size of pallets? _____
   What is the final use or destination of received pallets? _____
   _____

   Where (company name and address) are the pallets disposed? _____
   _____

   Does your company pay for the disposal of the pallets and do you receive a revenue for the pallets?
   _____

   Does the hauler provide the transportation and in what condition are the pallets?_____
   _____

   How many pallets are disposed per month? (List the quantity of each size disposed.) _____
   _____

   Is there a special receptacle for pallets only? (Please describe its dimensions.) _____

6. How are your supplies received, stored, and distributed throughout the facility?_____
   _____

7. How many personnel are utilized in the shipping, receiving, and distribution of supplies?_____
   _____

8. Please list any materials which are not handled by the conventional material handling system:
   _____
   _____

   How are these unusual materials handled? _____
   _____

9. Describe the extent to which returnable containers, cardboard recycling, and other waste reduction
   measures are used during material handling: _____
   _____

10. Is plastic returnable dunnage ever disposed of by your company? _____
    Where is it disposed? _____
    Who is the manufacturer (name and address) of the plastic dunnage? _____
    How much is disposed of per month, and please list the specification of the dunnage? (size, plastic type,
    etc.) _____

Figure 8.7　Existing records review worksheet—page 4.

11. How much office paper is purchased per month? Designate paper type and quantity in pounds and designate whether any of it is recycled or recycled content: _____
_____

12. Is cardboard recycled by your company? State recycler name and address: _____
_____
Does your company receive revenue from cardboard recycling? If yes, provide receipts from the recycler equivalent to 1 year. _____
Is there a designated receptacle for cardboard only? Please describe its dimensions and the frequency it is pulled: _____

13. Is wood recycled or disposed by your company? _____
Does your company receive revenue from wood recycling? If yes, provide receipts from the recycler equivalent to 1 year. _____
Is there a designated receptacle for wood only? Please describe its dimensions: _____

14. How do material handling and purchasing units communicate with one another? _____
_____
Please describe any coordination efforts between purchasing and material handling units to minimize waste: _____

Figure 8.8   Existing records review worksheet—page 5.

Section 4: Packaging and Shipping

1. Are empty containers returned to the supplier?   ☐ Yes   ☐ No
2. Do you buy pallets for shipping?   ☐ Yes   ☐ No
   If yes, how many pallets do you purchase per weeks or per month?
   _____
3. How is your final product packaged?_____
   _____
   What material do you use as packaging aids?_____
   _____
   What is the average weight of your final package for shipping? _____
4. Are any of the packing materials recycled products?   ☐ Yes   ☐ No
   If yes, please list: _____
5. What percentage of shipped product is returned by the customer? _____
   What percentage of these customer returns must be disposed of? _____
   Approximately what percentage of product by weight is shipped to the following range of distances?
             0–50 miles: _____
             50–100 miles: _____
         more than 150 miles: _____

Figure 8.9   Existing records review worksheet—page 6.

Once you have determined the beginning and ending activity steps, start mapping what is done between the two. Some key points to process mapping include

- ■ Keep it simple.
- ■ Start at a high level first.
- ■ Involve the people closest to the process.
- ■ Walk through the process yourself.

Section 5: Solid Waste

1. Please describe any current difficulties or obstacles in waste handling and collection: _____
_____
2. Please list the major components and annual component volumes or weight of your solid waste (e.g., wood, cardboard, mixed office paper): _____
3. How many in-plant receptacles do you have for trash? _____
Please list each end container for disposing waste. End containers include (but are not limited to) open top containers, compactors, and gaylords.

	Capacity	Times pulled (per week/month)	Rental cost (per month)	Cost per pull
End receptacle	_____	_____	_____	_____
End receptacle	_____	_____	_____	_____
End receptacle	_____	_____	_____	_____
End receptacle	_____	_____	_____	_____
End receptacle	_____	_____	_____	_____
End receptacle	_____	_____	_____	_____

On what day of the week are your end receptacles emptied? _____
4. Who is your solid waste hauler(s)? _____
What is our monthly solid waste pickup fee? _____
How are you billed for your solid waste pickup?

☐ By weight
☐ By volume

5. Is each department responsible for the cost of the waste they generate? ☐ Yes ☐ No

Figure 8.10   Existing records review worksheet—page 7.

- ■ Think end to end.
- ■ Work with a small group of three to seven people. A larger group can make the activity unwieldy.

Following are the three general steps to create a process map:

1 Begin by stating your intention to create a process map for activities completed between start (step A) and stop (step B).
2 Document the activities that people do between steps A and B on self-adhesive notes. Attach the notes to a wall, flip chart, or whiteboard in the order in which they are normally completed. Begin your documentation at a high level and then move into additional process maps that provide greater detail.
3 Document the process maps. Schedule your next meeting to review the information. Make sure that you verify and clarify the activities and their owners. Also, look for any immediate opportunities to create quick wins.

In order to identify areas for improvement, processes must be broken into subprocesses. A subprocess operates at a more detailed level than a core process and gets into who does what and why on a daily basis.

Section 6: Waste Reduction Activity

1. Please describe your current recycling efforts:_____
   _____

2. What percentage of waste is currently recycled in your facility?
   _____
   What is your current recycling revenue? _____

3. Do you reuse any scrap materials internally?   ☐ Yes   ☐ No
   If yes, please describe: _____
   _____

4. Please list your material recyclers: _____
   _____

5. Are there hand dryers in the restroom? _____
   How many cases of paper towels are used per month? _____
   What is the cost per case?_____
   How many rolls (or packs) of paper towels per case?_____
   How much does each (or roll/pack) weigh? _____
   How many restrooms are in the facility? _____
   List each restroom with the number of wash basins in the restroom: _____
   _____

6. Does your company use disposable cloth rags or reusable rags on the shop floor? If yes, continue. If no
   go on to the next question.
   For disposable rags, list the size and weight of the rags, the quantity used per month and the cost to
   purchase the rags. _____
   For reusable rags, list the company (launderer) name and address, and the terms of the contract. List
   how many rags are laundered per month, and how many have to be repurchased. Provide receipts from
   the launderer equivalent to 1 year. _____
   _____

7. If paper towels are used on the shop floor, please designate how floor paper towels are different than
   paper towels used in the restrooms/break room: _____
   How many cases of paper towels are used per month? _____
   What is the cost per case?_____
   How many rolls (or packs) of paper towels per case?_____
   How much does each (or roll/pack) weigh? _____

8. Does your company provide gloves for use by employees?_____
   What is the composition of the gloves (leather, canvas, etc.)? _____
   What is the quantity of gloves used per month? _____
   Are the gloves disposed or laundered and reused? _____
   If laundered, what is the average number of wears per glove? _____
   What are the repair and laundering costs per month? Provide receipts from the launderer equivalent to
   1 year. _____
   If disposed, what is the monthly cost for gloves? _____

9. Are aluminum pop cans generated at your company? _____
   Are the cans recycled? _____
   How many pounds are recycled or disposed per month? _____
   How and by whom are recycled aluminum cans handled?_____
   Is there any aluminum process scrap generated at your company? If so, how many pounds per month?
   _____
   What revenue does your company receive from recycling aluminum? Provide receipts from the recycler
   equivalent to 1 year:

Figure 8.11   **Existing records review worksheet—page 8.**

10. Is there a cafeteria or break room in your facility? If yes, are paper towels used here?
_____

    Does your company provide napkins, plastic flatware, cups, etc. in the break room? _____

    Are there vending machines in the breakroom? If yes, do they offer food in plastic or plastic clamshell containers? _____
11. Does your company use a sweeping compound for floor spills? _____
    What is it called and the name and address of the manufacturer of the sweeping compound? _____
    _____
    How many units (bag/barrel, etc.) of sweeping compound are purchased per month? _____
    What is the weight and dimensions of the unit? _____
    Is there a reason why this sweeping compound has been chosen (such as safety)? Explain.
    _____
    _____
    How is the sweeping compound disposed? _____
    What are the contaminants in the used sweeping compound? _____
12. Please describe any building features which will favor or hinder recycling and waste reduction efforts (e.g., loading docks, waste collection system): _____
    _____
13. Where are potential spaces for location of recycling containers and waste reduction equipment within your facility? (e.g., housekeeping closets, next to existing receptacles, near loading docks)
    _____
14. Describe any special concerns which might affect implementation and/or continuation of waste reduction activities: _____
    _____

Figure 8.12   Existing records review worksheet—page 9.

1. Does your company have a written environmental policy?   ☐ Yes   ☐ No
   If yes, does this policy include solid waste prevention?   ☐ Yes   ☐ No
2. Is there a solid waste prevention program in place?   ☐ Yes   ☐ No
   Is there a quality management program in place?   ☐ Yes   ☐ No
   Is there an employee involvement program to improve operations? ☐ Yes   ☐ No
   If yes, please explain your employee involvement program: _____
   _____

3. What are the priority areas where solid waste prevention efforts should be focused? _____
   _____
   _____

4. Please indicate any barriers to the procurement of recycled materials:
   ☐ Inadequate performance
   ☐ Corporate policy
   ☐ Engineering specifications
   ☐ Price
   ☐ Other _____

5. Please indicate any current procurement policies used by your company:
   ☐ Recycled content products
   ☐ Reusable containers
   ☐ Recyclability of materials concerns
   ☐ Environmental friendliness
   ☐ Minimum packaging
   ☐ Other _____

Comments:

Figure 8.13   Existing records review worksheet—page 10.

SUMMARY		
CATEGORY	NO.	DIST.
OPERATIONS		
TRANSPORTS		
INSPECTIONS		
DELAYS		
STORAGES		
TOTALS		

PROCESS DESCRIPTION:
COMPANY/LOCATION:
RECORDED BY:

DATE:

STEP	OPERATION	TRANSFER	INSPECTION	DELAY	STORAGE	DESCRIPTION OF METHOD	DISTANCE	CYCLE TIME
	○	⇨	□	D	▽			
	○	⇨	□	D	▽			
	○	⇨	□	D	▽			
	○	⇨	□	D	▽			
	○	⇨	□	D	▽			
	○	⇨	□	D	▽			
	○	⇨	□	D	▽			
	○	⇨	□	D	▽			
	○	⇨	□	D	▽			
	○	⇨	□	D	▽			
	○	⇨	□	D	▽			
	○	⇨	□	D	▽			
	○	⇨	□	D	▽			
	○	⇨	□	D	▽			
	○	⇨	□	D	▽			
	○	⇨	□	D	▽			
	○	⇨	□	D	▽			
	○	⇨	□	D	▽			
	○	⇨	□	D	▽			
	○	⇨	□	D	▽			

Figure 8.14　**Process flowchart form.**

Core processes must be broken down into enough detail to understand, monitor, manage, and analyze performance. As a general rule, processes must be described using the three general steps to create a process map dissussed in the previous paragraph before improvement teams are able to deal with them adequately. Figure 8.14 shows a data collection form that can be used when creating the process map and flowchart.

Figure 8.15 shows the process charts that were created during the case study example.

The primary purpose of the production analysis is to determine annual production rates (or capacities) and related cyclical patterns. These production rates will relate directly to the amount of waste generation by each cycle, whether the cycle is a week, month, or quarter. Figure 8.16 shows a capacity model worksheet that could be used to aid in this process. To calculate the capacity, the following information will be needed:

■ Annual forecast
■ Cycle time

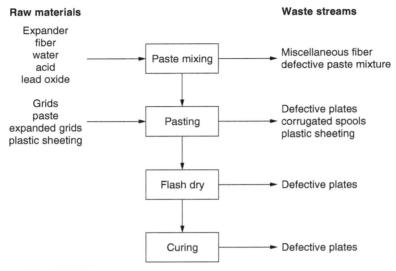

Figure 8.15   Process map example.

■ Efficiency (versus cycle time)
■ Time available
■ Utilization (uptime of the machine)

For the case study example, the company was capable of producing 25,000 batteries per year, with a cycle time of 8.4 minutes per battery over 250 production days per year operating in two 7-hour shifts (8-hour shifts minus 1 hour for breaks).

Company Name
Title (capacity model)
Location (city, state)

Date:

Production Parameters	
Days per year	233
Hours per shift	8

Production area		Time Required						Time Available			
Item #	Operation	Annual Forecast (units)	Daily Forecast (units)	Cycle Time (minutes)	Parts per Cycle (pcs/cycle)	Efficiency (percent)	Hrs Reqd Per Day (hrs/day)	Shifts Per Day (shifts/day)	Utilization (percent)	Hrs Avail Per Day (hrs/day)	Loading (percent)
1.0	LATHE	100,000	429	1.10	2	90%	4.37	2	90%	14.40	30.36%
2.0											
3.0											
4.0											
5.0											
6.0											
7.0											
8.0											
9.0											
10.0											
11.0											

Figure 8.16   Production analysis example.

8.5   Step 4: Solid Waste Sorts

Solid waste sorts provide detailed data regarding the composition of an organization's waste stream. Via the data collected from the solid waste sorts, the organization's waste stream can be characterized into the various materials that comprise the entire stream, including the annual amounts generated and the percentage that each component contributes to the entire waste stream. These data are invaluable when evaluating cost-effective methods or process changes to divert these components from landfills. The waste sort itself is affectionately referred to as dumpster diving, since the team will physically collect, sort, and weigh a representative sample of the organization's waste. The remainder of this section discusses the process to conduct a waste sort, including the required preparation, tools required, a step-by-step guide, and also provides data collection forms.

8.5.1   PREPARATION

Before scheduling the actual waste sort, some preparation work is required, including a team meeting to discuss several items to ensure a smooth project. Specifically, the following items should be addressed and finalized prior to the sort:

- Hold a team meeting
    - Create a waste assessment plan and assign roles and facility areas of responsibility
    - Conduct waste sort training
    - Review a layout of the facility
- Gather the needed tools and equipment
- Assign data collection roles to temporary team members
- Prepare the data collection forms
- Hold a custodial staff meeting
- Disseminate employee messaging regarding the sort and what to expect

The team meeting prior to the waste sort should be held to get the team on the same page. The primary outcomes from this meeting are training and creating a waste sort plan. The training should be conducted by the technical expert and focus on the use of tools and data collection form. The waste sort plan should assign team members to the various areas of the facility to conduct data collection effectively and efficiently. The equipment required for the waste sort includes

- Gloves
- Yard sticks
- Plastic bags
- Scales
- Clipboards (with the data collection form)

Based on the size of the facility, it may be necessary to recruit additional support to collect data during the waste audit. If additional help is used, these individuals should receive the same training as the core team. An approach that seems to work well is to assign one temporary team member from each work unit within the facility. For example, assign a supervisor or shop floor worker from each production work area such as the metal stamping unit, the paint shop, and the accounting offices (work units will differ by business type). The advantages of assigning temporary team members within each work unit include faster and more accurate data collection. As a member of the work unit in which data are being collected, the temporary team member will possess specialized knowledge on the waste generation types and amounts.

The data collection form is the most important document of the waste sort. The data collected with the form will be used to extrapolate the annual generation for the facility so care should be taken when collecting the data to ensure accuracy. At a minimum, the required information on the data collection form is

- The date the data were collected
- The team members collecting the data (this is very useful if follow-up or clarification is needed)
- Work unit and location of the waste receptacle
- Source of the waste (previous operation or supplier)
- Disposal method (baler, compactor, recycler, uncompacted dumpster)
- Size of the container in cubic feet (can be derived from length, width, and height measurements)
- The container type (desk side, recycling bin, dumpster)
- Percent full
- Times emptied (per day, week, or month)
- Container contents and percent of each component
- Condition of material (loose, compacted, baled)
- Notes and comments that may be useful when analyzing the data (including names and contact information for work units members with specialized knowledge of the waste or generation levels)

Figure 8.17 shows a template of the data collection form.

A meeting should be held with the maintenance/custodial supervisor and custodial team to clarify the scope and support required for the waste sort. On the day of the waste sort, the custodial staff should be instructed not to empty any waste containers so that a full day's data can be collected. Notes and messages should be placed in the custodian's work and break areas the day before and the day of the waste sort to ensure that all waste is accounted and measured. During the meeting, the custodial team should be made aware of the goal of the project and the purpose of the waste sort. Oftentimes, the custodial team has value-added suggestions and comments that will improve the study, such as the areas to target and high-volume waste streams.

Finally, several days before the waste sort, messaging in the form of e-mails, service talks, and postings in common areas should be disseminated to all facility employees discussing the details of the waste sort. The messaging should include the date, the purpose, and the process of the waste sort.

Solid Waste Analyses and Minimization Project Waste Sort Data Collection Form			
Unit name		Date of study	
Unit location		Performed by	
Source		Disposal method	
Waste Composition			
Container type		Percent full	
Container location		Times emptied	hour/day/week
Container size		Percent recycled	
Container Contents			
Item	**Percent of Container**	**Item**	**Percent of Container**
Aerosol cans		NCR (carbonless paper)	
Aluminum		Newsprint (ONP)	
Boxboard		OCC	
Carbon paper		Paper	
Cardstock		Paper cups	
Cloth rags		Polyethylene	
Computer printout (CPO)		Polypropylene	
Envelopes		Paper towels (wash/breakroom)	
Film plastic		Paper towels (shop floor)	
Floor sweepings		Plastic banding	
Food waste		Plastic bottles	
Glass		Plastic clamshell containers	
Gloves—leather		Plastic cups	
Gloves—rubber		Safety masks	
Gloves—cloth		Shredded plastic	
Greenbar		Steel	
Kraft paper		Styrofoam	
Label backing		Tape	
Labels		Terry sleeves	
Metal		Wire (list metal type if known)	
Metal banding		Wood	
Miscellaneous (Describe in Notes)		VCI paper	
Notes			

Figure 8.17   Waste sort data collection form.

May 15th

Memo to All Employees:

On May 22nd the solid waste analysis and minimization team will conduct a waste audit at our facility between 2 p.m. and 5 p.m. The purpose of this waste audit is to determine the amount and composition of our waste stream to help us become more environmentally friendly by reducing, reusing, and recycling. On the day of the assessment, please do not empty any waste containers in your work unit. Also, if you have any ideas or suggestions to help us become more environmentally friendly, please share them with the assessment team on May 22nd.

Thank you for your support.

President and CEO

Figure 8.18   Sample service talk.

This will help ensure a smooth waste sort by informing employees to keep all trash available to measure and prevent data collection delays resulting from the team continually being asked who they are and what they are doing. In addition, the messages could solicit suggestions for improvement and employee comments. By sending the messages earlier, it will give the employees time to think about waste generation in their work units and provide more value-added feedback. Figure 8.18 displays a sample service talk that could be used to inform all employees.

8.5.2 WASTE SORT GUIDE

The waste sort itself is straightforward; every individual waste and recycling receptacle in the facility is analyzed and a data collection form is completed. During the walk-through, the assessment team will gather raw data on what they observed as well as record the location and the main purpose of each dumpster. The team also examines the company's main production, the waste flow in the facility, number of waste streams in the facility, and draws a process and material flow diagram. Ideally, if a layout of the facility is available, it may be used to segment the plant to collect data for the team members. In general, a team of two people is recommended for each mini team. This will allow one person to measure and sort and the other to record the data. Figure 8.19 shows a sample of a completed form.

8.6   Step 5: Data Analysis

The primary outcome of the analysis phase is the annualized waste generation baseline data for the facility. This baseline data should be broken down by the component waste stream (paper, metal, etc.), work unit of generation, and how the component waste stream is currently handled (landfill, recycled, burned, etc.). Each component should

Solid Waste Analyses and Minimization Project Waste Sort Data Collection Form			
Unit name	*Door assembly*	Date of study	*5/5/2007*
Unit location	*Column A7*	Performed by	*Bill K. and Brad C.*
Source	*Door preassembly*	Disposal method	*Uncompacted dumpster*
Waste Composition			
Container type	*Round 55 gallon*	Percent full	*85%*
Container location	*Assembly bench #4*	Times emptied	*1 per day*
Container size	*55 gallon*	Percent recycled	*0%*
Container Contents			
Item	**Percent of Container**	**Item**	**Percent of Container**
Aerosol cans	10%	NCR (Carbonless paper)	
Aluminum		Newsprint (ONP)	5%
Boxboard		OCC	
Carbon paper		Paper	10%
Cardstock	25%	Paper cups	
Cloth rags		Polyethylene	20%
Computer printout (CPO)		Polypropylene	
Envelopes		Paper towels (wash/break room)	
Film plastic	10%	Paper towels (shop floor)	
Floor sweepings	5%	Plastic banding	20%
Food waste		Plastic bottles	
Glass		Plastic clamshell containers	
Gloves—leather		Plastic cups	
Gloves—rubber	5%	Safety masks	
Gloves—cloth		Shredded plastic	
Greenbar		Steel	
Kraft paper		Styrofoam	
Label backing		Tape	
Labels		Terry sleeves	
Metal		Wire (List metal type if known)	
Metal banding		Wood	
Miscellaneous (Describe in Notes)		VCI paper	
Notes			

Figure 8.19   Completed waste sort data collection form.

be given in terms of weight (tons per year) and volume (cubic yards per year). The key questions that are answered from this analysis are

- What are the waste streams generated from the facility and how much?
- Which processes of operations do these waste streams come from?
- Which waste streams are classified as hazardous and which are not? What makes them hazardous?
- What are the input materials used that generate the waste streams of a particular process or facility area?
- How much of a particular input material enters each waste stream?
- How much of a raw material can be accounted for through fugitive losses?
- How efficient is the process?
- Are any unnecessary wastes generated by mixing otherwise recyclable hazardous materials with other process wastes?
- What type of housekeeping practices are used to limit the quantity of wastes generated?
- What types of process controls are used to improve process efficiency?
- How much money is the company paying to dispose of solid waste, and how much revenue does it generate from the sale of recyclable materials?

To answer these questions and generate the baseline data, the existing records collected, the data gathered during the waste audit, and the team member knowledge will be used. Additional data collection or verification may be required during the analysis portion. The first step in the data analysis process is to summarize the collected data in one document. The most efficient manner to accomplish this is with the use of a linked spreadsheet. The data collected from the waste audit and records review are input into a linked spreadsheet to extrapolate annual generation and break down disposal costs and revenues. The information entered into this spreadsheet includes the data collected from each collection form during the waste audit. After the data are entered, the spreadsheet can be created to automatically calculate annual generation based upon predetermined density factors. The spreadsheet in Fig. 8.20 provides a format to enter these data and includes common density factors.

After all the data have been entered into the spreadsheet, annual generation in terms of weight and volume can be tabulated. To estimate the annual waste stream in terms of both weight and volume is very important because, in general, waste haulers charge based on volume (cubic yards that fill a dumpster) and processors pay for recyclable materials based on weight (tons in a bale). Following is an example calculation for cardboard (OCC).

A company reported using a dumpster of 12 yd^3 that was used exclusively for compacted OCC (cardboard) that was emptied two times per month by a recycling vendor. These data were converted into annual tonnage using the following equation:

$$\text{Tons per year} = (\text{dumpster size in cubic yards}) \times (\text{times emptied per month})$$
$$\times (12 \text{ months per year})$$
$$\times (\text{EPA average material density} - \text{tons/cubic yards})$$

$$\text{OCC} = (12 \text{ yd}^3) \times (2/\text{month}) \times (12 \text{ months/year}) \times (0.45 \text{ tons/yd}^3)$$
$$= 129.6 \text{ tons of OCC per year}$$

Stream or Container #1

Adjust density assumptions based on observations Density Factors (lb/cy)				Waste Material	Percent of Container or Stream	Material Density lb/cy	Recycle Percent	Recycled Volume CY/yr	Recycled Weight tons/yr	Disposed Volume CY/yr	Disposed Weight tons/yr
Loose	Stack	Comp	Baled	**Paper**							
200	350	500	700	Newsprint							
60	200	350	650	Corrugated cardboard							
60	350	500	700	Kraft paper							
60	200	450	450	Waxed cardboard							
				Office Paper							
300	450	650	750	Computer printout (CPO)							
300	450	650	750	White ledger (WL)							
250	450	650	750	Mix office pap (MOP)							
300	400	600	650	Ground wood CPO							
250	400	600	650	Cardstock/folders							
200	350	600	600	Box(gray)board							
300	550	650	700	Books							
300	600	700	800	Magazines & glossy							
150	400	550	650	Tissue paper							
250	400	500	600	Other mixed paper							
Loose	Stack	Comp	Grnd	**Wood**							
200	300	700	500	Pallets							
300	600	700	500	Lumber, etc.							
	700		500	Logs, etc.							
Loose	Stack	Comp		**Compostables**							
400		1000		Yard waste**							
150	200	400		Brush							
300		600		Soiled paper							
850		1200		Food waste							
500		800		Other compostables							
Loose	Stack	Baled	Grnd	**Metals**							
150	1000	1600	2000	Ferrous scrap							
50	400	500	400	Aluminum							
150		16	2000	Other nonferrous **							
Estimate by Conditions				**Plastics**							
15			100	PS foam							
150			500	PS rigid							
100			400	PE (nonfilm/nonfoam)							
100			400	PP (nonfilm/nonfoam)							
150			500	PVC (nonfilm/nonfoam)							
20			300	Foams							
25			1100	Film plastic							
				Other plastics							
Loose			Baled	**Textiles**							
200			400	Wool/cotton/linen							
100			400	Synthetic fabrics/fibers							
			Bulk	**Other Materials**							
			1200	Sludge							
				Miscellaneous							
**density varies greatly											
				TOTAL							

Figure 8.20 Data entry form template.

TABLE 8.2 ANNUAL WASTE STREAM EXAMPLE	
CURRENT WASTE DISPOSITION (LB)	
Disposed	1,029,670
Recycled	616,317
Generated	1,645,987
Recycling rate: 37.4%	

This leads to the next step, which entails calculating the waste removal costs and revenue generated by each waste component. The data to calculate this are taken from solid waste billing statements (collected during the existing records review) and the extrapolated waste stream. From the case study below is a typical summary (Table 8.2) for the extrapolated waste stream and disposal cost analysis.

At the time of the waste audit the company was paying approximately $75 per ton for disposal for a total cost of just over $38,900 per year. Table 8.3 summarizes production waste. The column heading titled "Annual Nonhazardous Generation" refers to the materials found in containers labeled "Nonhazardous." Likewise, "Annual Hazardous Generation" refers to materials labeled "Hazardous." Note that uncoated lead accounts for more than a quarter of production waste.

Table 8.4 illustrates nonhazardous and hazardous waste generated outside of production areas, but does not include waste generated in administrative areas.

Table 8.5 illustrates the administrative waste generated at the company. Office paper accounts for nearly 9 percent of the company's total waste stream comprised of production, nonproduction, and administrative waste. The types of office paper generated are summarized in Table 8.5.

The results of the research show that the company is currently generating approximately 10.1 lb of waste per employee on a daily basis. This figure does not include in-process recycled waste, such as uncoated lead, recycled OCC, and other hazardous by-products from the production process.

After analyzing the overall waste stream, the team determined that there are a number of major waste streams by weight. These 11 major streams and their subsequent generation are shown in Table 8.6; this represents the baseline data for the facility.

About 85 percent of waste generated by the plant can be categorized into one of the 11 major waste streams. In Sec. 8.7 alternative disposal and recycling systems for the recyclable waste streams will be discussed. Recycling waste steam material can result in reduced disposal costs and increased revenue due to the sale of recyclable material.

TABLE 8.3 PRODUCTION-RELATED WASTE

MATERIAL	ANNUAL NONHAZARDOUS GENERATION (LB)	ANNUAL HAZARDOUS GENERATION (LB)	PERCENTAGE OF PRODUCTION-RELATED WASTE STREAM
Uncoated lead	417,026	20,430	31.2
Oxide inserts	84,514	–	6.0
Polypropylene components	103,432	–	7.4
Miscellaneous	17,242	5993	1.7
Gloves	5109	3065	0.6
Plastic gloves	1088	2492	0.3
Safety masks	123	–	0.0
Terry sleeves	2886	–	0.2
Floor sweepings	18,185	72,867	6.5
OCC*	275,512	26,532	21.6
Crepe paper	–	15,169	1.1
Boxboard	10,803	6810	1.3
Kraft paper	4457	1430	0.4
Labels	8804	–	0.6
Label backing	28,569	5993	2.5
Wood	73,447	6027	5.7
Metal banding	30,959	575	2.3
Metal	6547	–	0.5
Rubber-coated metal	1894	–	0.1
Wire	1313	–	0.1
Filter material	16,845	–	1.2
Film plastic	12,079	6766	1.3
Mixed plastic	3787	1502	0.4
Plastic banding	409	737	0.1
Shredded plastic	43	–	0.0
Cloth rags	4668	5857	0.8
Paper towels	47,076	13,943	4.4
Tape	–	23,859	1.7
Aerosol cans	–	3933	0.3
Total	1,176,817	233,979	100.0
Employee generation	8.59 lb. per employee per day		

*The generation for OCC does not include the 733,333 lb that are currently recycled by a local vendor.

TABLE 8.4   NON-PRODUCTION-RELATED WASTE

MATERIAL	ANNUAL NONHAZARDOUS GENERATION (LB)	ANNUAL HAZARDOUS GENERATION (LB)	PERCENTAGE OF PRODUCTION-RELATED WASTE STREAM
Food waste	44,636	–	44.5
Paper cups	21,851	–	21.8
Aluminum	15,256	–	15.2
Aseptic packaging	14,299	–	14.2
Plastic cups	2259	–	2.3
Glass	1328	–	1.3
Plastic clamshell containers	198	–	0.2
Styrofoam	149	–	0.1
Plastic bottles	71	358	0.4
Total	100,047	358	100.0
Employee generation	0.62 lb per employee per day		

TABLE 8.5   ADMINISTRATIVE WASTE

MATERIAL	ANNUAL NONHAZARDOUS GENERATION (LB)	ANNUAL HAZARDOUS GENERATION (LB)	PERCENTAGE OF ADMINISTRATIVE WASTE STREAM
Paper	87,140	6670	5.66
NCR	20,188	–	1.23
Computer printout (CPO)	14,126	–	0.86
Mixed office paper (MOP)	8044	–	0.49
Cardstock	3816	–	0.23
Greenbar paper	1203	–	0.07
Envelopes	1073	–	0.06
Carbon paper	776	–	0.05
Newsprint (ONP)	972	776	0.11
Total	137,338	7446	8.76
Employee generation	0.92 lb. per employee per day		

TABLE 8.6   BASELINE DATA—MAJOR COMPONENTS GENERATED WITH DISPOSAL COSTS

MATERIAL	ANNUAL GENERATION (LB)	PERCENTAGE OF TOTAL MATERIAL STREAM	ANNUAL DISPOSAL COST	ANNUAL RECYCLING REVENUE
Uncoated lead	437,456	26.58	$10,453	
OCC	303,521	18.44	$3232	$8327
Administrative waste (paper)	137,338	8.34	$5189	–
Polypropylene battery components	103,432	6.28	$3908	–
Floor sweepings	91,052	5.53	$3440	–
Oxide inserts	84,514	5.13	$3193	–
Wood	79,474	4.83	$3003	–
Paper towels	61,019	3.71	$2305	–
Food waste	44,636	2.71	$1686	–
Label backing	34,562	2.10	$1306	–
Metal banding	31,535	1.92	$1191	–
Total	1,408,539	85.57	$38,905	$8327
Employee generation		8.96 lb per employee per day		

8.7   Step 6: Identify Major Waste Minimization Opportunities

After the baseline data have been calculated, the assessment team can begin to investigate individual components in the waste stream that should be targeted for reduction, reuse, or recycling. A useful method to accomplish this task is to conduct a Pareto analysis. A *Pareto analysis* is a statistical technique in decision making that is used for the selection of a limited number of tasks that produce the significant overall effect. Pareto analysis is a formal technique useful where many possible courses of action are competing for your attention. In essence, a problem solver estimates the benefit delivered by each action, then selects a number of the most effective actions that deliver a total benefit reasonably close to the maximal possible one. The analysis uses the Pareto principle—a large majority of effects, in this case waste generation are produced by a few key causes, in this case waste components. The Pareto principle is also known as the 80/20 rule, in that 80 percent of the effects are caused by 20 percent of the causes. The idea is to identify the 20 percent significant waste components that generation 80 percent of the total waste and then target the 20 percent significant causes for waste minimization. Table 8.7 and Fig. 8.21 display the Pareto analysis for the recyclable material within the waste stream for the case study company.

TABLE 8.7   RECYCLABLE MATERIAL WITHIN THE WASTE STREAM

MATERIAL	ANNUAL GENERATION (LB)	EMPLOYEE GENERATION (LB/DAY)	PERCENTAGE OF TOTAL MATERIAL STREAM
Uncoated lead	437,456	2.784	26.58
OCC	303,521	1.931	18.44
Polypropylene battery components	103,432	0.658	6.28
Paper	87,140	0.873	5.29
Wood	79,474	0.506	4.83
Label backing	34,562	0.220	2.10
Metal banding	31,535	0.201	1.92
NCR	20,188	0.128	1.23
Film plastic	18,845	0.120	1.14
Boxboard	17,613	0.112	1.07
Aluminum	15,256	0.097	0.93
Computer printout (CPO)	14,126	0.090	0.86
Mixed office paper (MOP)	8044	0.051	0.49
Metal	6547	0.042	0.40
Kraft paper	5887	0.037	0.36
Mixed plastic	5289	0.034	0.32
Cardstock	3816	0.024	0.23
ONP	1748	0.011	0.11
Glass	1328	0.008	0.08
Wire	1313	0.008	0.08
Greenbar	1203	0.008	0.07
Plastic banding	1146	0.007	0.07
Envelope	1073	0.007	0.07
Plastic bottles	429	0.003	0.03
Total	1,200,971	7.960	72.96

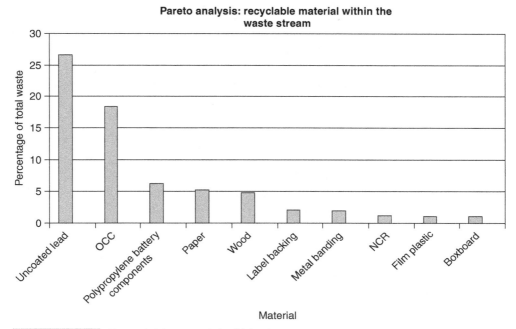

Figure 8.21 Recyclable material within the waste stream.

From the table and graph, it is straightforward to identify the largest waste stream contributors that are recyclable and should be targeted for reduction. In the case study, the largest recyclable contributors to the waste stream are uncoated lead, cardboard (OCC), polypropylene battery components, paper, and wood. These five components represent 84.2 percent of the recyclable waste stream and 61.4 percent of the total waste stream and should serve as the starting point for the waste assessment team when developing waste minimization strategies and alternatives.

8.8   Step 7: Determine, Evaluate, and Select Waste Minimization Alternatives

Once the major waste streams have been quantified, the team can begin to develop alternatives to minimize the solid waste and move closer to the ultimate goal of the project. In this phase of the project, the team identifies alternatives to minimize the major components of the waste stream and evaluates the economic and operational feasibility while rating each alternative on its ability to achieve the waste minimization goal. This section covers the process to generate, screen, and select waste minimization alternatives. In addition, a comprehensive list of common materials that can be reduced, reused, or recycled and a list of common waste minimization alternatives is also provided.

8.8.1 GENERATING ALTERNATIVES

The alternatives are based on the existing records review, the waste audit results, and the analysis phase. Various methods and tools are available to develop the initial list of alternatives. The environment in which these alternatives are created should be done in one that encourages creativity and free thinking by the team. Following is a suggested list of methods to identify and create these alternatives:

- Discussions with trade associations
- Discussions with plant engineers and operators
- Internet and literature reviews
- Information available from federal, state, or local governments
- Discussions with equipment manufacturers or vendors
- Discussions with environmental or business consultants
- Brainstorming
- Benchmarking

Trade associations generally provide assistance and information about environmental regulations and the various tools and techniques to address these issues. The information is usually industry specific and at times is free of charge. The National Association of Manufacturers (NAM), the National Association of Purchasing Management (NAPM), and the American Plastics Council (APC) are some very good examples of trade associations.

Discussions with plant employees are often very low cost and reliable method to develop alternatives to minimize solid waste. Employees are very familiar with a facilitie's processes and operations. In addition to generating feasible options, it also aids in fostering employee buy-in for the program and increases support. Establishing focus groups, town hall meetings, service talks, facility postings, or one-on-one interviews are methods to gather feedback from employees.

Literature and Internet reviews are another low-cost method to generate waste minimization alternatives. These include Internet searches, journal reviews, technical magazine reviews, and government reports. These sources describe similar waste components that other organizations have successfully minimized and the methods, tools, and equipment used. Often a company does not need to reinvent the wheel if another company successfully minimized a similar waste stream and reported the results. Many articles also provide contact information of the authors to gather further information. Some examples of popular environmental magazines and journals are *Resource, Recycling and Conservation, The Journal of Solid Waste Technology and Management, Resource Recycling*, and *The Journal of Cleaner Production.*

The government is also a great source of no-cost information and guidebooks. The Environmental Protection Agency (EPA) and local governments have developed programs that include technical assistance and information on industry-specific waste minimization tools and techniques. The EPA has published the *Waste Minimization Opportunity Manual* and the *Business Guide for Reducing Solid Waste.* Both of these are available at no cost.

Equipment manufacturers and vendors are another source of information to generate alternatives. The downside to this source is a bias toward the vendor's own products. Nonetheless, this source is a good method to identify equipment-related options and provide installation assistance.

Consultants are another resource to generate alternatives. A consultant with environmental experience and industry-specific knowledge can be a very valuable asset to an organization's waste minimization efforts. The downside is the added cost to contract with a consultant, because the information will rarely be given free of charge.

Team brainstorming sessions are another alternative. Brainstorming is a group creativity technique designed to generate a large number of ideas for the solution of a problem. Throughout the early stages of the project, the team will have of answer several what, why, and how questions. Brainstorming can accomplish this goal. Following is a list of general brainstorming rules:

- Collect as many ideas as possible from all participants with no criticisms or judgments made while ideas are being generated.
- All ideas are welcome, no matter how silly or far out they seem. Be creative. The more ideas the better because at this point you don't know what might work.
- Absolutely no discussion takes place during the brainstorming activity. Talking about the ideas will take place after brainstorming is complete.
- Do not criticize or judge. Don't even groan, frown, or laugh. All ideas are equally valid at this point.
- Do build on others' ideas.
- Do write all ideas on a flipchart or board so the whole group can easily see them.
- Set a time limit (i.e., 30 minutes) for the brainstorming.

Benchmarking may also be used to generate ideas. Benchmarking is defined as the concept of discovering what is the best performance being achieved, whether within the team's company, by a competitor, or by an entirely different industry. Benchmarking is an improvement tool whereby a company measures its performance or process against other companies' best practices, determines how those companies achieved their performance levels, and uses the information to improve its own performance. Data for benchmarking can be achieved from company visits, annual reports, or Internet searches. Often times, team members have colleagues in other companies and industries who can provide these data. Data can also be collected from regional committees. For example, in northwest Ohio, a multiorganization recycling committee was formed by a local nonprofit group that meets monthly to discuss environmental challenges and successes. At the monthly meeting, lunch is provided and new members are welcome at no cost. Also at the meeting, each member provides a status update of his or her company's environmental efforts, process changes, and issues with which the company could use assistance.

Each waste minimization alternative should be documented to facilitate the screening and review process. A simple spreadsheet that records the name of the alternative, a brief description, the author with contact information, and sources (if any) is a very useful tool. Figure 8.22 is a sample spreadsheet that may be used.

Potential Waste Minimization Options

	Alternative	Description	Author	Contact Information (phone/e-mail)	Sources or Additional Information
1					
2					
3					
4					
5					
6					
7					
8					
9					
10					
11					
12					
13					
14					
15					
16					
17					
18					
19					
20					

Figure 8.22   Alternative generation worksheet.

8.8.2   IDENTIFYING MARKETS FOR RECYCLABLE MATERIALS

As part of this process, identifying markets for recyclable materials is a very important step in implementing a waste minimization alternative. Each state and local area has recycling brokers for most wastes. These companies can be located through phone calls or Internet directories or the state environmental agency. In California, for example, a booklet called *The California Waste Exchange* can be obtained from the State Department of Toxic Substance Control.

Once the information has been collected, it will need to be analyzed and summarized into a useable format for evaluation. The remainder of this chapter will provide the process to perform this analysis and provide several examples.

8.8.3   COMMON WASTE MINIMIZATION ALTERNATIVES (MATERIALS AND METHODS)

This section provides a brief list of common solid waste minimization alternatives that many companies have successfully implemented. When considering alternatives to minimize waste, the solid waste minimization hierarchy should be considered. Consideration should first be given to reducing the waste (find a process or purchasing change to prevent generating the waste in the first place); next, reuse the waste item; and next recycle the waste item; then finally, dispose at the landfill. Source reduction can be accomplished through process modifications, technology changes, input material changes, or product changes. The following list provides commonly applied waste reduction and reuse opportunities that many companies have successfully implemented. Most of these are relatively low cost and are considered the low-hanging fruit, or simple to launch:

- *Office paper*—Many easy options exist to reduce office paper usage, including implementing an organization-wide double-sided copying policy (set the defaults of copiers and printers to print double-sided), reuse old paper as scratch paper, put company bulletins in electronic form (e-mail), centralize files to reduce redundant copies, save files electronically versus hard copy, and donate old magazines to hospitals or other organizations.
- *Packaging*—Order merchandise in bulk, work with suppliers to minimize packing materials, establish a reuse policy for cardboard boxes, implement returnable containers, and reuse shredded newspaper as packing material.
- *Equipment*—Use rechargeable batteries, reuse old tires for landscaping or pavement, install reusable filters, donate old furniture, and sell obsolete equipment and computers.
- *Organic waste*—Compost yard trimmings, choose low-maintenance landscape designs, and use mulching mowers.
- *Inventory/purchasing*—Set up an area in the facility where employees can exchange used items, purchase more durable products, order in bulk to reduce packaging supplies, and use a waste exchange program.

Before discussing the alternatives, a brief synopsis of common recyclable materials is provided:

- Paper products (high-grade paper, newspaper, computer printout, colored ledger, corrugated cardboard, envelopes)
- Glass (separated by color)
- Plastics (virtually all plastic is recyclable, and classified by one of seven types)
- Metals (most metals are recyclable, including both ferrous and nonferrous metals, aluminum, steel, tin cans, and bimetal containers)
- Compost materials (food scraps and yard trimmings)
- Batteries (lead acid and household)

Purchasing options fall into special category because they can encompass prevention, reuse, and recycling alternatives, but are specifically related to purchasing changes. These include buying supplies with reduced packaging to careful inventory control to avoid over ordering and possibly throwing away perishable items. In addition, the need for favoring products made with recycled content also may be noted. The team may need to have meetings with suppliers and vendors to discuss viable options that would meet organizational criteria. After the team has identified opportunities to purchase recycled products and products that can aid the company in reducing waste, each item should be rated in terms of availability and cost. Comparisons should also be made in terms of the life cycle of the product, not initial cost only.

For example, a picnic table made of recycled plastic costs more than a wood table, but the plastic table lasts up to four times longer. Similarly, while reusable products may cost more to purchase initially, they often save money over time by avoiding frequent purchases of single-use items. The comparison process is discussed in greater detail in Secs. 8.8.4 and 8.8.5.

Finally, there are several great and company-tested methods to encourage employee participation that may be included in the list of waste minimization alternatives. The goal is to develop an ongoing effort to increase and sustain employee participation. Communication is a critical ingredient involved in all the methods, which include

■ Holding regular environmental meetings with representatives from each department to discuss possible program changes on a regular basis.
■ Holding monthly or quarterly company-wide luncheons to promote and recognize environmental efforts of the organization, departments, and individuals.
■ Publicizing program changes and achievements in the company newsletter.
■ Announcing special events in memos and paycheck stuffers.
■ Rewarding employees for program involvement.

8.8.4   SCREENING ALTERNATIVES

The process of creating waste minimization alternatives can generate hundreds of options. It would be very time consuming for the team to conduct detailed financial and operational feasibility evaluations on each option. A quick screening process can help to quickly identify the options worthy of full evaluation and possible inclusion in the waste minimization program. Additionally, noneffective options can be weeded out, saving the team time and money in the evaluation process. An effective screening process should be based on the original goals of the project and at a minimum should examine the following:

■ Expected solid waste reduction (tons per year)
■ Expected start-up costs
■ Impact on waste removal costs (dollars per year)
■ Impact on purchasing costs (dollars per year)
■ Impact on employee moral
■ Ease of implementation

The team should keep in mind that the goal of the screening process is to quickly identify options worthy of further analysis. A weighed scoring system can be developed to consistently rank each alternative in an objective manner. A quality deployment function, such as the *House of Quality* is an excellent tool to accomplish this evaluation. The *House of Quality* is a graphic tool for defining the relationship between an organization's desires and its capabilities. It utilizes a planning matrix to relate the organizational wants (for example, solid waste reduction and cost performance) to how the waste minimization program will or can meet those wants (for example, process changes or recycling efforts). It looks like a house with a correlation matrix as its roof, and organizational wants versus waste minimization options as the house. The House of Quality can also increase cross-functional integration within organizations using it, especially among marketing, engineering, and manufacturing.

The basic structure is a table with whats as the rows on the left and hows as the columns. Rankings based on the whys and the correlations can be used to calculate priorities for the hows. House of Quality analysis can also be cascaded, with hows from one level becoming the whats of a lower level; as this progresses the decisions get closer to the engineering/manufacturing details.

Before proceeding with the screening process, the team should decide on the evaluation criteria (the whats) and weighting system. A scale of 1 to 10 for weighting each criterion is recommended. These weightings should be determined by the team, project manager, facility manager, or a combination of all three. The evaluation criteria should be directly related to the overall goals of the project, such as

- Reduction in waste amounts
- Reduction in waste toxicity
- Reduction in waste disposal costs
- Reduction in purchasing costs
- Revenue generation potential
- Low start-up costs
- Productivity improvements
- Quality improvement
- Ease of implementation
- Impact on employee morale
- Impact on organization image
- Impact on safety
- Other factors as determined by the team

Once these criteria have been created, the team should rank them on a scale of 1 to 10 based on importance. For example, waste reduction in waste amounts could received an importance rating of 10 (meaning it is highly important) versus low start-up costs receiving an importance rating of 2 (meaning that start-up costs are of low importance and not a major factor in the decision process). These criteria should then be placed into the column headers of the spreadsheet with importance weightings in parentheses. Figure 8.23 provides an example.

Once the criterion and importance ratings have been established, the team should list each alternative in the rows under the option column. In the row for each alternative, the team should place a rating score corresponding to the level of which the alternative meets the criterion, with 0 being no impact and 10 being great impact. For example, if the team is considering the purchase of a cardboard baler, the reduction in waste amounts could be significant, so the team could rate it an 8, but in the start-up cost criterion, the team could rate it lower, such as a 1, due to the high implementation cost. Once each alternative is rated, the ratings should be multiplied by the importance factor and each row should be summed. This score will allow the team to objectively screen each alternative. Once all of the alternatives are listed and scored,

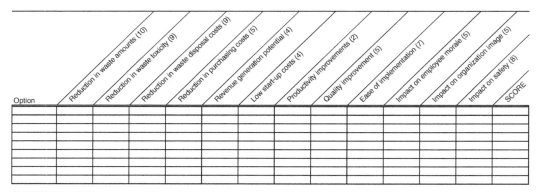

Figure 8.23 Alternative screening worksheet.

the team can screen them based on their total score. Alternatives with higher total scores pass the screening process and become eligible for further evaluation. To determine the cut-off point, depending on time and money resources, the team could set the threshold at a specific point value, accept the top 20 percent, or accept the top 10 for further analysis. When first starting a solid waste minimization program it is recommended that the team select the top third of all alternatives for further screening to compensate for estimation error. Figure 8.24 is an example of a completed screening worksheet.

8.8.5 ANALYZING AND SELECTING ALTERNATIVES

After trimming down the list of alternatives via the screening process, the remaining alternatives should be further analyzed to determine the best fit for the organization to minimize solid waste and hence include in the program. The analysis process focuses on identifying the benefits, costs, and drawbacks of each alternative. To accomplish this, each alternative is evaluated on

Option	Reduction in waste amounts (10)	Reduction in waste toxicity (9)	Reduction in waste disposal costs (9)	Reduction in purchasing costs (5)	Revenue generation potential (4)	Low start-up costs (4)	Productivity improvements (2)	Quality improvement (5)	Ease of implementation (7)	Impact on employee morale (5)	Impact on organization image (5)	Impact on safety (8)	SCORE
Purchase OCC Baler	8	0	7	0	5	1	0	0	4	4	4	5	275
Copy default to double sided	8	0	5	8	0	10	5	0	9	5	4	0	323

Figure 8.24 Completed alternative screening worksheet.

■ Impact on the program goal
■ Technical feasibility
■ Operational feasibility
■ Economic feasibility
■ Sustainability
■ Organizational culture feasibility

The key outcome of this phase is to fully document, analyze, and arrive at a final acceptance decision for each alternative. To accomplish this, process flowcharts are analyzed, the annualized amount of waste generated is determined, a complete feasibility analysis is conducted (including technical, operational, and organizational), a cost justification study is completed, feedback is collected and analyzed, and finally a decision is made regarding each alternative. This section provides a complete discussions and documentation of each alternative that will be used in the implementation phase if the alternative is accepted. During this process the team must keep a clear understanding of the overriding goals of the waste minimization project. For example, the relative importance of reducing costs versus minimizing environmental impact. Some alternatives may require extensive analysis, including gathering additional data from vendors or analyzing market trends for recyclable material commodity markets. The first consideration when evaluating alternatives is its impact on the goals of the project established in the first phase of the project. These goals can range from solid waste generation to the cost-benefits associated with waste minimization. When considering alternatives, the hierarchy of solid waste management should be kept in mind (discussed in Chap. 1). Efforts should first be made to reduce waste generation, next to reuse waste materials, then recycle waste material (in and out of process), and finally dispose the waste in a landfill. The idea behind the hierarchy is to engineer methods to eliminate the generation of a waste stream altogether, and hence eliminate the need to manage the solid waste stream via recycling or landfill disposal. Alternatives should be separated into different categories to aid with this process. The categories are (based on the hierarchy)

■ Waste prevention alternatives
■ Reuse alternatives
■ Recycling alternatives
■ Composting alternatives

The evaluation process itself consists of seven steps to rate each alternative. The process is completed sequentially and after each step, the alterative is either accepted and moved to the next phase or rejected and the analysis is terminated. If the alternative does not meet thresholds or feasibility tests it is eliminated from further review to save the team time and resources. The alternative should still be kept on file in the event technical or organizational changes render the option feasible. The seven steps of the evaluation process are as follows:

1 Fully describe each alternative in terms of the equipment, raw material, process, or purchasing additions or modifications

2 Calculate the annualized waste reduction impact in terms of tons per year and whether the waste reduction is source reduction, reuse, or recycling

3 Compile and analyze the process flowcharts that create the waste stream

4 Conduct feasibility analyses (technical, operational, and organizational)

5 Conduct a cost justification for each alternative (payback, internal rate of return, and net present value)

6 Gather feedback from all stakeholders

7 Approval and sign off from the waste minimization team and executives

Technical and operational feasibility is concerned with whether the proper resources exist or are reasonably attainable to implement a specific alternative. This includes the square footage of the building, existing and available utilities, existing processing and material-handling equipment, quality requirements, and skill level of employees. During this process, product specifications and facility constraints should be taken into account. Typical technical evaluation criteria include

- Available space in the facility
- Safety
- Compatibility with current work processes and material handling
- Impact on product quality
- Required technologies and utilities (power, compressed air, and data links)
- Knowledge and skills required for operating and maintaining the alternative
- Additional labor requirements
- Impact on product marketing
- Implementation time

When evaluating technical feasibility, facility engineers or consultants should be contacted for input. In addition it is also wise to discuss the technical aspects with workers directly impacted by the change, such as production and maintenance. If an alternative calls for a change in raw materials, the effect on the quality of the final product must be evaluated. If an alternative does not meet the technical requirements of the organization, it should be removed from consideration. From a technical standpoint, the three areas that require additional evaluation are

1 Equipment modifications or purchases

2 Process changes

3 Material changes

If an alternative involves an equipment modification or purchase, an analysis on the equipment should be conducted. The team should investigate whether the equipment is available commercially and then contact the manufacturer for more information. Performance of the machine should also be addressed, including cost, utility requirements, capacity, throughput, cycle time, required preventative maintenance, space requirements, and possible locations in the facility in which the equipment could be installed. In addition, whether production would be affected during installation, should also be evaluated. The vendor or manufacturer could provide more information regarding potential shutdowns. Required modifications to work flow or production procedures should be analyzed as well as required training and safety concerns related to the equipment purchase or modification. From an operational standpoint, attention should be given to how the alternative will improve or reduce productivity and labor force reductions or increases.

If a waste minimization alternative involves a process change or a material change, the affected areas should be identified and feedback gathered from the area managers, employees, maintenance staff, and engineers (if applicable). With the process changes, training requirements should also be discussed. Also, the impacts on production, material handling/storage, and quality should be addressed. A material-testing program is highly recommended for new items with which the engineering team may not be familiar, to analyze quality and throughput impacts. A design of experiment (DOE) that tests the changes versus the current material is an excellent method to gauge impacts. A DOE is the design data gathering tests where variation is present, whether under the full control of the experimenter or not. Often the experimenter is interested in the effect of some process or intervention, such as using a new raw material, on some outcome such as quality.

From an economic standpoint, traditional financial evaluation is the most effective method to analyze alternatives. These measures include the payback period, (discounted payback period), internal rate of return, and net present value for each alternative. If the organization has a standard financial evaluation process, this should be completed for each alternative. The accounting or finance department would have this information. To perform these financial analyses, revenue and cost data must be gathered and should be based on the expectations for the alternatives. This is more complicated than it sounds, especially if a project will have an impact on the number of required labor hours, utility costs, and productivity, not to mention initial investments. A comprehensive estimation of the cost impacts (revenues and costs) per year over the life of the alternative is required to begin the analysis. The first step of the economic evaluation process is to determine these costs. These costs include capital costs (or initial investment), operating costs/savings, operating revenue, and salvage values for each waste minimization alternative.

Capital costs are the costs incurred when purchasing assets that are used in production and service. Normally they are non-reoccurring and used to purchase large equipment such as a baler or plastic grinder. Capital costs include more than just the actual cost of the equipment; they also include the costs to prepare the site for production. Following is a brief list of typical capital costs; also known as the initial investment:

- Site development and preparation (including demolition and clearing if needed)
- Equipment purchases, including spare parts, taxes, freight, and insurance
- Material costs (piping, electrical, telecommunications, structural)
- Building-modification costs (utility lines, construction costs)
- Permitting costs, building inspection costs
- Contractor's fees
- Start-up costs (vendor, contractor, in-house)
- Training costs

After the initial investment has been calculated, the reoccurring costs, savings, and revenues from the waste minimization alternative must be determined. The concept is to reduce waste disposal and raw material costs based on the implementation of the alternative under analysis. For example, if a company considers the installation of a cardboard baler, the annual operating costs of the baler (such as labor and utilities), the annual cost savings from reduced disposal costs, and the revenue from the sale of the baled cardboard must be considered. Reducing or avoiding present and future operating costs associated with solid waste storage and removal are critical elements of the solid waste minimization process. Due to increased solid waste disposal costs (around $30–$80 per ton in the United States); many companies are finding that the cost of waste management has become a significant factor in their cost structures. Some common reoccurring costs include

- *Reduced solid waste disposal costs*—Waste generation is reduced or diverted to recycling streams, resulting in less waste being sent to the landfill for disposal and lower hauler charges. These include disposal fees, transportation costs, and predisposal treatment costs.
- *Input material cost savings*—Options that reduce scrap, reduce waste, or increase internal recycling tend to decrease the demand for input materials.
- *Changes in utility costs*—Utility costs may increase or decrease depending on the installation, modification, or removal of equipment.
- *Changes in operating and maintenance labor/benefits*—An alternative may increase or decrease labor requirements and the associated benefits. This may be reflected as changes in overtime hours or as the number of employees.
- *Changes in operating and maintenance supplies*—An alternative may result in increased or decreased operating and maintenance supply usage.
- *Changes in overhead costs*—Large projects may increase or decrease these values.
- *Changes in revenues for increased (or decreased production)*—An alternative may result in an increase in the productivity of a unit. This will result in changes in revenue.
- *Increases in revenue from by-products*—An alternative may generate a by-product that can be sold to a recycler or sold to another company as raw material. This will result in increased revenue.

It is suggested that savings in these costs be taken into consideration first, because they have a greater impact on the project economics and involve less effort to estimate reliably. The remaining elements usually have a smaller impact and should be included in an on-needed basis or to fine-tune the analysis.

A project's profitability is measured by estimating the net cash flows each operating year over the life of the project. A net cash flow is calculated by subtracting the cash outlays from the cash incomes starting in year zero (the year the project is initiated). Figure 8.25 is an example of a cash-flow diagram.

If a project does not have an initial investment, the project's profitability can be judged by whether an operating cost savings occurred or not. If such a project reduces overall operating costs, it should be implemented. For example, suppose an organization

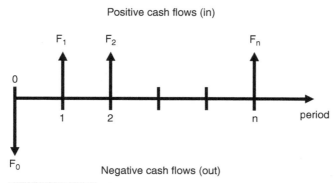

Figure 8.25   **Cash-flow diagram example.**

currently recycles plastics and metals. If the organization currently ships comingled plastics and metals to a recycling processor, a process change could be implemented that requires employees to separate plastics from metals before shipment. There is little to no initial investment with the example, but there will added labor costs for separation versus additional revenue generated by the finer sort to the processor. If the additional revenues outweigh the additional costs, the alternative should be implemented.

For projects with significant initial investments or capital costs, a more detailed profitability analysis is needed. The three standard measures of profitability are

1 Payback period
2 Internal rate of return (IRR)
3 Net present value (NPV)

The *payback period* for a project is the amount of time it requires to recover the initial cash outlay for the project. The formula for calculating the payback period on a pre tax basis in years is

$$\text{Payback period} = \frac{\text{capital investment}}{\text{annual operating cost savings}}$$

For example, suppose a manufacturer installs a cardboard baler for a total cost of $65,000. If the baler is expected to save the company $20,000 per year, then the payback period is 3.25 years. Payback period is typically measured in years. However, some alternatives may have payback periods in terms of months. Many organizations use the payback period as a screening method before conducting a full financial analysis. If the alternative does not meet a predetermined threshold, the alternative is rejected. Payback periods in the range of 3 to 4 years are usually considered acceptable for low-risk investments. Again, this method is recommended for quick assessments of profitability. If large capital expenditures are involved, it should be followed by a more strenuous financial analysis such at the IRR and NPV.

The *internal rate of return (IRR)* and *net present value (NPV)* are both discounted cash-flow techniques for determining profitability and determining if a waste minimization alternative will improve the financial position of the company. Many organizations use these methods for ranking capital projects that are competing for funds, such as the case with the various waste minimization alternatives. Capital funding for a project can depend on the ability of the project to generate positive cash flows beyond the payback period to realize an acceptable return on investment. Both the IRR and NPV recognize the time value of money by discounting the projected future net cash flows to the present. For investments with a low level of risk, an after tax IRR of 12 percent to 15 percent is typically acceptable.

Each cash inflow/outflow is discounted back to its present value (PV). Then they are summed. The equation for NPV is

$$NPV = \sum_{t=0}^{N} \frac{C_t}{(1+r)^t}$$

where t = the time of the cash flow
N = the total time of the project
r = the discount rate (the rate of return that could be earned on an investment in the financial markets with similar risk)
C_t = the net cash flow (the amount of cash) at time t (for educational purposes, C_0 is commonly placed to the left of the sum to emphasize its role as the initial investment)

The IRR is a capital budgeting metric used by firms to decide whether they should make investments. It is an indicator of the efficiency of an investment, as opposed to NPV, which indicates value or magnitude. The IRR is the annualized effective compounded return rate that can be earned on the invested capital; that is, the yield on the investment.

A project is a good investment proposition if its IRR is greater than the rate of return that could be earned by alternate investments (investing in other projects, buying bonds, even putting the money in a bank account). Thus, the IRR should be compared with any alternate costs of capital including an appropriate risk premium.

Mathematically the IRR is defined as any discount rate that results in a net present value of zero for a series of cash flows. In general, if the IRR is greater than the project's cost of capital, or hurdle rate, the project will add value for the company. The IRR is determined by calculating the interest rate (r) when NPV equals zero or the equation and solving for r

$$NPV = \sum_{t=0}^{N} \frac{C_t}{(1+r)^t} = 0$$

TABLE 8.8 NET PRESENT VALUE ANALYSIS	
YEAR	CASH FLOW
0	$65,000
1	$20,000
2	$20,000
3	$20,000
4	$20,000
5	$20,000
6	$20,000
7	$20,000
8	$20,000
9	$20,000
10	$20,000
MARR	15.0%
IRR	28.2%
NPV	$30,761

Most spreadsheet programs typically have the ability to automatically calculate IRR and NPV from a series of cash flows. Table 8.8 is an example applying these financial evaluation concepts. Returning to the baler example discussed previously, recalling an initial cost of $65,000 and $20,000 in annual savings and assuming a baler life span of 10 years and an organization minimum attractive rate of return (MARR) of 15 percent. The MARR is the minimum return on a project that a manager is willing to accept before starting a project, given its risk and the opportunity cost of forgoing other projects. The MARR is calculated by the management working with the finance department and typically ranges between 3 to 15 percent for most organizations. Table 8.8 shows the cash flows, IRR, and NPV.

As shown in the last two rows, the IRR is 28.2 percent and the NPV is nearly $31,000 at an MARR of 15 percent. The fact that the IRR is greater than the 15 percent MARR and the NPV is positive indicates that the project is a good financial decision.

Waste minimization alternatives should also be evaluated based on sustainability and the cultural fit within the organization. *Sustainability* is defined as an organization's investment in a system of living, projected to be viable on an ongoing basis that provides quality of life for all individuals of sentient species and preserves natural ecosystems. Sustainability in its simplest form describes a characteristic of a process or state that can be maintained at a certain level indefinitely. The term, in its environmental usage, refers to the potential longevity of vital human ecological support systems, such as the planet's climatic system, systems of agriculture, industry, forestry, fisheries, and the systems on which they depend. In other words, the waste minimization alternatives should be evaluated based on how well they meet this definition, such that the alternative can be sustained without large amounts of effort or additional resources and continue to protect the environment. Often, this will be related to the culture of the organization. Criteria commonly used to evaluate the sustainability of an alternative include

- Dealing transparently and systemically with risk, uncertainty, and irreversibility
- Ensuring appropriate valuation, appreciation, and restoration of nature
- Integrating environmental, social, human, and economic goals in policies and activities
- Equal opportunity and community participation/sustainable community
- Conserving biodiversity and ecological integrity
- Ensuring intergenerational equity
- Recognizing the global integration of localities
- Committing to best practice
- No net loss of human capital or natural capital
- The principle of continuous improvement
- The need for good governance

When an alternative involves working with a recycler or commodity broker, there are several key questions to ask potential candidates to determine the best fit for the organization. These questions include

- What types of materials does the company accept and how must they be prepared?
- What contract terms does the buyer require?
- Who provides the transportation?
- What is the schedule of collections?
- What are the maximum allowable contaminant levels and what is the procedure for dealing with rejected loads?
- Are there minimum quantity requirements?
- Where will recyclable material be weighed?
- Who will provide containers for recyclables?
- Can escape clauses be included in the contract?
- Be sure to check references.

In a similar vein, when working with equipment vendors, there are a several key questions to ask

- What is the total cost of the equipment including freight and installation?
- What are the building requirements and specifications for the equipment (compressed air, electricity, space, minimum door widths)?
- Is a service contract included in the purchase price or is there an additional charge?
- Do you offer training to the employees, engineers, and maintenance employees who will be working with the equipment? If so, is there a charge?

- What is the process if the equipment malfunctions and the company needs support? Is there a representative available 24 hours per day? What is the charge for these visits?
- Do you offer an acceptance test process to ensure that the equipment operates within the promised specifications (capacity and cycle time)?
- What is the required installation time and must production be shut down?

The worksheets in Figs. 8.26 through 8.34 may be used to evaluate each waste minimization alternative. The first worksheet is a cover page that provides a general description of the alternative. Based on the responses to the questions on the cover sheet one of the four additional worksheets is attached:

- Worksheet A: Equipment purchases or modifications
- Worksheet B: Raw material changes
- Worksheet C: Process changes
- Worksheet D: Purchasing changes

The worksheets are designed to work together to completely describe and evaluate each alternative. If the alternative is approved, the form also provides sign offs for the

Waste Minimization Alternative Feasibility Analysis Worksheet
Cover Sheet

The purpose of this worksheet is to evaluate the feasiblity of alternatives that have passed the screening process

Company name: _____
Location: _____
Date: _____

Alternative description:

Alternative tracking number: date and two-digit number
Work unit of departments affected:

Source or contact person:
Contact information:

Hierarchy of waste minimization (check the ONE that applies):

	Waste reduction
	Waste reuse
	Recycling
	Other

Does the alternative involve (check the ONE that applies):

	Equipment purchases of modifications> **Please attach worksheet A**
	Raw material changes> **Please attach worksheet B**
	Process changes > **Please attach worksheet C**
	Purchasing changes> **Please attach worksheet D**

Figure 8.26 *Feasibility analysis coversheet.*

Waste Minimization Alternative Feasibility Analysis Worksheet A
Equipment Purchases or Modifications

The purpose of this worksheet is to evaluate the feasibility of alternatives that have passed the screening process.
All grayed boxes should be completed before completing the Approval Process in Step 6.

Company name: _____
Location: _____
Date: _____
Alternative description: _____
Alternative tracking number: _____

PASS

Step 1: Estimate annual waste reduction (tons per year) and attach supporting documents   material

tons/year
cy/year

Current annual disposal costs $

Step 2: Attach process flowcharts that generate waste stream
Step 3: Feasibility analysis

Technical

Does the equipment exist (vendor or manufacturer)?   **Yes**   **No**
Description of machine

Vendor name
Vendor contact information

Estimate machine and installation cost $
Implementation time (days)   days

Required utilities
Power   **Yes**   **No**
Compressed air   **Yes**   **No**
Data link   **Yes**   **No**
Other

Compatibility with current work processes and material handling

Training concerns
Skill level required to operate equipment
Skill level required to maintain equipment

Space requirements
Space required for machine and staging   ft²
Available space in the facility   ft²
Proposed locations

Operational
Machine cycle time   minutes
Machine capacity   units
Labor impacts (additional work hours)
Supervisory needs
Maintenance needs
Productivity impacts
Safety concerns and impacts
Product quality impact
Additional labor requirements

Organizational
Impact on sales
Impact on marketing
Impact on employee morale
Impact on corporate image
Impact on supply chain

Figure 8.27   Feasibility analysis worksheet A—page 1 of 2.

Waste Minimization Alternative Feasibility Analysis Worksheet A (Page 2 of 2)
Equipment Purchases or Modifications

Step 4: Cost analysis

Machine costs

Machine cost	$
Site development	$
Material costs	$
Building modification costs	$
Permit and inspection costs	$
Contractor fees	$
Start-up costs	$
Initial training costs	$

Operating expenses

Utility cost impacts	$
Input material changes	$
Labor cost impacts	$
Supervision cost impacts	$
Maintenance cost impacts	$
Operating and maintenance supply impacts	$
Changes in overhead costs	$

Operating savings and revenue

Reduced solid waste disposal costs	$
Revenues from increased sale of recyclable material	$
Revenues from the sale of by-products	$

Life of machine or product (whichever is shorter)   years

Total initial investment	$
Annual operating savings	$
Payback period	$
Net present value (NPV)	$
Internal rate of return (IRR)	
Organization minimum attractive rate of return (MARR)	

Companies to purchase recycled material

$ per ton $

Exchange options

Step 5: Feedback analysis

Feedback from operators

Feedback from management

Feedback from maintenance

Feedback from finance

Step 6: Approval

Waste minimization team leader

	name	date
Manager, maintenance	name	date
Manager, operations	name	date
Manager, finance	name	date
CEO	name	date
	name	date

Figure 8.28   Feasibility analysis worksheet A—page 2 of 2.

Waste Minimization Alternative Feasibility Analysis Worksheet B
Raw Material or Modifications
The purpose of this worksheet is to evaluate the feasibility of alternatives that have passed the screening process.
All grayed boxes should be completed before completing the Approval Process in Step 6.

Company name: _____
Location: _____
Date: _____
Alternative description:_____
Alternative tracking number: _____   **PASS**

Step 1: Estimate annual waste reduction (tons per year)   material
    and attach supporting documents   tons/year
                              cy/year
    Current annual disposal costs $
Step 2: Attach process flowcharts that generate waste stream
Step 3: Feasibility analysis
    Technical

	Yes	No
Does the material exist and is available		
Compatibility with current work processes and material handling		

    Operational
        Labor impacts (additional work hours)
        Supervisory needs
        Maintenance needs
        Productivity impacts
        Safety concerns and impacts
        Product quality impact
        Additional labor requirements
    Organizational
        Impact on sales
        Impact on marketing
        Impact on employee morale
        Impact on corporate image
        Impact on supply chain

Figure 8.29   **Feasibility analysis worksheet B—page 1 of 2.**

leadership team to indicate review and acceptance. Continuing with the case study, the following list shows alternative evaluations that were selected with accompanying data. Also included are

- The cost analyses
- A list of accepted recycling strategies
- An overview of nonrecyclable items
- Proposed changes to waste handling practices
- Suggestions to the office paper-recycling program
- Corporate waste minimization strategy

8.9   Case Study—Potential Recycling Strategies

The waste assessment team has identified 11 major waste streams produced by the manufacturer. Of these 11 streams, 8 of the streams are potentially recyclable. At present, two of these waste streams are being recycled. These major streams are the old

Waste Minimization Alternative Feasibility Analysis Worksheet B (Page 2 of 2)
Raw Material or Modifications

Step 4: Cost analysis
    New material costs
        Previous material cost per load $
        New material cost per load $
        Material cost differential $
        Annual loads purchased $
        Annual material cost differential $

    Operating expenses
        Utility cost impacts $
        Labor cost impacts $
        Supervision cost impacts $
        Maintenance cost impacts $
        Operating and maintenance supply impacts $
        Changes in overhead costs $

    Operating savings and revenue
        Reduced solid waste disposal costs $
        Revenues from increased sale of recyclable material $
        Revenues from the sale of by products $

    Total initial investment $
    Annual operating savings $
    Payback period $
    Net present value (NPV) $
    Internal rate of return (IRR)
    Organization minimum attractive rate of return (MARR)

    Companies to purchase recycled material
    $ per ton $
    Exchange options

Step 5: Feedback analysis

    Feedback from operators

    Feedback from management

    Feedback from maintenance

    Feedback from finance

Step 6: Approval

    Waste minimization team leader
    Manager, maintenance   name   date
    Manager, operations   name   date
    Manager, finance   name   date
    CEO   name   date
        name   date

Figure 8.30   Feasibility analysis worksheet B—page 2 of 2.

corrugated cardboard and the mixed office paper. By separating the potentially recyclable material from the waste stream, the company could significantly reduce both the amount of material going to the landfill as well as disposal costs. The following sections provide alternatives for dealing with the recyclable materials not currently recycled. The 11 major waste streams are mentioned in Table 8.9:

Waste Minimization Alternative Feasibility Analysis Worksheet C
Process Changes

The purpose of this worksheet is to evaluate the feasibility of alternatives that have passed the screening process.
All grayed boxes should be completed before completing the Approval Process in Step 6.

Company name: _____
Location: _____
Date: _____
Alternative description:_____
Alternative tracking number: _____

PASS

Step 1: Estimate annual waste reduction (tons per year) and attach supporting documents

    material

    tons/year

    cy/year

    Current annual disposal costs $

Step 2: Attach process flowcharts that generate waste stream
Step 3: Feasibility analysis
    Technical
        Compatibility with current work processes and material handling

    Operational
        Labor impacts (additional work hours)
        Supervisory needs
        Maintenance needs
        Productivity impacts
        Safety concerns and impacts
        Product quality impact
        Additional labor requirements
    Organizational
        Impact on sales
        Impact on marketing
        Impact on employee morale
        Impact on corporate image
        Impact on supply chain

Figure 8.31   Feasibility analysis worksheet C—page 1 of 2.

8.10   Case Study—Major Waste Streams

The following is an analysis of the major streams and/or their components.

8.10.1   UNCOATED LEAD

Lead is the largest component of the company's waste stream. Over 437,000 lb of lead leave the facility in its waste stream every year. This figure does not include much of the lead waste and lead-contaminated waste sent to the recyeling processor. Again, more accurate figures were unavailable to the waste assessment team due to the confidentiality of the contract between the manufacturer and the recycling processor.

While there are plant rules regarding employees eating on the production floor, food containers and waste were found in both the hazardous and nonhazardous containers. Not only is eating on the floor a health risk, it also is an added expense for the

Waste Minimization Alternative Feasibility Analysis Worksheet C (Page 2 of 2)
Process Changes

Step 4: Cost analysis

New material costs

Previous material cost per load	$
New material cost per load	$
Material cost differential	$
Annual loads purchased	$
Annual material cost differential	$

Operating expenses

Utility cost impacts	$
Labor cost impacts	$
Supervision cost impacts	$
Maintenance cost impacts	$
Operating and maintenance supply impacts	$
Changes in overhead costs	$

Operating savings and revenue

Reduced solid waste disposal costs	$
Revenues from increased sale of recyclable material	$
Revenues from the sale of by products	$

Total initial investment	$
Annual operating Savings	$
Payback period	$
Net present value (NPV)	$
Internal rate of return (IRR)	
Organization minimum attractive rate of return (MARR)	

Companies to purchase recycled material	
$ per ton	$
Exchange options	

Step 5: Feedback analysis

Feedback from operators

Feedback from management

Feedback from maintenance

Feedback from finance

Step 6: Approval

	name	date
Waste minimization team leader		
Manager, maintenance	name	date
Manager, operations	name	date
Manager, finance	name	date
CEO	name	date
	name	date

Figure 8.32   Feasibility analysis worksheet C—page 2 of 2.

Waste Minimization Alternative Feasibility Analysis Worksheet D
Purchasing Changes

The purpose of this worksheet is to evaluate the feasibility of alternatives that have passed the screening process.
All grayed boxes should be completed before completing the Approval Process in Step 6.

Company name: _____
Location: _____
Date: _____
Alternative description:_____
Alternative tracking number: _____

PASS

Step 1: Estimate annual waste reduction (tons per year) and attach
supporting documents
material
tons/year
cy/year

Current annual disposal costs $
Step 2: Attach process flowcharts that generate waste stream
Step 3: Feasibility analysis
Technical
Does the material exist (vendor or manufacturer)? | Yes | No |
Compatibility with current work processes and material handling

Operational
Safety concerns and impacts
Product quality impact
Additional labor requirements

Organizational
Impact on sales
Impact on marketing
Impact on employee morale
Impact on corporate image
Impact on supply chain

Figure 8.33   Feasibility analysis worksheet D—page 1 of 2.

company since it is assumed that the cost to dispose waste via the recycling processor
is more than disposing waste through the waste hauler.

Since lead waste and lead-contaminated waste constitutes such a large portion of
Manufacturer' waste stream, process improvements geared toward significantly reducing the amount of lead disposed should be heavily investigated.

Visual inspections during walkthroughs discovered that nonhazardous and hazardous
materials were frequently mishandled. For example, a large percentage of apparently
nonhazardous materials were found in hazardous waste containers. This material could be
disposed more economically through conventional means. Research determining
which materials need to be disposed via the smelter should be performed. Employee
training discussing proper lead disposal should follow up this research.

Some alternative regional smelters interested in recycling lead are listed in
Table 8.10.

8.10.2 CARDBOARD (OCC)

Over 1,036,854 lb of corrugated cardboard is generated yearly. The company captures
approximately 733,333 lb or 71 percent of this waste stream and it is recycled by Lake
Erie Recycling. However, there is opportunity to capture more of the OCC and

Waste Minimization Alternative Feasibility Analysis Worksheet D (Page 2 of 2)
Purchasing Changes

Step 4: Cost analysis

New material costs

Previous material cost per load	$
New material cost per load	$
Material cost differential	$
Annual loads purchased	$
Annual material cost differential	$

Operating expenses

Utility cost impacts	$
Labor cost impacts	$
Supervision cost impacts	$
Maintenance cost impacts	$
Operating and maintenance supply impacts	$
Changes in overhead costs	$

Operating savings and revenue

Reduced solid waste disposal costs	$
Revenues from increased sale of recyclable material	$
Revenues from the sale of by products	$

Total initial investment	$
Annual operating savings	$
Payback period	$
Net present value (NPV)	$
Internal rate of return (IRR)	
Organization minimum attractive rate of return (MARR)	

Companies to purchase recycled material	
$ per ton	$
Exchange options	

Step 5: Feedback analysis

Feedback from operators

Feedback from management

Feedback from maintenance

Feedback from finance

Step 6: Approval

Waste minimization, team leader		
Manager, maintenance	name	date
Manager, operations	name	date
Manager, finance	name	date
CEO	name	date
	name	date

Figure 8.34 Feasibility analysis worksheet D—page 2 of 2.

increase the manufacturer's revenue for the waste stream. Currently, direct and indirect labor personnel put OCC in trash containers throughout the plant. Large wooden or metal gaylords could be placed next to the trash containers to collect OCC. Material handlers could then take the OCC to the baling area on a regular basis, such as once

TABLE 8.9   WASTE HANDLING OPTIONS FOR THE 11 MAJOR WASTE STREAMS

WASTE STREAM	POUNDS/YEAR GENERATED	WASTE HANDLING OPTIONS
1. Uncoated lead	437,456	Smelter—if hazardous
2. OCC*	1,036,854	Currently recycled
3. Polypropylene battery components	103,432	Recyclable
4. Paper	137,338	Recyclable
5. Floor sweepings	91,052	Smelter—if hazardous
6. Oxide inserts	84,514	Smelter—if hazardous
7. Wood	79,474	Recyclable
8. Paper towels	61,020	Reducible
9. Food waste	44,636	Nonrecyclable
10. Label backing	34,562	Possibly recyclable
11. Metal banding	31,535	Recyclable

per shift or daily. An analysis of the available floor space for the gaylord and the OCC generation rate would have to be conducted in order to determine the best process.

OCC prices are quoted per ton. Prices vary greatly by region and service provider. Lake Erie Recycling pays the manufacturer market value. According to *Recycling Times*, September 29 *Markets Page*, the current rate for OCC in this area is $15.00 to $90.00. Therefore, up to an additional $10,243 in revenue could be achieved by the manufacturer. This number does not include any potential cost savings resulting in the diversion of OCC waste from the compactor waste stream.

8.10.3   BATTERY CASES AND BUTTONS

The company annually disposes of about 52 tons of polypropylene from discarded battery cases and buttons. This figure is based on the trash sorts performed by the waste assessment team and equates to 8667 lb per month. A breakout of the nonhazardous polypropylene components found in the waste stream is shown in the following table. This plastic could be ground and recycled for revenue. This plastic, in the form of grind or flake, could be sold to the company's Kentucky plant at $0.30 per pound thereby generating revenue of $31,003 per year. Costs would be cut for the overall corporation, since utilizing the reground plastic would reduce the virgin plastic purchasing costs.

Table 8.11 is a cost-benefit analysis for purchasing a grinder and operating it in-house. The waste assessment team used a figure for discarded polypropylene of 5497 lb per month rather than the 8667 lb per month found from trash sorts. This figure was found on the recycling processor receipts and was used in order to be conservative in the analysis.

TABLE 8.10 SAMPLE LIST OF SMELTER VENDORS	
Louis Padnos Iron & Metal Co. P.O. Box 1979 Holland, MI 49422-1989 Phone: (616) 396-6521 Fax: (616) 396-7789 Web: www.macatawa.org/~padnos	Contact: Bob Klink Provide transportation, recycle, and disposal of lead and lead contaminated materials. Can accommodate necessary quantities
Northern Mountain State Metals, Inc. 1823 Morgantown Ave. Fairmont, WV 23554 Phone: (304) 367-1207 Fax: (304) 367-1208 Web: www.mountain.net/~nmsm	Contact: Tim Gooch
A & B Recycling 212 First Street Ft. Oglethorpe, Georgia 30742 Phone: (708) 866-7098 Fax: (708) 866-7068	Contact: Lamar Bearden
American Waste Transport & Recycling, Inc. 421 Jamaica Dr. Cherry Hill, NJ 08002 Phone: (800) 673-7998 Fax: (609) 985-1778 Web: www.amwaste.com/#lead	Contact: Bruce Levin Provide transport, recycle, and disposal of lead waste including lead paint waste, lead waste debris, and batteries
Edelstein Recyclers 1320 Lagrange Toledo, OH 43608 Phone: (419) 241-5274 Fax: (419) 241-2191	
R. J. Brown Phone: (419) 241-5274	
Environmental Recyclers Phone: (419) 269-1443	

Other benefits of purchasing and installing a grinder exist in addition to the financial advantage. By grinding the polypropylene, the volume of material is reduced which lowers the transportation costs and also increases the value of the material to the customer since further processing will not be needed.

Another option for the polypropylene is to have an outside vendor recycle it. Findlay Foam, a local vendor, has expressed interest in the battery cases and buttons. Some vendors interested in recycling polypropylene are listed in Table 8.12 along with their conditions. All prices are given in dollars per pound.

TABLE 8.11 GRINDER COST ANALYSIS	
Annual estimated polypropylene generation	$65,964
Estimated grinder cost	$25,000
Annual deducted labor expense	$11,400
Annual maintenance expense	$1500
Annual additional labor expense	$9120
Annual additional electrical expense	$75
Annual cost of lost space	$484
SL depreciation (7 years)	$3571
Tax rate	35.0%
Annual revenue after depreciation & taxes	$14,257
MARR	15.0%
IRR	54.3%

8.10.4 PAPER

The manufacturer generates 99 tons of paper per year. Included in this waste stream are mixed office paper, paper towels, NCR paper, computer printout paper, old newsprint, greenbar computer paper, envelopes, and carbon paper. Approximately 48.5 tons of this stream is recyclable and up to 30.5 tons could be eliminated by paper towel alternatives as illustrated in Sec. 8.10.8. Of the recyclable amount, a minimal amount of office paper is collected and recycled by the waste hauler at charge of $10 per month. However, the company is not paid for their waste paper. There are paper

TABLE 8.12 SAMPLE LIST OF PLASTIC VENDORS	
Findlay Foam	**Marsh Plastics, Inc.**
1831 East Manhattan Blvd.	4043 Maple Road
Toledo, OH 43608	Amherst, NY 14226
727-8090	(716) 834-6500
Contact: Jim Johns	$0.10 (dependent upon plastic packaging)
Alternative Plastic Services	**CSI**
200 Brown Street	(800) 933-6889
Lawrenceburg, IN 47025	$0.04–$0.05
(800) 219–8734	
$0.05–$0.07	**Mr. Plastic Recyclers, Inc.**
(they can only pick up once a month)	(614) 229-8488
	$0.11–$0.18

TABLE 8.13   PROPOSED OFFICE-PAPER RECYCLING AND SOURCE-REDUCTION STRATEGIES

MATERIAL	ANNUAL NONHAZARDOUS GENERATION (LB)	INCREASE RECYCLING ACTIVITY TO (PERCENTAGE)	REDUCTION COST SAVINGS (DISPOSAL & REVENUE) (DOLLARS)	TOTAL WASTE RECYCLED (CURRENT & INCREASE)
Paper	87,140	50	1545	43,570
NCR	Not recyclable			
Computer printout (CPO)	14,126	85	850	12,007
Mixed office paper (MOP)	8044	85	309	6837
Cardstock	3816	85	147	3244
Greenbar	1203	85	46	1023
Envelopes	1073	85	41	912
Carbon paper	Not recyclable			
Newsprint (ONP)	972	85	35	826
Total	116,374		2973	68,419

collection boxes located near the copiers and printers and scattered in various offices. However, less than 40 percent is currently recycled.

Therefore, one of the best strategies for recycling some waste paper is to fully implement an office-paper recycling program. Table 8.13 illustrates the potential for recycling based upon the current office paper generation at the company. A description of how to successfully implement such a program is described in Sec. 8.10.14. Paper recyclers are also listed in the mentioned section.

8.10.5   FLOOR SWEEPINGS

There are 91,052 lb or over 45.5 tons of floor sweepings generated on the plant floor each year. There are two types of floor sweepings used in the facility: one is chemical based, the other is sawdust. It is not possible to recycle either type of floor sweepings due to lead contamination on the plant floor. However, it may be possible to reduce the amount of floor sweepings that are sent to the smelter at an unknown cost.

To calculate the possible reduction four steps must be followed:

1 Determine the volume of the two sweeping compounds needed to clean up a standard or representative spill. Specify the compound that uses the lowest volume to clean up the average spill as C_l, and the other compound as C_h, and the volumes determined to clean up the spill as V_l and V_h. Then calculate the weight fraction of each type of sweeping compound used annually and designate them as W_l and W_h. These are found by dividing the respective annual weights by the total annual weight.

2 Calculate the current total volume of sweepings generated annually, V_T, by summing annual volumes for each compound, V_L and V_H. These volumes are calculated by multiplying the weight fraction by the total weight (91,052 lb) and dividing by the density.

3 Calculate the lowest possible annual total volume, V_{LT} by adding V_L plus V_H and multiplying by V_l and dividing by V_h.

4 Determine the possible volume reduction for the floor sweeping, R, by taking the difference between the total volume, V_T and the lowest total volume, V_{LT}. Also consider whether the compounds are needed to perform specific jobs and the cost and benefit of changing the current situation.

8.10.6 OXIDE INSERTS

There are 84,514 lb or over 42 tons of oxide inserts in the waste stream each year. The oxide insert waste is generated in the COS area of the plant. Due to lead contamination oxide inserts are considered hazardous materials, non-recyclable and therefore must be sent to the smelter. An updated process analysis of the carbon oxide sulphide (COS) area may highlight opportunities to decrease insert process scrap.

8.10.7 WOOD

There are 79,447 lb of wood going into the waste stream each year. Wood is reusable and recyclable and should be handled in such a way that it is beneficial to both the company and the environment. The waste assessment team observed that the wood consists of particleboard, pallets, and assorted broken pieces. The wood is collected by BFI and most likely chipped and sold as environmentally friendly mulch. The market for chipped wood is saturated and therefore particleboard and broken pieces cannot be sold for revenue. However, a market remains for standard-sized pallets in good condition. Table 8.14 is a list of vendors who purchase used, standard-sized pallets.

Unfortunately, wood was not separated into the three components (particleboard, broken pieces, and pallets) and therefore it is not possible to estimate the disposal-cost savings and revenue generation from selling the pallets.

8.10.8 PAPER TOWELS

Currently, the manufacturer purchases 36 cases of restroom paper towels per month. These are used to supply the seven restrooms within the facility. The rolls are purchased at $34.50 per case of 12, or $2.88 per roll. This results in a cost of $1242 per month and $14,904 per year spent on paper towels. Wall-mounted hand dryers were researched to determine whether or not it would be beneficial to install hand dryers instead of paper

TABLE 8.14 SAMPLE LIST OF WOOD RECYCLING VENDORS	
Binker Geo Wooden Pallets Contact Name: Mr. Craig Binker 28961 Oregon Road Perrysburg, Ohio 43551 Tel: (419) 666-3185	Final terms upon inspection
Pallets and Containers Corp. of America 901 Buckingham Toledo, Ohio 43607 Tel: (419) 255-1256 Fax: (419) 241-2833	Final terms upon inspection 3 ft × 3 ft only size accepted @ $0.50 apiece
Superior Pallet 2445 Hill Avenue Toledo, Ohio 43165 Tel: (419) 382-0693 Fax: (419) 536-4741	Final terms upon inspection Will only pay for 48 in × 40 in with four-way entry

towel dispensers. Not only does warm air drying help to eliminate viruses and bacteria from washed hands, but it is also environmentally and economically beneficial.

Each case of paper towels weighs 29 lb, equating to 6.3 tons of paper towels per year. Note that the waste assessment team's trash sorts indicated that about 30.5 tons of paper towels are consumed per year. While most of these towels are not used for drying hands, there still may be additional opportunities for reducing paper towel usage. This should be investigated.

The World Dryer Corporation was contacted at the following location:

World Dryer Corporation

Scott Kerman, ext. 115

1-800-323-0701

5700 McDermott Drive

Berkeley, IL 60163

Tel: (708)449-6950

Fax: (708)449-6958

Table 8.15 is a cost-benefit analysis for purchasing hand dryers. This analysis assumes a 100 percent hand dryer usage rate and does not account for increased electrical usage due to the hand dryers.

As stated, the manufacturer uses 6.3 tons of paper towels per year, which translates to 106 trees and 18.8 yd^3 of landfill space. Purchasing hand dryers would completely

TABLE 8.15 HAND DRYER COST COMPARISON		
	MODEL A SERIES (PUSH BUTTON)	**MODEL XA5 SERIES (AUTOMATIC)**
Cost of purchase	$302.50 per dryer	$346.50 per dryer
Savings per year	$14,904	$14,904
Payback period	One dryer for six restrooms, two dryers for men's restroom: 1.95 months One dryer for five restrooms, two dryers for women's restrooms, four dryers for men's restroom: 2.68 months One dryer for five restrooms, four dryers for women's restrooms, eight dryers for men's restroom: 4.14 months	One dryer for six restrooms, two dryers for men's restroom: 2.23 months One dryer for five restrooms, two dryers for women's restrooms, four dryers for men's restroom: 3.07 months One dryer for five restrooms, four dryers for women's restrooms, eight dryers for men's restroom: 4.74 months

eliminate paper towel waste and subsequently reduce the amount of trees and landfill space consumed each year by the manufacturer. Also, using hand dryers decreases the spread of viruses and bacteria that are able to survive on paper towels.

8.10.9   FOOD WASTE

Food waste and food-contaminated waste will not be discussed because it is not recyclable. However, the manufacturer may want to consider reducing food packaging waste through its vending machine selections.

8.10.10   LABEL BACKINGS

Approximately 17 tons of label backings enter the company's waste stream each year from the shipping area. According to the company's personnel, this label backing has been recycled in the past. After several inquires, the identity of the past recycler is still unknown. Previous research by the waste assessment team into the reuse/recycling of label backing has not yielded any local alternatives. Plastics recyclers may be interested in label backing; however, local recyclers have not shown interest at this time. Further research could be done to determine if this material is reclaimable or can be eliminated altogether.

8.10.11   METAL BANDING

There are 31,535 lb of metal banding going into the company's waste stream each year. OmniSource recycles the metal banding disposed of by the company. However, there may be opportunity for the manufacturer to sell the metal banding to a metal recycler rather than paying to have it hauled away. The company would need to determine the material composition of the banding in order to investigate metal recyclers.

8.10.12   NON-MAJOR RECYCLABLE WASTE STREAMS

The following is a list of recycling opportunities for waste streams that are not one of the 11 major waste streams.

Used beverage containers (pop cans)   The company generates about 4645 lb of aluminum pop cans per year. Aluminum pop cans are easily recyclable. Another benefit of recycling pop cans is that proceeds from recycling could be donated to a public relations project. Current market value for aluminum is $0.60 per pound, according to the Global Recycling Network. This would create annual revenue of about $2787 from selling the cans.

In order to ease the maintenance and storage of pop cans at the facility, a can crusher could be installed. The waste assessment team performed an economic analysis to determine if this was cost justifiable. Table 8.16 illustrates the result.

Unfortunately, purchasing a can crusher is not currently cost justifiable for the company because the annual revenue after depreciation and taxes is negative.

An aluminum can recycling program can still be instituted at the company by setting up areas to collect cans and contracting with a local recycler. Some aluminum recyclers are shown in Table 8.17 along with their terms.

Film plastic (low density polyethylene)   The thick plastic wrap used for shipping is potentially recyclable. The company currently disposes approximately 9.4 tons of plastic wrap per year. Previously, attempts had been made to recycle the plastic

TABLE 8.16 ECONOMIC ANALYSIS OF INSTALLING A CAN CRUSHER	
Annual weight of aluminum cans	4645 lb
Estimated crusher cost	$15,000
Annual deducted labor expense	$2280
Annual maintenance expense	$1000
Annual additional labor expense	$5244
Annual additional electrical expense	$90
Annual cost of lost space	$198
SL depreciation (7 years)	$2143
Tax rate	35.0%
Annual revenue after depreciation & taxes	–$202

TABLE 8.17   SAMPLE LIST OF PAPER RECYCLERS

Browning-Ferris Industries		As a customer, the company will be
1301 E. Alexis	Phone: 729-2273	charged an additional $4 to
Toledo, OH 43614	Fax: 848-8100	current rate
Edelstein Recycling Center		
1320 Lagrange	Phone: 241-5274	
Toledo, OH 43508	Fax: 241-2191	
Lake Erie Recycling		$0.43/lb (total collection weight
1011 Malzinger	Phone: 726-4446	under 200 lb)
Toledo, OH 43612	Fax: 726-4808	$0.45/lb (total collection weight
		over 200 lb)
Omni Source		$0.45/lb, will provide pickup for
5000 N. Detroit Ave.	Phone: 537-9400	more than 40,000 lbs
Toledo, OH 43607		
State Paper		
1118 W. Central Ave.	Phone: 243-5567	
Toledo, OH 43610		

wrap. To date, this effort has been unsuccessful because of a 4 in by 11 in paper label adhered to each piece of plastic. Most recyclers will not accept plastic wrap with labels. Three alternatives exist for alleviating this problem:

1 First option would be to hire an outside firm to remove the labels and ship the plastic wrap to recyclers. Lott Industries would be suitable for this type of work. For more information, contact the Lott Industries sales office. Note that this option still requires disposal of the labels.
2 Second option would be to replace the paper labels with a plastic label that would not have to be removed prior to recycling. It is critical that the plastic label is the same type of plastic as the plastic wrap; otherwise it will have to be removed.
3 Third option would be to remove the label with in-house labor and ship the plastic wrap to recyclers. Note this option also requires disposal of the labels.

Further research to determine the most economic option for the manufacturer should be conducted. However, the manufacturer should base their decision on the ease of implementation, functionality, and cost. Please see Table 8.12 for a list of plastic recyclers.

8.10.13   PROPOSED CHANGES TO CURRENT WASTE HANDLING PRACTICES

Given current observations from the waste composition in conjunction with the potential waste reduction figures, the waste disposal needs of the company should be evaluated from the standpoint of the capacity of its waste collection system.

Since much of the company's disposal is done on an on-call basis, the hoppers are most likely full at disposal. It is critical that this practice remains in order to most economically dispose of material. However, scheduled pickups are generally less expensive than on-call pickups. This would require the company to adequately estimate frequency of pickup. Volumes should be monitored from time to time to ensure that the hoppers are full when serviced. Also consider negotiating scheduled pickups with the waste haulers and requiring them to note the percentage the hopper is full per pickup on the invoices.

Also, investigation of source reduction and in-plant reuse of materials might provide the means to further reduce disposal costs. Currently, the company has three end-disposal receptacles (plus a collection of drums and OCC bales). The capacity requirements of these containers and alternatives for end-disposal receptacles are presented below.

1 *Steel hopper*—OmniSource Corporation provides a 20-yd³ hopper for steel free of rental charge. OmniSource pulls the hopper on an on-call basis about 3 to 4 times per year. The charge for pulling the hopper is $105 per pull minus the market value of the steel. There is also a hopper for old machinery that is provided and pulled by OmniSource. However, this hopper is said to be temporary. After investigating other possible avenues of steel removal, it became apparent that OmniSource remains the ideal steel-removal service for the company's current situation. All either do not handle scrap steel or require a faster turnaround than the manufacturer can provide. Six recyclers were contacted during this investigation.

2 *Wood hopper*—There is a 40-yd³ hopper for wood, rented from BFI for $50 per month. BFI pulls the hopper an average of 2 times per month at $260 per pull. If the manufacturer requires more than two pulls in a month, the charge further increases to $320 per pull. The hopper is kept free of most contaminants. Nails and screws are left in the wood and represent the only contaminants in the hopper.

3 *Compactor*—A 42-yd³ compactor for general waste is rented from BFI for $118.45 per month. BFI pulls the compactor 5 to 6 times a month on an on call basis for a charge of $297.93 per pull. The components of the general waste stream are food, pop cans, office paper, paper towels, boxboard, and others. Of the waste in the stream, office paper, aluminum pop cans, and plastic are readily recyclable. The paper towel volume can be reduced by installing electric hand dryers. Removing the office paper, aluminum cans, film plastic, and paper towels from the waste stream would reduce the number of pulls per month by 2.4 pulls, equating in a reduced pull charge of $715. Also, the company can contact another local vendor who provides general waste hauling services and compactor rental.

4 *Baled OCC*—OCC is currently baled and then recycled through Lake Erie Recycling. The terms are that in exchange for the OCC, Lake Erie provides 60 marshmallow boxes which are worth $5.00 per box to the manufacturer, plus the revenue obtained from the sale of the baled OCC at market price. Lake Erie picks up the OCC generated by the manufacturer. The pickups are on-call when 40 bales (1100 lb each) are ready. This generally occurs every 2 to 3 weeks.

5 *Drums (lead waste)*—Many of the materials generated as waste on the plant floor are sent to a smelter due to the lead contamination inherent in the process. These materials are packaged in 55-gallon steel drums before being transported to the smelter. The drums are piled up outside the rear of the building and are then taken in truckloads to the smelter. The Horwitz & Pintis Company supplies the drums. Additional steel drum suppliers are listed below.

From visual inspections made during walkthrough, it was found that nonhazardous and hazardous materials are incorrectly disposed of in inappropriate containers. A large percentage of apparently nonhazardous materials were found in hazardous waste containers. This material could be disposed of more economically in the general trash since the waste assessment team assumes that it costs the manufacturer more to dispose waste via the independent waste hauler.

Further research should be conducted to determine which materials actually need to be disposed of through the smelter.

8.10.14 IMPLEMENTING A SUCCESSFUL OFFICE-PAPER RECYCLING PROGRAM

The company generates approximately 137,338 lb of nonhazardous paper waste annually and receives no revenue from recycling it. Therefore, it would be profitable to find a way to reduce the amount of paper generated from the facility and a way to recycle it for revenue.

Of the 137,338 lb of nonhazardous paper disposed annually, 95,184 lb are paper plus mixed office paper (MOP). In addition, there is also a significant amount of computer printout (CPO). Under the current arrangement, this paper waste, along with all other types of paper, is hauled by the waste hauler to a landfill, and no revenue from recycling is received. This translates to approximately $5446 per year in paper-disposal costs. Calculations for this figure are shown in Table 8.18. Table 8.19 illustrates the potential annual revenue by instituting a paper-recycling program.

The company began an office-paper recycling program about 1 year ago. BFI collects toters of mixed office paper for a total of $10 per month. No revenue is achieved in this program. Also, BFI believes that less than 3000 lb of mixed office paper is collected per month. If the company were to step-up the paper-recycling program, not

TABLE 8.18 ANNUAL COST OF PAPER DISPOSAL	
Total paper disposed of per year	137,338 lb
Size of BFI compactor	42 yd³
Cost per pull	$297.93
Hopper rental cost per month	$118.45
Approximate paper density	250 lb/yd³
Total paper disposed of per month	45.8 yd³
Resultant annual cost	$5446

TABLE 8.19 ESTIMATED POTENTIAL REVENUE FROM PAPER RECYCLING				
MATERIAL STREAM	ANNUAL RECYCLED (LB)	*CURRENT MARKET PRICE (PER TON) (MINIMUM)	*CURRENT MARKET PRICE (PER TON) (MAXIMUM)	POTENTIAL ANNUAL REVENUE
Paper plus MOP	95,184	$0.00	$30.00	$1427.76
Computer printout	14,126	$80.00	$110.00	$776.93
		Total estimated revenue		$2,204.69

only could they achieve potential revenue from the sale of the paper, but they would also cut disposal costs.

8.10.15 OFFICE-PAPER RECYCLING PROGRAM

An easy and cost-effective program that can be established in any business is an office-paper recycling program. Paper is relatively easy to collect but requires a cooperative effort in order to lower costs through proper waste management. A successful office-paper recycling program can provide a stimulus for a company's future efforts in the minimization of other costly waste materials such as wood, cardboard, and production waste. The following is an outline for instituting an office-paper recycling program at the company.

1 Identify an office recycling leader When establishing a new paper-recycling program, it is important to pick employees on a volunteer basis to lead the recycling program. This person is responsible for talking to coworkers about proper locations for different types of papers, identifying the kinds of paper that can be recycled in these containers, and concerns about the program. A union workplace should encourage union workers to get involved in this position.

2 Contact vendors Recycling vendors should be contacted to solicit current prices for the items in Table 8.20.

3 Create a plan of focus The recycling leader should decide which products the manufacturer should segregate. This should be based on estimates of which materials are generated in quantities worth segregating. We recommend that the manufacturer segregate office paper into three categories, if possible: white ledger, computer printout (CPO), and mixed office paper (MOP). These categories will supply the highest return on investment.

4 Investigate locations and types of recycling receptacles The recycling leader, with the help of coworkers, should identify the location and number of recycling receptacles needed. Disposal receptacles should be present at any location where

TABLE 8.20 PAPER PRICES AT TIME OF AUDIT			
TYPE	DESCRIPTION	CURRENT PRICE PER TON (MINIMUM)	CURRENT PRICE PER TON (MAXIMUM)
White ledger	White fibrous paper—segregated	$30.00	$90.00
Colored ledger	Colored fibrous paper—segregated	$10.00	$35.00
Mix office paper	All grades of office paper	$0.00	$30.00
Newspaper		$0.00	$10.00
Magazines		$0.00	$5.00

recyclable paper is being generated. This usually includes all desks, copiers, and printers. Once generation points are identified, estimate the volume of paper that will be generated at each point. This task provides information for purchasing recycling receptacles. Receptacles vary in size and appearance.

5 Choose a vendor   Factors to consider in choosing a vendor include

- *Convenience of service*—Will the recycler pick up the materials? Will pickup be as needed or as scheduled?
- *Provision of receptacles*—Will the recycler provide free, lease, or sell recycling storage receptacles?
- *Pricing*—Will the recycler pay for the material and continue to pick up the material during market downswings?

Depending on the company's motivation, the factors for choosing a vendor may be equally important. However, in reality, one factor often is sacrificed to accomplish another. For example, some vendors will provide recycling storage receptacles, will pick up material, but will not pay for the material.

6 Purchase recycling receptacles   The capacity of these receptacles should be relative to the paper generated at each location to avoid excess dumping. For inconspicuous locations, used cardboard boxes can be utilized as inexpensive recycling receptacles. For visible locations, attractive receptacles may be purchased. Table 8.21 is a list of suppliers of recycling receptacles.

The waste assessment team recommends the following standard purchases if manufacturer's specific needs do not dictate otherwise.

- A 11 in × 17 in stand-up cardboard box for each desk.
- A 14-gallon plastic bin for copiers.
- A 14-gallon plastic bin or stackable 11-gallon bin for the central segregation point.

TABLE 8.21   SAMPLE LIST OF RECEPTACLE SUPPLIERS

Bruce Mooney Associates, Inc.
1849 Fairhill Road
Allison Park, PA 15101
(800) 454-2686
FAX: (412) 367-1015
Cardboard desk side

Busch Systems International, Inc.
130 Saunders Road
Unit #7
Barrie, Ontario L4M 6E7
(800) 565-9931
FAX: (705) 722-8972
Plastic desk side and general

Fibrex, Inc.
3734 Cook Blvd.
Chesapeake, VA 23323
(800) 346-4458
FAX: (757) 487-5876
Plastic desk side and general

Jedstock, Inc.
P.O. Box 4405
Warren, NJ 07059-4405
(908) 754-0404
Fax: (908) 754-2247
Cardboard desk side
Plastic desk side and general

OTTO Industries, Inc.
12700 General Drive
P.O. Box 410251
Charlotte, NC 28241-0251
(800) 227-5885
FAX: (407) 588-5250
Plastic general

Recy-CAL Supply
40880B County Center Drive, Ste. P
Temecula, CA 92591
(800) 927-3873
FAX: (909) 695-5228
Cardboard desk side and general
Plastic desk side and general

R.P.I.
P.O. Box 1929
Andover, MA 01810
(800) 875-1735
FAX: (508) 475-1983
Plastic desk side and general

---

7 Implement the recycling program   Train personnel about the recycling program. Training should emphasize benefits, but should also point out costs and extra effort required by personnel to make the program successful. The basic program can be instituted as follows:

- Place recycling receptacles at each desk.
- Make users responsible for emptying each receptacle when it is full. Union cleaning crews should empty the boxes when applicable.
- Transport receptacle contents to a central location in the office where larger receptacles are located.
- Segregate paper by grade into the appropriate box.

8 Establish generation rates   The volume of each segregated bin multiplied by the number of pickups should be monitored to establish the generation rate for each type of paper. Once a rate is determined, emphasis should be on reducing the amount of paper used and monitoring compensation from the recycling vendor. Periodic evaluation of the program will allow it to evolve.

9 Employee involvement   Post container locations for segregated recycling bins clearly throughout the facility. Emphasize the importance of not contaminating a segregated product with food or other waste. Contaminated recyclable are sometimes treated as trash and landfilled. These policies should be regularly reinforced.

This paper collection program should be monitored periodically to ensure that all the staff members at the company are disposing of paper in the recycling bins. Also, find out if any changes or improvement in the collection and segregation method are necessary.

10 Monitor and reevaluate vendor services   This is an important on-going detail of any vendor/customer relationship. Customers must require vendors to provide adequate services for all types of materials receiving and disposal. This includes oil and solvents as well as solid waste and recycling services.

11 Reduce the amount of paper used   Once the paper-recycling program is in place, the disposal schedules become indicative of amount of paper being disposed. At this point, the company can set reduction goals for the future.

12 Other practices   There are several other practices that can help reduce the amount of office paper waste:

1 *Only place contaminated paper in the hazardous containers*—There was a significant amount of paper found in the hazardous containers. Any paper placed in these receptacles can not be recycled.
2 *Purchase recycled paper*—Due to the fluctuation of the market for recycled goods, it is important that companies strive to purchase goods manufactured from recycled products. Purchasing recycled products enables recyclers to pay higher prices for your reusable materials. This paper can be used in copiers, printers, and fax machines. Recycled content paper now performs as well as fresh paper products and should pose no problem in equipment.
3 *Make two-sided photocopies*—This can not only save paper, it can cut mailing costs and reduce the need for filing cabinets. If your department's copier doesn't have double-sided capability, suggest that it is upgraded.
4 *Limit the use of "post it" notes*—They are usually colored and hard to recycle.
5 *Use the unprinted side of paper*—Reuse computer paper that has print on one side. Either load this back into the printer or use it for handwritten notes.
6 *Circulate documents*—Rather than making multiple copies of a memo or document, circulate them to reduce paper.

7 *Utilize electronic mail*—If available, this option can greatly reduce waste paper volume by sending messages and files via computer to be viewed on a screen rather than on paper.

8.10.15 CORPORATE WASTE MINIMIZATION STRATEGY

Following is a list of helpful suggestions to assist in the institution of a corporate waste minimization/pollution prevention strategy:

- Educate employees about the overall benefit of a corporate waste minimization strategy. Benefits include
  - Reducing waste disposal costs
  - Improving corporate Image
  - Stimulating the growing market for recyclables
  - Increasing employee awareness that may carry over personal practices
- Develop written corporate source reduction policies.
- Monitor and understand solid waste generation rates.
- Involve all levels of employees in the waste reduction program.
- Post container locations for segregated recycling bins throughout the plant. Emphasize the concession areas. Contaminated recyclable are sometimes treated as trash and landfilled. Therefore, stress the importance of not contaminating a segregated product with food or other waste. These policies should be regularly reinforced.
- Buy recycled products whenever possible and inform employees of this practice.
- Keep records of your waste minimization measures and publicize successes.
- Reward employees by recognizing their efforts and success. Rewards may or may not be financial. A thank you is always appreciated.

The following example illustrates the global impact of a simple waste minimization program:

Every ton of recycled paper produced protects 17 trees, saves approximately 7200 gallons of water, conserves 4200 kWh of energy, preserves 410 gallons of fuel, and frees upto 3 yd^3 of landfill space.

8.11   Step 8: Documentation and Development of the Deployment Plan

The goal of this phase of the waste minimization process is to translate the list of accepted alternatives into an achievable implementation plan and to document the selected alternatives. The remainder of this section provides an overview of the deployment plan, a discussion regarding obtaining funds, and details regarding each of the 12 sections of the deployment plan. Finally, a template for the deployment plan is provided.

8.11.1   OVERVIEW OF THE DEPLOYMENT PLAN

A deployment plan is a comprehensive document that details the what, when, where, and how of each alternative. It serves as an implementation guide to aid the organization in achieving its waste minimization goals. The deployment plan describes the set of tasks necessary to implement a program such that it can be effectively transitioned within the organization. The deployment plan provides a detailed schedule of events, persons responsible, and event dependencies required to ensure successful cutover to the new system.

Deployment can impose a great deal of change and stress on the employees. Therefore, ensuring a smooth transition is a key factor in the implantation process. The deployment plan should minimize the impact of the cutover on the organization's staff, production system, and overall business routine. The waste minimization team leader usually is responsible for creating the deployment plan and it is implemented by the entire organization, so communication and feedback is critical in the development process. The waste minimization deployment plan consists of 10 sections:

1 Cover page with official approvals
2 Overview
3 Assumptions, dependencies, constraints
4 Operational readiness
5 Timeline for implementation
6 Training and documentation
7 Notification of deployment
8 Operations and maintenance plan
9 Contingency plan
10 Appendices

8.11.2   OBTAINING FUNDING

Many waste reduction alternatives involve cost savings via cost reductions and process efficiencies, not the generation of additional revenue. This can pose a problem for the waste minimization team if they are competing with a limited amount of capital funds to support the proposed alternatives. If the organization is more focused on the creation of future revenues, cost savings projects may not get the same attention. To remedy this, an understanding of the approval authority for capital projects within the organization can be a big help in determining the best route to seek funds. For example, within larger organizations, smaller projects within a $5000 to $10,000 range can generally be approved by a local manager whereas larger projects (over $10,000) must be approved at a regional or vice president level.

Oftentimes, organizations evaluate projects via a committee. An understanding of who is on this committee and the process that they use to rate projects can help to maximize the alternatives chances for success. Meeting with the committee members and discussing the merits of the project in order to gather their initial feedback can go a long way in improving the submission to the committee. Some key selling points to the committee may include

■ Past experience in the field by the waste minimization team.
■ What the market and the competition is doing.
■ How the implementation program fits into the company's overall business strategy.
■ Advantages of the proposal in relation to competing requests for capital funding (i.e., environmental impacts).

Even when a project promises a high internal rate of return, some organizations will have difficulty raising funds internally for the initial investment. In this case, the organization should examine options for external funding. The two options generally considered are private sector funding and government assisted funding. Private sector funding includes bank loans. Government funding may be available in some cases. It is usually worthwhile to contact the state's Department of Commerce or the federal small business administration for information regarding loans for pollution prevention or waste reduction projects. Some states can provide technical and financial assistance as well. For example, in Ohio, the Department of Natural Resources has historically sponsored grants to private companies to reduce waste. Other potential government contacts include

■ The Environmental Protection Agency
■ The U.S. Department of Natural Resources
■ The U.S. Green Building Council
■ The U.S. Department of Development
■ State environmental protection agencies
■ Local waste management districts

8.11.3 CONTENTS OF THE DOCUMENTATION AND DEPLOYMENT PLAN

This section provides an in-depth discussion of the contents of the deployment plan. One of the most important components of the plan is the cover sheet, which contains the official signatures of the executive management team indicating full approval of the deployment plan.

Cover sheet   In addition to the official approvals, the cover page should include the company name, the title of the program (such as "The Solid Waste Minimization Deployment Plan"), the date the report was written, and the author or department that created the report. If the organization uses specialized tracking or identification numbers for projects, these should be included too.

Overview This section provides the purpose, business context, and project summary in an executive summary format. It identifies the purpose of the deployment plan and its intended audience, describes the business processes that will be modified as a result of the deployment, and provides a summary of the plan. It also includes an overview of activities necessary to get the program launched into the business environment such as installation, configuration, and initial operational activities. It also includes details regarding the location that the assessment was conducted, the dates that it was conducted, the names of the individuals conducting the audit, and a map of the facility.

Assessment findings and recommendations This section provides a summary of the principle findings, recommendations, and observations. This section discusses the data collected during the assessment in terms of waste stream composition, volumes, and disposal costs. It also provides a listing of the approved alternatives for implementation.

Assumptions, dependencies, and constraints This section describes the assumptions about the current organizational capabilities and the day to day operations of the program. In addition it describes the dependencies that can affect the deployment of the program, such as working within the constraints of housekeeping or waste-removal contracts and the factors that limit the ability to deploy the program.

Operational readiness This section describes the preparation required for the site on which the program and alternatives will operate. It defines any changes that must occur to the operational site and specifies features and items that should be modified to adapt to the program alternatives. It also describes the method for use in assessing deployment readiness and identifies the configuration audits and reviews to be held after the program is tested and accepted and before the program or equipment is installed in the production environment.

Timeline for implementation This section describes the timetable for the implementation of each waste minimization project or program. It serves as the control document to facilitate communication within (departments) and outside the organization (suppliers and contractors). A Gantt chart is an excellent diagram to include in this section as it displays the order or precedence of events and the percent completion to the established timeline. Figure 8.35 shows an example.

Training and documentation This section describes the plans for preparing and conducting training for the purpose of training all stakeholders regarding program or process changes. It also identifies and describes each document that will be produced for the purpose of aiding in implementation, support, or use of the new programs. The section should include the activities needed to create each document.

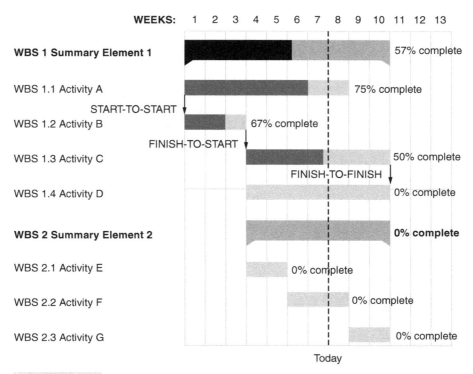

Figure 8.35   Sample implementation timeline.

Notification of deployment   This section describes the method of notifying all stakeholders of the successful release of all waste minimization programs and identifies stakeholders and groups requiring notification.

Operations and maintenance plans   This section describes the maintenance and operations activities for each program or piece of equipment. For example, preventative maintenance schedules should be included for each new piece of equipment.

Contingency plan   This section describes the contingency plan to be executed if problems occur during deployment activities. A contingency plan is devised for specific situations if or when things do not occur as expected or circumstances change. Contingency plans include specific strategies and actions to deal with specific variances to assumptions resulting in a particular problem, emergency, or state of affairs. They also include a monitoring process and "triggers" for initiating planned actions. They are very useful to help governments, businesses, or individuals to recover from serious incidents in the minimum time with minimum cost and disruption.

Appendices   This section contains all relevant appendices related to the project. The alternative evaluation sheets should be included in this section.

8.11.4 DEPLOYMENT PLAN TEMPLATE

Following is a template that may be used to create the deployment plan:

DEPLOYMENT PLAN

[Agency/Organization Name]
[PROJECT NAME]

VERSION: [VERSION NUMBER]   **REVISION DATE: [DATE]**

Approver Name	Title	Signature	Date

Section 1. Overview

1.1 Purpose
Identify the purpose of the deployment plan and its intended audience.

1.2 Business Context
Describe the business processes that will be modified as a result of the deployment specified in the deployment plan.

1.3 Summary
Provide a summary of the deployment plan. Include an overview of activities necessary to get the product into a production environment such as installation, configuration, and initial operational activities.

Section 2. Assessment Findings and Recommendations

2.1 Overview of Major Waste Streams
Describe the major waste streams in terms of annual generation and disposal costs including charts and graphs.

2.2 Recommendation and Approved Alternatives
List the approved alternatives and attach the feasibility analysis sheet with all approval signatures.

Section 3. Assumptions, Dependencies, Constraints

3.1 Assumptions
Describe the assumptions about the current capabilities and use of the program when it is released to production.

3.2 Dependencies
Describe the dependencies that can affect the deployment of the program.

3.3 Constraints
Describe factors that limit the ability to deploy the program.

Section 4. Operational Readiness

4.1 Site Preparation
Describe the preparation required for the site on which the system will operate. Define any changes that must occur to the operational site and specify features and items that should be modified to adapt to the new product. Identify the steps necessary to assist the customer in preparing the customer's designated sites for installation of the accepted programs.

4.2 Assessment of Deployment Readiness
Describe the method for use in assessing deployment readiness.

4.3 Product Content
Identify the configuration audits and reviews to be held after a product is tested and accepted and before the product is installed in the production environment.

4.4 Deviations and Waivers
Provide additional information regarding any waivers and deviations from the original software requirements specification.

Section 5. Timeline for Implementation

Describe activities for a phased function rollout or a phased user base rollout.

Section 6. Training and Documentation

6.1 Training
Describe the plans for preparing and conducting training for the purpose of training all stakeholders on the use of the product.

6.2 Documentation

6.2.1 Documents
Identify and describe each document that will be produced for the purpose of aiding in installation, support, or use of the product.

6.2.2 Documentation Activities
Describe activities required to develop each document.

Section 7. Notification of Deployment

Describe the method of notifying all stakeholders of the successful release of a product. Identify stakeholders and groups requiring notification.

Section 8. Operations and Maintenance Planning

Describe the maintenance and operations activities for the product.

Section 9. Contingency Plan

Describe the contingency plan to be executed if problems occur during deployment activities.

Section 10. Appendices

Include any relevant appendices.

8.12   Implementation and Execution of the Solid Waste Minimization Plan

A well-developed deployment plan based on viable options will yield poor results if the plan is not executed properly. There is no such thing as overcommunication when it comes to rolling out a new project or program. The three key components of a successful implementation and execution are following the deployment plan, communication, and recognizing the need to adjust in certain circumstances.

To facilitate the communication process, at a minimum, weekly progress meetings should be held with all key stakeholders. These meetings should focus on the status of each project versus the timeline and established goals. An agenda and the project timeline should be prepared in advanced and serve to lead the discussion. The task leader (as determined in the deployment plan) should take the lead role in discussing the status of each project or program. Any obstacles or delays should be discussed so that the team may determine solutions.

During the deployment process it is critical not to overwhelm employees with process changes. Effort should be taken to ensure that all employees are aware of upcoming changes, timelines, and the reasons behind the change. This can be accomplished with service talks, postings, or newsletters in paychecks. All three options may be used to ensure that the message is heard and that employees are not confused and buy in to the programs.

In general, less effort is required for operational and process changes. These options can usually be implemented in a much quicker fashion than equipment or material changes. Following is a general outline of the scope of an implementation effort:

- Approve the project or program
- Finalize the specifications and design for each alternative
- Submit and gather bid requests and quotes (if necessary)
- Complete and submit a purchase order
- Receive and install the equipment
- Finalize operating and maintenance procedures
- Train affected employees
- Start the project or program
- Complete regulatory inspections
- Track implemented project cost savings and waste reductions

8.13 Validate the Program versus Goals

Many companies require a validation process to ensure that projects and programs have met the goals that were set at the onset of the project. This includes validating the project or program was installed at or below cost, that it is operating within the expense and revenue limits, and that it is achieving the waste reduction goals. Even if an organization does not require a validation process, it can be a very valuable tool for future planning processes to identify were estimation errors occurred and take effort to correct them. Alternatives that do not meet the established goals or performance expectations may require rework or modifications. It is also critical to store warranties and contracts from vendors prior to the installation of the equipment. Also, the experience gained in implementing an option at one facility can be used to reduce the problems and costs of implementing options at subsequent facilities.

An alternative performance analysis should be completed for each equipment, process, or material change. The analysis provides a standardized method to compare project performance against estimates in terms of

- Project duration
- Implementation cost
- Operating expenses and revenue
- Waste reduction volume
- Cycle time and productivity
- Product or process quality
- Safety

It is useful to emphasize that the purpose of the worksheet is not a "gotcha game," but a method to improve future project estimates and learn from mistakes if applicable. In terms of project duration, the alternatives should be evaluated based on the time required to implement the alternative versus the original estimate. Explanations should be provided for large deviations, such as "an additional 2 weeks required to obtain building permits." Actual implementation cost should be analyzed versus estimates, in addition to operating expenses (including labor, materials, and utilities), revenue generation (from the sale of recyclable material), and cost savings from process changes or waste hauling costs. The waste reduction volume should be evaluated in a similar manner. For example, if the purchase of a cardboard baler was expected to reduce cardboard waste by 20 tons per week and the baler is only reducing the waste stream by 12 tons, a root cause analysis should be conducted to explore and improve the deviation. Any cycle time or productivity deviations from the original estimates should also be explored. These deviations could have a very negative effect on the organization's profitability and in most cases are very closely watched by upper management. The same goes for product quality. Finally, any safety concerns should be addressed immediately. A walk though by the team leader, safety captain, and area supervisor can quickly identify and resolve these issues. To aid in the validation process a worksheet is provided in Fig. 8.36. The validation process should be performed within 4 to 8 weeks of implantation.

8.14   Monitor and Continually Improve Performance

After the waste minimization program has been implemented and validated it must be monitored on a periodic basis to ensure that it is still performing as planned and to make any necessary adjustments. This includes monitoring the waste reduction amounts and operational, and financial performance versus the goals. In addition, emphasis should be placed on continuous improvement to enhance current waste reduction programs and to identify new opportunities. It may be beneficial to conduct period waste assessments, facility walkthroughs, or employee interviews by the original waste reduction team to accomplish these goals. When evaluating the program it is important to

- Keep track of program success and to build on past successes.
- Identify new ideas for waste reduction.
- Identify areas needing improvement.
- Document compliance with state or local regulations.
- Determine the effect of new additions to the facility or program.
- Keep employees informed and motivated.

In addition, consider reviewing the organization's waste removal receipts and purchasing records on at least a quarterly basis to ensure that the waste minimization

Waste Minimization Validation Worksheet

Company name: _____

Location: _____

Date: _____

Alternative description: _____

Alternative tracking number: _____

	Estimated	Actual	% Difference	Comments
Project Duration				
Space Requirements				
Implementation Cost				
Machine costs				
Machine cost				
Site development				
Material costs				
Building modification costs				
Permitting and inspection costs				
Contractor fees				
Start up costs				
Initial training costs				
Operating expenses				
Utility cost impacts				
Input material changes				
Labor cost impacts				
Supervision cost impacts				
Maintenance cost impacts				
Operating and maintenance supply impacts				
Changes in overhead costs				
Operating savings and revenue				
Reduced solid waste disposal costs				
Revenues from increased sale of recyclable material				
Revenues from the sale of by products				
Waste Reduction Volume				
Cycle Time				
Product Quality				
Safety				

Figure 8.36 Waste minimization validation worksheet.

program is working. New product or process changes should also be evaluated at the onset to ensure that the design minimizes environmental impact. This is easily accomplished by adding waste minimization review to the new product or process checklist or standard operating procedure.

The waste minimization program is a continuing process versus a one-time project. Generally, the first waste assessment and implemented alternatives will target only the high-volume waste streams. Once these high-volume streams have been reduced, the team can focus on smaller volume waste streams. From a systems standpoint, the ultimate goal of the team is to minimize all input materials into the facility and by-products generated by the facility. The frequency at which the additional waste assessments are conducted will depend on the budget of the company. In general, organizations that conduct assessments 1 to 4 times per year have achieved paybacks. In addition, if there are special circumstances that indicate the need for further review, a waste assessment should be conducted. These special circumstances include

■ A change in raw material or product requirements
■ Higher waste management costs
■ New regulations
■ New technology
■ A major event with undesirable environmental consequences (such as a major spill)

To be truly effective, an organizational culture of waste minimization must be fostered within the organization. Executive management must ensure this through repeated communications and acknowledgements for success stories from individuals or business units. This will make waste minimization an integral part of the organization's operations.

9

TRAINING AND IMPLEMENTATION

9.1 Introduction

Training is a critical element of any successful program launch. Training teaches all affected employees the "how" of each process change. This is in contrast to an education program that teaches the "why" behind a process change. Both are very necessary ingredients, but training's main thrust is the execution system. This training should be centered on the employees actually performing the waste minimization processes. The training should focus on an understanding of the basic principles behind the waste minimization efforts and the wise application and integration of these tools with other and current techniques. A just-in-time training approach is usually most effective so that employees will be applying the training very shortly after learning the concepts. A word should also be said regarding the delivery of the training. Poor delivery of training material can turn even the best material into a boring exercise that fails to impart useful know-how. The true test of effective training is not an enthusiastic student evaluation, but rather the student's ability to perform new tasks effectively on the job.

9.2 Strategy

The key strategy to training involves organizing and delivering material in such a way that trainees can immediately apply the concepts. The training will be fresh in their minds and their level of excitement with the waste minimization project will be at its highest. On-the-job application is also highly recommended at the shop floor level. For example, if a new cardboard baler is installed, the operators should be given a detailed training session on the machine's operation, a standard operating procedure, a safety talk, and a list of contacts in the event the machine malfunctions or if they have improvement ideas. One-on-one coaching should also be made available as needed.

During the training session, example problems and trainee exercises may be beneficial to allow the trainees to work together to solve waste minimization problems in their work units and identify new waste reduction opportunities. A group discussion can also be a useful component of the training exercise to gather feedback and employee perceptions. The trainees should be made aware of the key metrics of the waste reduction program. Specifically the environmental and economic impacts in their work units should be presented and discussed. Many employees have a desire to help the environment and by discussing the tons of solid waste that can be reduced per year by applying the waste minimization programs and process changes, the employees can gain a sense of contribution and motivation. By discussing the financial benefits of waste minimization to the organization the employees can be made aware that the programs will enhance job security by improving the bottom line of the organization. The tracking and monitoring methods of the waste reduction programs should also be discussed. Trend charts that display waste reduction and economic performance versus expectations should be presented. The location of these charts and how to read them should also be provided. Suitable locations include common areas such as lunch rooms or informational boards and should be in line with organizational policies. Figure 9.1 shows a sample performance chart that may be used as a template.

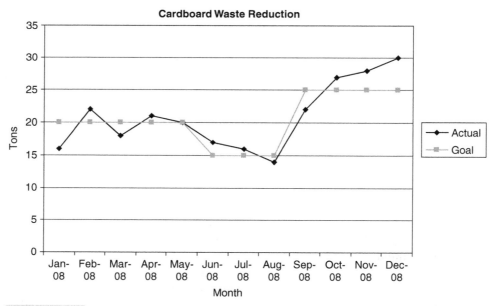

Figure 9.1   Sample performance-tracking chart.

In conjunction with the training session, the waste minimization and management team should carefully consider a certification or recognition process. Most trainees will be more motivated if there is some type of reward process; a simple thank you from the unit supervisor, a catered lunch, door prizes, or monetary prizes for meeting established goals. Employees may gain a stronger sense of team work if a certification process is involved with new equipment deployment. For example, if a new cardboard baler is installed, the work unit can be provided with a certification checklist. If all items on the checklist are met, the process can be officially certified and the work unit can be recognized and/or presented with an award mentioned above. Common elements of the checklist may include

■ Adherence to safety standards
■ Adherence to the standard operating procedure
■ Housekeeping and cleanliness
■ Achievement of waste reduction goals over a 3-month time frame
■ Achievement of cost-reduction/revenue goals over a 3-month time frame

9.3   Agendas for Training

A typical training agenda includes a 1-day workshop for the affected employees that discusses the new process changes, equipment, the goals of the program, tracking methods, continuous improvement for the waste minimization program, and the need for a waste reduction mindset to change the organization culture. Below is a list of lessons learned that may be useful when developing agendas:

■ Ensure integrity of the measurement and tracking system.
■ There is no such thing as overcommunication.
■ Ensure a feedback mechanism to strengthen the program and foster employee buy-in.
■ Offer employees suitable rewards for meeting the program goals (some organizations prefer time off, bonus checks, luncheons, or even a donation made in the organization's name to an environmental group or charity).
■ Existing databases are inadequate for the job.
■ Avoid diversions.

Table 9.1 is a sample agenda that may be used as a template.

TABLE 9.1 SAMPLE TRAINING AGENDA	
TIME	**AGENDA ITEM**
9:00 a.m. to 9:30 a.m.	Registration and continental breakfast
9:30 a.m. to 9:45 a.m.	Welcoming address by CEO
9:45 a.m. to 10:15 a.m.	Waste reduction program overview and goals
10:15 a.m. to 11:30 a.m.	Overview of program changes and new procedures
11:30 a.m. to 12:30 p.m.	Catered lunch
12:30 p.m. to 1:00 p.m.	Tracking reports and monitoring processes
1:00 p.m. to 1:30 p.m.	Feedback and continuous improvement
1:30 p.m. to 2:00 p.m.	Rewards program
2:00 p.m. to 3:30 p.m.	Facility tour and on the job training for new processes
3:30 p.m. to 4:00 p.m.	Question session and closing remarks

A separate 2-day training session/workshop should be held for executives and managers to gain a better understanding of the program and the facilitation process. The program should emphasize the manager's role in coaching employees and the execution and monitoring of the system.

10

COMMUNICATING AND
LEVERAGING SUCCESS

Oftentimes, the waste minimization team will discover that the various departments in the facility are working on similar projects and often had the same frustrations. For example, two departments may be recycling aluminum cans, but not utilizing the same vendor or method, reducing leverage against the vendor. There can be a lack of synergy between departments and no established method for them to communicate. The need for mentoring and communicating lessons learned can be satisfied in a variety of ways. Many organizations have been successful in accomplishing this goal by creating a Web site or intranet to discuss the current waste minimization projects, goals, and results. The internal Web site can contain

- A list of current projects and results
- A project-selection process that maps steps
- Templates for project proposals
- Templates for project reports
- Presentations completed by the waste reduction team and management
- Lessons learned
- Contests for projects of the month

Coordination of the waste minimization projects and Web site is critical to the communication efforts; if the Web site is not maintained, it is better not to use one. The waste minimization team leader or a dedicated executive leader is the best individual to monitor and control this Web site and facilitate communication. If this function is overlooked, an organization can waste time and resources in completing similar projects and constantly reinventing the wheel.

A key component of the communication process is project reports and status updates to upper management. These reports should be created in a one-page executive summary format. An effective report can consist of a few good tracking charts and associated text description. A periodic review process is a good idea as well.

Waste Minimization Status Report
Executive Summary

Company Name:_____   Team Leader:_____
Location: _____   Team Members: _____
Date: _____
Tracking Number: _____

Project Description

| |
| |

Waste Reduction Goal (tons/quarter)
Actual Waste Reduction (tons/quarter)
Percent Difference

Reasons for Differences (if needed)

| |

Corrective Action Plan (if needed)

| |
| |

Cost Reduction Goal ($/quarter)
Actual Cost Reduction ($/quarter)
Percent Difference

Reasons for Differences (if needed)

| |

Corrective Action Plan (if needed)

| |
| |

Please include year-to-date waste and cost reduction graphs

Figure 10.1   Executive summary report example.

An established process with periodic written evaluations for every accepted project is an efficient method to brief upper management. An appropriate time frame should be established for these periodic reports, usually quarterly or monthly will suffice. A standardized template should be developed to report on each project. This will make it much easier for upper management to evaluate the projects and for comparison purposes. Graphs and charts often tell the best story on each project. Figure 10.1 shows a template that may be used for these periodic reports.

11

THE ROLE OF CREATIVITY, INVENTION, AND INNOVATION IN SOLID WASTE MINIMIZATION

Creativity, invention, and innovation are necessary skills within an organization to create a sustainable and continuously improving waste minimization program. The executive team must create an environment that fosters these skills and rewards employees who exhibit them. A culture of innovation may also have positive effects of the organization in other ways and lead to competitive advantages in terms of cost reduction and product or service offerings. This chapter gives ideas and concepts regarding the creation of a culture and infrastructure that foster creativity.

Creativity is defined as is a mental process involving the generation of new ideas or concepts, or new associations of the creative mind between existing ideas or concepts. *Invention* is the creation of new and useful things. *Innovation* is transforming an invention or idea into a profitable good or service. EPA has placed a huge emphasis on the need for innovation. In 2002, EPA laid out a strategy for achieving better environmental results through innovation. This important framework recognizes that environmental problems are becoming increasingly complex, and that new ways of thinking and doing are needed to fully address them. Developed by EPA's Innovation Action Council, the strategy is based upon previous experiences with innovative approaches, careful consideration of recommendations from outside policy groups, and discussions with the states and a variety of stakeholders about the best ways to achieve environmental progress. It focuses EPA on finding solutions to a set of priority problems, developing new tools and approaches that can enhance problem solving, strengthening the innovation partnership with states and tribes, and fostering innovation through EPA's culture and organizational systems.

There are numerous methods to enhance innovation at any organization and to improve processes and products or services. The two areas that an organization has control over when it comes to creativity are creating an environment that fosters creativity and hiring creative employees.

The environment plays a critical role in an employee's level of creativity. Much research has been conducted in this field. One notable study from Teresa Amabile published in *Research in Organizational Behavior* in 1988 titled "A Model of Creativity and Innovation in Organizations" analyzed the variables that motivate or inhibit the employees of an organization from being creative. Some of the motivators (as could be related to solid waste management) included

- Freedom and control over work
- Good management that sets goals, prevents distractions, and is not overly strict
- Sufficient resources
- Encouragement of new ideas
- Collaboration across divisions
- Recognition of creative work
- Sufficient time to think
- Challenging problems
- Sense of urgency in getting things done

On the contrary a list of inhibitors include:

- Lack of freedom
- Red tape and bureaucracy
- Insufficient resources
- Unwillingness to risk change
- Apathy
- Poor communication
- Defensiveness in an organization
- Poor rewards
- Time pressures
- Critical, unrealistic, or inappropriate evaluation

The management team should do their best to exude the motivators and minimize the inhibitors or demotivators. The overarching concept is to create an environment where innovative and creative employees can flourish.

Creating and establishing a creative environment is not enough to ensure an innovative organization. It is one piece of the puzzle; the other is hiring creative employees. The next area that fosters creativity relates to the traits and skills of the individuals within the organization. The four critical traits of a creative individual are

1 Intelligence
2 Knowledge
3 Personality
4 Motivation

A minimum level of intelligence is necessary for an individual to be creative. It is important to comment that intelligence is a necessary, but not a sufficient trait for someone to be creative. An individual can be very intelligent as rated by an IQ test, but lacks a creative mind. Knowledge on the other hand is having the skills in the area to understand the critical issues and constraints of a situation. For example, an accountant may have difficulty analyzing the composition of a raw material for potential changes, but an engineer would have the training and skill sets to understand such a problem and the possible ramifications.

The next trait is personality. This refers to the way that an individual interacts with his or her work environment. In 1995 Sternberg and Lubart identified six personality traits of creative individuals. These traits are

1 Perseverance to face obstacles
2 Willingness to take sensible risks
3 Willingness to grow
4 A tolerance for ambiguity
5 Openness to experience
6 Belief in self and courage of one's convictions

When interviewing potential employees, several questions should be geared around these traits to identify candidates with creative potential. These same questions can be used when selecting the waste minimization team.

Finally, motivation refers to the driving force that leads someone to action. Simply stated, a creative individual must be motivated to create and apply new ideas to reduce solid waste. The challenge of the waste minimization project may be enough to keep the team motivated, but other common motivators that lead to creative thinking include

- Need for contributing to the greater good
- Need for power (including resume building for promotions)
- Need for association with others
- Need for achievement

The waste minimization team structure and communication should apply several or all of these concepts to keep the team motivated and creative. Books can also serve as an excellent motivation tool. There are many great books that provide techniques and processes to enhance motivation in an organization. The waste minimization team leader may consider purchasing a common book related to innovation that all the team members can read and discuss. The team could then apply the concepts and ideas provided in the book. Some suggestions are

1 *Innovation: The Five Disciplines for Creating What Customers Want* by Curtis R. Carlson and William W. Wilmot (Crown Publishing Group, 2006).
2 *The New Age of Innovation: Driving Cocreated Value through Global Networks* by C. K. Prahalad and M. S. Krishnan (McGraw-Hill, 2008).
3 *Innovation to the Core: A Blueprint for Transforming the Way Your Company Innovates* by Peter Skarzynski and Rowan Gibson (Harvard Business School Press, 2008).

4 *The Art of Innovation: Lessons in Creativity from IDEO, America's Leading Design Firm* by Tom Kelley with Jonathan Littman (Currency/ Doubleday, 2001).

5 *The Ten Faces of Innovation: IDEO's Strategies for Defeating the Devil's Advocate and Driving Creativity throughout Your Organization* by Tom Kelley with Jonathan Littman (Currency/Doubleday, 2005).

6 *Innovation and Entrepreneurship* by Peter F. Drucker (Butterworth Heineman, 2007).

7 *Making Innovation Work: How to Manage It, Measure It, and Profit from It* by Tony Davila, Marc J. Epstein, and Robert Shelton (Wharton School Publishing, 2005).

8 *Harvard Business Review on Innovation* by Clayton M. Christensen, Michael Overdorf, Ian MacMillan, and Rita McGrath (Harvard Business School Press, 2001).

9 *The Myths of Innovation* by Scott Berkun (O'Reilly, 2007).

10 *Ten Rules for Strategic Innovators: From Idea to Execution* by Vijay Govindarajan and Chris Trimble (Harvard Business Press, 2005).

11 *Innovation Nation: How America Is Losing Its Innovation Edge, Why It Matters, and What We Can Do to Get It Back* by John J. Kao (Free Press, 2007).

12 *The Game-Changer: How You Can Drive Revenue and Profit Growth with Innovation* by A. G. Lafley and Ram Charan (Crown Business, 2008).

13 *Diffusion of Innovations*, 5th ed., by Everett M. Rogers (Simon & Schuster, 2003).

Part 3

SOLID WASTE GENERATION

MODELING: CHARACTERIZATION,

PREDICTION, AND EVALUATION

INTRODUCTION

12.1 Model Overview

Concepts for the integrated environmental model developed for this research were created to accomplish the research goal of better predicting and evaluating the solid waste generation quantities of businesses and government agencies. The underlying hypotheses for the model concepts were also developed to guide the research process. This chapter discusses the intended functions of the proposed model and the theoretical structure of the model concepts hypotheses. This research involved examining the feasibility of the model concept to determine if such a model with the specified functions could be established. The model concepts were important to the research process because their successful development would create new knowledge in the field and provide insights into individual company waste generation.

12.2 Functions of the Proposed Model

This research investigated the development of a model to better understand solid waste generation and recycling rates of individual U.S. companies and government agencies. To begin the model development process, the proposed functions of the model were established to guide the research. Research was conducted to develop the model with two primary functions:

1 Prediction of the solid waste generation quantities and waste stream compositions of individual companies and government agencies.
2 Standardized solid waste generation evaluation of individual companies and government agencies with statistical performance parameters.

The performance parameters work as a report card to monitor and control solid waste generation of individual companies. The performance indices offer a uniform comparison basis and provide insights into waste generation of U.S. industries.

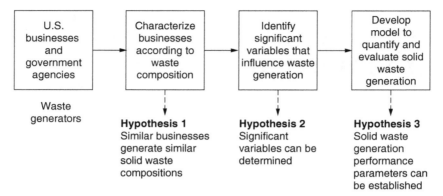

Figure 12.1 **Model concepts and hypotheses.**

12.3 Theoretical Structure of the Model Concepts and Hypotheses

A theoretical structure for the proposed model was established based on the intended model functions and theoretical concepts. To acheive the goal, related hypotheses were developed to build the evaluation model. Three concepts were interrelated to develop the proposed model and hypotheses required to accomplish the research goal. These concepts are listed below and can be seen in Fig. 12.1.

1 Group businesses that generate similar waste stream compositions.
2 Identify significant variables that aid in the prediction of the solid waste rates from businesses and government agencies that generate similar compositions.
3 Develop and validate a model to evaluate solid waste generation.

13

RESEARCH METHODOLOGY

This chapter discusses the methodology that was used to create the waste estimation model discussed in the previous chapter. The methodology applied the scientific method to investigate the proposed model concepts and theories discussed in the previous chapter. The methodology was designed to incorporate the positive aspects of previous studies and to overcome many of the shortcomings to lead to new knowledge. An overview of the research methodology follows:

1 Determine types of data required for the model concept and hypotheses.
2 Collect data.
3 Characterize businesses according to waste stream compositions.
    a. Conduct multivariate cluster analysis on mean material composition percentages using SIC codes as initial grouping basis (ANOVA and F-tests).
    b. Perform mean and variance hypotheses tests to validate results (F-tests and Bartlett tests).
    c. Analyze and interpret the results.
4 Identify dominant variables that influence waste quantities.
    a. Conduct multivariable stepwise regression analysis.
    b. F-tests, R^2 analyses, t-tests to determine significant variables.
    c. Validate regression model assumptions.
    d. Analyze and interpret the results (including correlation analysis).
5 Develop evaluation parameters for solid waste generation.
6 Integrate the model and develop the program for the model.
7 Validate the model.
    a. Evaluate and analyze additional data collected from other methods versus the model.
    b. Conduct case studies using the regression models and performance parameters.
8 Discuss summary of findings and conclusions.

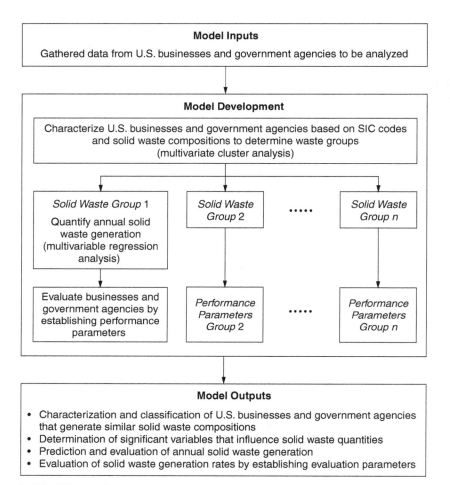

Figure 13.1   Methodology framework.

Figure 13.1 provides a visual representation of the methodology framework. This diagram emphasizes the analyses and concepts that were integrated to develop the environmental model, as discussed in the previous chapter.

14

DATA COLLECTION

14.1   Introduction

A data collection procedure was established to determine the required data types and quantities of data to collect. The data collection procedure focused on efficient and effective gathering of the required data. This chapter provides a detailed discussion of the data collection conducted for creating the waste generation model. The remaining sections of this chapter discuss

- The required data types to develop the integrated model
- Data collection process and survey instrument
- The data collection results and discussion
- Existing data collection discussion

14.2   Required Data Types

To begin the data collection process, the types of data required to develop the integrated environmental waste generation model were determined. Since the model investigates the solid waste generation of individual companies and government agencies, annual solid waste records of individual companies and other relevant data were identified for collection. The dependent variable to be investigated was annual solid waste generation of U.S. businesses and government agencies. Specifically, the following data was collected:

- Total annual solid waste generation (tons)
- Major materials disposed (type and tons)
- Major materials recycled (type and tons)

Specific solid waste material types that were investigated are

- Aerosol cans
- Aluminum (cans, scrap, sheet, other)
- Batteries (lead acid and household)
- Biomedical (hazardous)
- Concrete and cement
- Composite construction and demolition debris (concrete, wood, etc.)
- Fabric
- Food waste
- Glass (bottles and scrap)
- Lamps and ballasts
- Metal scrap (ferrous and nonferrous)
- Nonhazardous chemicals
- Old corrugated containers (cardboard)
- Paint
- Paper (newspaper, white ledger, mixed office paper, etc.)
- Plastics (No. 1, PETE; No. 2, HDPE; No. 3, PVC; No. 4, LDPE; No. 5, PP; No. 6, PS; No. 7, Other)
- Rubber
- Sludge
- Stone, clay, or sand
- Used oil
- Wood

The independent variables hypothesized to predict the annual solid waste generation of U.S. businesses and government agencies is displayed in the following list. A brief note on the reason each potential independent variable was selected is provided in the list.

- *Standard Industry Classification (SIC) code*—Collected to classify each business process and aid in the prediction of waste stream composition (a complete list can be found in Sec. 14.2.1 of this chapter).
- *Brief description of the business*—Collected to supplement SIC code and classify businesses based on processes, this information was also useful to determine SIC codes not supplied or identify incorrect SIC codes.
- *Number of employees at the facility*—Collected to predict the poundage of solid waste streams generated.
- *Working days at the facility per year*—Collected to predict the poundage of solid waste streams generated.
- *Overall recycling percentage for the facility*—Collected to examine the effects for waste reduction on overall waste reduction (derived from reported data).
- *Landfill disposal cost per ton*—Collected to predict the poundage of solid waste streams generated (in terms of dollars per ton to dispose).

■ *ISO 9000 registration (if yes, date received)*—Collected to predict the poundage of solid waste streams generated.

■ *ISO 14000 registration (if yes, date received)*—Collected to predict the poundage of solid waste streams generated.

■ *U.S. state of operation*—Collected to examine the effects of varying environmental regulations among states.

The sources from which the variables were identified are as follows:

■ Literature review (related previous studies on aggregate waste generation)
■ Survey of the industry (consulted leading U.S. waste management experts)
■ The researcher's experience and previous research

The potential independent variables were believed to aid in the prediction of solid waste because of their relationship to waste compositions and quantities. Many were obvious, such as number of employees and business type, while others, like environmental certifications and landfill-disposal costs, were hypothesized based on information from the literature review.

SIC codes were used as the basis to define business functions for each company. SIC codes were chosen as the basis to group companies to characterize solid waste because

■ Most U.S. environmental regulators use this classification system when reporting solid waste data.
■ SIC code classification has been used in previous waste generation studies (as indicated in the literature review).
■ Broad business function–based system (99 classifications of companies)
■ Mutually exclusive classifications
■ Researched and developed by the U.S. government

14.2.1 SIC CODE GROUPINGS

The SIC system is a series of numbers, each ranging from 0 to 9, used to label industries. Primary industries use either 0 or 1, secondary industries use 1, 2, or 3, and tertiary industries use numbers 4 through 9. As more digits are added to the number, the classification becomes more specific. For example, the SIC code 8 refers to the service industry, 82 refers to educational services, and 829903 refers to music and drama schools (Encarta, www.encarta.msn.com, retrieved January 9, 2001).

The SIC system was a useful indicator for a given company's solid waste generation rates and composition. The SIC codes were useful in grouping companies into categories for this research. Following is a complete listing of all Standard Industrial Classification (SIC) codes developed by the U.S. Department of Labor to classify industries (U.S. Department of Labor, www.dol.gov, retrieved May 4, 2002):

Division A: Agriculture, forestry, and fishing
    Major group 01: Agricultural production crops
    Major group 02: Agricultural production livestock and animal specialties
    Major group 07: Agricultural services
    Major group 08: Forestry
    Major group 09: Fishing, hunting, and trapping

Division B: Mining
    Major group 10: Metal mining
    Major group 12: Coal mining
    Major group 13: Oil and gas extraction
    Major group 14: Mining and quarrying of nonmetallic minerals, except fuels

Division C: Construction
    Major group 15: Building construction general contractors and operative builders
    Major group 16: Heavy construction other than building construction contractors
    Major group 17: Construction special trade contractors

Division D: Manufacturing
    Major group 20: Food and kindred products
    Major group 21: Tobacco products
    Major group 22: Textile mill products
    Major group 23: Apparel and other finished products made from fabrics
    Major group 24: Lumber and wood products, except furniture
    Major group 25: Furniture and fixtures
    Major group 26: Paper and allied products
    Major group 27: Printing, publishing, and allied industries
    Major group 28: Chemicals and allied products
    Major group 29: Petroleum refining and related industries
    Major group 30: Rubber and miscellaneous plastics products
    Major group 31: Leather and leather products
    Major group 32: Stone, clay, glass, and concrete products
    Major group 33: Primary metal industries
    Major group 34: Fabricated metal products
    Major group 35: Industrial and commercial machinery and computer equipment
    Major group 36: Electronic and other electrical equipment and components
    Major group 37: Transportation equipment
    Major group 38: Measuring, analyzing, and controlling instruments
    Major group 39: Miscellaneous manufacturing industries

Division E: Transportation, communications, electric, gas, and sanitary services
    Major group 40: Railroad transportation
    Major group 41: Transit and interurban highway passenger transportation
    Major group 42: Motor freight transportation and warehousing
    Major group 43: U.S. postal service

Major group 44: Water transportation
Major group 45: Transportation by air
Major group 46: Pipelines, except natural gas
Major group 47: Transportation services
Major group 48: Communications
Major group 49: Electric, gas, and sanitary services

Division F: Wholesale trade
Major group 50: Wholesale trade—durable goods
Major group 51: Wholesale trade—nondurable goods

Division G: Retail trade
Major group 52: Building materials, hardware, and garden supply
Major group 53: General merchandise stores
Major group 54: Food stores
Major group 55: Automotive dealers and gasoline service stations
Major group 56: Apparel and Accessory stores
Major group 57: Home furniture, furnishings, and equipment stores
Major group 58: Eating and drinking places
Major group 59: Miscellaneous retail

Division H: Finance, insurance, and real estate
Major group 60: Depository institutions
Major group 61: Non-depository credit Institutions
Major group 62: Security and commodity brokers, dealers, exchanges, and services
Major group 63: Insurance carriers
Major group 64: Insurance agents, brokers, and service
Major group 65: Real estate
Major group 67: Holding and other investment offices

Division I: Services
Major group 70: Hotels, rooming houses, camps, and other lodging places
Major group 72: Personal services
Major group 73: Business services
Major group 75: Automotive repair, services, and parking
Major group 76: Miscellaneous repair services
Major group 78: Motion pictures
Major group 79: Amusement and recreation services
Major group 80: Health services
Major group 81: Legal services
Major group 82: Educational services
Major group 83: Social services
Major group 84: Museums, art galleries, botanical and zoological gardens
Major group 86: Membership organizations

Major group 87: Engineering, accounting, research, management, and services
Major group 88: Private households

Division J: Public administration
Major group 91: Executive, legislative, and general government, except finance
Major group 92: Justice, public order, and safety
Major group 93: Public finance, taxation, and monetary policy
Major group 94: Administration of human resource programs
Major group 95: Administration of environmental quality and housing programs
Major group 96: Administration of economic programs
Major group 97: National security and international affairs
Major group 99: Non-classifiable establishments

14.3   Data Collection Process

A seven-step process was utilized to collect the required data. The data was directly collected using a national survey that was conducted in August 2002 through September 2004 of manufacturing companies, service providers, and government agencies located in the United States. The seven-step process was developed by Creative Research Systems and revised in November of 2001 (www.surveysystem .com). Below is an overview of the seven-step survey process:

1 Establish the goals of the survey.
2 Determine the sample and sample size.
3 Develop the sampling methodology.
4 Create the questionnaire.
5 Pretest the questionnaire.
6 Distribute the survey (data collection).
7 Analyze the data (survey results).

The following subsections expand on each step of the survey process and describe how each was applied for this research.

14.3.1 GOAL OF THE SURVEY

A goal of the survey was to balance cost of collecting the data versus the required accuracy. This was balanced by the number of observations included for the survey. If a sample is too large, time, talent, and money are wasted (the cost of collecting the data increases as the number of surveys increases). Conversely, if the number of observations included in the sample is too small, inadequate information may be collected for the population (the data may not be accurate enough to base any decisions or create any models).

14.3.2 POPULATIONS, SAMPLES, AND SAMPLING METHODOLOGY

A simple random sampling procedure was used for this survey. Each of the 99 SIC code groups represented a different population (N_i), where i represents the SIC code group (1–99). Each business group (i) represents a different population (N_i) because each business grouping generates different types of waste. For example, a metal-stamping plant primarily generates materials such as metal scrap, wastewater, and pallets; whereas a commercial printing business primarily generates paper waste and ink waste. The underlying assumption is that similar businesses produce similar types of waste.

A list of U.S. companies was complied for each SIC code group and each company was given an ordered number. The ordered list was created from an Internet data warehouse that stores U.S. corporate information (www.superpages.com). This Internet warehouse utilizes a directory format and offers extensive search options by business category. A random number generator was then used to select the needed number of surveys for each population. Each organization was contacted via Internet, phone, or personal visits to determine whether the company was willing to participate in the survey. After confirmation was received, each firm was sent a survey by mail, fax, or e-mail (according to the respondents' preference).

14.3.3 SURVEY DESIGN

A three-page survey form was developed to collect information from companies regarding their waste generation rates (based on the data requirements). Human factor concepts were utilized to develop a short and easy to understand survey form. The form was printed on a bright colored paper to attract attention. The survey form format was based on similar surveys conducted by two solid waste management districts in Ohio. One survey form was developed and used by the Green County solid waste management district (courtesy of Mr. Robert Snow). This survey form was used for over 190 successfully completed mailings/returns. The second survey form was developed by Fayette/Highland/Pickaway/Ross solid waste management district (courtesy of Ms. Lorna Abbott) and used for over 130 successfully completed mailings/returns. The survey form was pretested and finalized before it was sent to all the sample groups. The finalized survey form is displayed in Fig. 14.1.

14.3.4 PRETESTING

The survey instrument developed for this research was pretested to gather feedback and critique before the finalized survey was mailed. The survey was pretested on a group of 15 industrial contacts and members of the Working Council for Employee Involvement (WCEI) recycling committee. The WCEI recycling Committee consists of environmental representatives from 28 Northwest Ohio corporations. The survey was sent to each previously contacted member requesting he or she complete the survey and provide positive and negative feedback on

#_____

Recycling at Your Facility

Please identify the tons per year of each of the following materials recycled at your facility.

Materials Recycled	Tons Per Year	Materials Recycled	Tons Per Year
Aluminum		Paper	
1. Foil/sheet		1. Newspaper	
2. Cans		2. Office	
3. Other		3. Computer	
Aerosol cans		4. Mixed	
		5. Corrugated containers (cardboard)	
Ash (including fly ash)		6. Other (specify below)	
Batteries		_____	
1. Lead acid		Plastics (including shrink wrap)	
2. Other		1. PETE	
Concrete		2. HDPE	
Fabric		3. Vinyl (PVC)	
Gas/oily rags		4. LDPE	
Glass		5. Polypropylene	
1. Brown/amber		6. Polystyrene	
2. Clear		7. Other (specify below)	
3. Other (specify below)		_____	
_____			
Lamps (bulbs and ballasts)		Rubber	
Metals		Sludge/waste water	
1. Ferrous		Stone/clay/sand	
2. Nonferrous (excluding aluminum)		Used oil	
Paint		Wood	

If any other materials are recycled at your facility, please enter in the table below the material and annual tonnage recycled:

Other Material Recycled	Tons per Year

Upon completion, please return this survey in the self-addresses stamped envelope.

Thank you for your help in this survey !

Figure 14.1 Recycling survey form.

a separate sheet. The main points emphasized from the feedback and integrated into the finalized survey were

- Added more space for respondents to enter information.
- Added an additional comments section.
- Several respondents noted that for large corporations, difficulty might arise in determining annual sales and profits.
- Several respondents noted confidentially issues might reduce the sales and waste disposal costs information from being released by respondents.

Before mailing, the survey was also reviewed by Mr. James Walters, manager of the Lucas County Solid Waste District. Mr. Walters is one of the top solid waste management leaders in the United States. He spent over 20 years in the automobile-manufacturing industry in the field of waste management. Mr. Walters suggested several value-added changes to the survey and gave it his approval as a useful and effective tool to collect information.

14.3.5 DISTRIBUTION

As discussed in the sampling methodology section, an ordered list and a random number generator were used to select companies that would receive a survey. An ordered list of companies in each business group was compiled and a random number generator (Microsoft Excel) was applied to select the required companies for each SIC code business grouping.

Each selected company was contacted by phone, fax, or e-mail to determine if they were willing to participate in the survey. This method was used to increase the response rate. Surveys were also sent to the researcher's industrial contacts to increase response rate. The companies willing to participate were asked their contact information and mailed the survey. Companies declining to participate were deleted from the ordered list and the random number generator was rerun to achieve the required sample size for each business group. These companies were contacted by phone, fax, or e-mail and asked if they were willing to participate in the survey. This process was iterated until all 1500 surveys were mailed proportionality to each SIC code group.

14.4   Data Collection Results and Discussion

Over 4100 businesspersons were contacted via phone, fax, or e-mail to participate in the survey. Most were contacted by phone (75 percent), followed by e-mail (20 percent) and fax (5 percent). Of that total, approximately 35 percent agreed to participate in the survey and 65 percent declined or did not return phone calls or e-mails. A total of 1500 surveys were mailed to companies and 438 were returned properly completed, from 65 of the

99 SIC code groups. Surveys that were missing information, such as company name or number of employees were cross-referenced with a number designated to each survey. Once the company was determined from this survey number, the number of employees was collected from the Harris Registry and corporate information from various Internet sources (www.cnn.money.com). Surveys that were incomplete were discarded (6.8 percent or 30 surveys were discarded). The following is a summary of the figures:

- A total of 4135 businesses were contacted and requested to participate.
- Only 1500 companies (36.3 percent of those contacted) verbally agreed to participate in the survey.
- A total of 468 companies (31.2 percent of those who agreed to participate) returned the survey.
- Only 438 (29.2 percent of those who agreed to participate) companies correctly returned the survey with useable data.

Of the 4135 companies initially contacted, 438 correctly returned the survey, resulting in an overall response rate of 10.6 percent. This is a very common response rate for a "cold mailing," in which random surveys are mailed without initially contacting potential respondents. The use of phone calls, e-mails, and faxes before mailing the survey aided to reduce the percentage of nonresponders. Prescreening potential respondents with phone calls, e-mails, or faxes increased the response rate of the actual mailed survey by nearly 20 percent (from 10.6 to 29.2 percent). The pre-screening process eliminated companies that would definitely not respond to the survey and identified a pool of potential companies that would be more likely to complete the survey, hence increasing the response rate. The number of surveys mailed and successfully returned from each SIC code group can be seen in Fig. 14.2.

Waste material generation given in terms cubic yards was converted into annual tonnage using the average conversion factors developed by the EPA. An example of a conversion is discussed in Sec. 8.6. This method was applied to convert all generation rates given in volume to weight.

Potential sources of error in the data collected are

- Inaccurate data generated from the surveys (including the solid waste management district surveys and the national survey).
- Data collected may not accurately describe the population. Companies that are more conscientious about environmental initiatives and protection may be more likely to submit results.
- Some respondents had difficulty separating facility financial data from corporate data, but made attempts to do so. Surveys with incomplete financial data were discarded or the company was contacted and asked for an estimate.
- Some respondents were unwilling to release financial data (including annual waste costs) on the survey. Some of the existing data did not include these figures either. Landfill disposal charges in the area were used as a substitute. This information was easily attainable from the companies and various Internet sources, including the Solid Waste Association of North America (www.swana.org).

SIC Group Number	Description	Number of Firms Analyzed	SIC Group Number	Description	Number of Firms Analyzed
01	Agricultural production crops	6	56	Apparel and accessory stores	7
02	Agricultural production livestock	5	57	Home furniture and furnishings	8
07	Agricultural services	2	58	Eating and drinking places	13
08	Forestry	3	59	Miscellaneous retail	8
10	Metal mining	3	60	Depository institutions	8
12	Coal mining	3	61	Non-depository credit institutions	6
15	Building construction and general contractors	8	62	Security and commodity brokers	6
16	Heavy construction	7	63	Insurance carriers	7
17	Construction special trade contractors	6	64	Insurance agents, brokers, and service	9
20	Food and kindred products	8	65	Real estate	8
22	Textile mill products	8	67	Holding and other investment offices	6
23	Apparel and other finished products	5	70	Hotels and other lodging places	7
24	Lumber and wood products	7	72	Personal services	8
25	Furniture and fixtures	7	73	Business services	12
26	Paper and allied products	6	75	Automotive repair, services, and parking	9
27	Printing, publishing, and allied industries	8	76	Miscellaneous repair services	8
28	Chemicals and allied products	5	78	Motion pictures	6
29	Petroleum refining and related industries	6	79	Amusement and recreation services	4
30	Rubber and miscellaneous plastics products	5	80	Health services	14
33	Primary metal industries	4	81	Legal services	7
34	Fabricated metal products	6	82	Educational services	8
35	Industrial and commercial machinery	9	83	Social services	6
36	Electronic and other electrical equipment	8	84	Museums and zoological gardens	4
37	Transportation equipment	14	86	Membership organizations	6
38	Controlling instruments	8	87	Engineering and management services	5
39	Miscellaneous manufacturing industries	7	91	Executive and legislative government	5
42	Motor freight transportation and warehousing	5	92	Justice, public order, and safety	6
50	Wholesale trade-durable goods	8	93	Public finance, taxation, and monetary policy	4
51	Wholesale trade-non-durable goods	7	94	Administration of human resource programs	5
52	Building materials and hardware	8	95	Administration of environmental quality	4
53	General merchandise stores	7	96	Administration of economic programs	3
54	Food stores	9	97	National security and international affairs	3
55	Automotive dealers and service stations	10			

Figure 14.2 Survey data collected by SIC group.

■ Hazardous wastes were not reported by many manufacturing respondents. These companies view this data as sensitive, and since it is illegal to dispose of hazardous wastes at landfills, these types of wastes were not included as responses in most cases. Generally hazardous wastes compose a very low percentage of the total waste stream.

Table 2.2 displays the EPA volume to mass conversion factors for common waste components. The conversion factors were obtained from EPA (www.epa.gov/epaoswer/non-hw/recycle/recmeas/, retrieved July 10, 2007).

The following densities in Table 14.1 were experimentally determined at the University of Toledo from 1998 through 2007. The densities are used during the data analysis phase of waste assessment to convert volumes.

TABLE 14.1   ECML VOLUME TO WEIGHT CONVERSION FACTORS

MATERIAL	CONDITION (LEVEL OF COMPACTION)	DENSITY CONVERSIONS FACTOR
Mixed office paper	Loose in desk trash bin	55 lb/yd^3
Old corrugated containers (cardboard)	Not broken down and uncompacted	45 lb/yd^3
Food waste	Loose in desk trash bin	68 lb/yd^3
Old news print	Loose in desk trash bin	325 lb/yd^3
Paper towels	Loose in restroom trash bin	20 lb/yd^3

14.5   Existing Data Collection

Existing raw data was identified for this research and provided valuable direction for the additional data collection. The existing data was primarily utilized to validate the integrated model. These existing raw data sources contained 1282 solid waste generation and recycling records of U.S. companies. These data sources were identified from previous research conducted (Franchetti, 1999), professional waste management contacts, information gathered during the literature review, and internet searches. The existing records and studies are from the Waste Minimization Research Project (Lucas County, Ohio) and similar research projects, manufacturing company records, and data collected from the U.S. government. Much raw data existed on waste generation quantities of individual companies, but little information. Many state and local environmental agencies collected waste generation data to use for various aggregate reports. These reports examined waste generation for entire geographical regions, not individual companies. Limited research has been conducted to compile and analyze waste generation for individual companies. No sources were found that archived and consolidated this data from the various sources. Most government agencies collected the individual company data and used it solely to establish aggregate waste rates for reporting requirements. Few in-depth statistical analyses were conducted on the data by the government agencies. The individual data was stored in files and rarely used for other purposes. The existing data sources are listed below

■ Lucas County's (Ohio) Waste Minimization Research Project solid waste assessments records (24 companies in total).
■ Mahoning County's (Ohio) Industrial Waste Minimization Project, which conducts solid waste for manufacturers (46 companies in total).
■ U.S. governmental records and offices, specifically
   ■ Solid waste management district offices located across the United States (1,060 companies in total); many of the districts have conducted industrial waste assessments throughout their respective areas.

■ The U.S. Environmental Protection Agency on the local, state, and federal level (132 companies in total).

Specific data collection methods were applied for each data source, depending on the nature of the source.

Initially, interviews with waste-hauling company management and recycling collector management were planned for this research. These data sources were not available from these companies due to confidentiality and customer protection. As a substitute, phone and personal interviews were conducted with solid waste management districts across the United States.

In conclusion, the data collection phase of this research was successful; sufficient data was collected to begin modeling. Several issues were identified from the collection process and the actual data. First, many government agencies collected the required data to build the waste evaluation models, but did not use it for that purpose. These government agencies could achieve significant improvements in the reliability of the studies they conduct if they worked together and shared information. A common database would aid in storing and disseminating such information. Also identified from the government agencies were a wide variety of data collection methods and analysis techniques to accomplish the same goals. A common goal was to estimate the annual generation amounts for a specific region of the United States. Nonstandardized methods were used to collect the data and nonscientific analyses techniques were used to estimate these aggregate totals. This creates comparison discrepancies when different analysis techniques are used. A standardized approach to estimate generation amounts coupled with government agencies sharing information would reduce this problem. The model developed for this research applied a scientific, standardized approach to estimate and predict solid waste generation rates and will also aid in reducing this problem.

14.5.1 THE WASTE MINIMIZATION RESEARCH PROJECT (LUCAS COUNTY, OHIO)

Significant research was conducted at the University of Toledo regarding solid waste estimation. This research provided valuable insights into the nature of solid waste generation of individual companies. From 1996 through August of 2002 the Environmentally Conscious Design and Manufacturing Lab (ECDML) located at the University of Toledo, College of Engineering has conducted 24 solid waste assessments for manufacturing and service companies in Lucas County, Ohio. The largest research project at the ECDML was the Waste Minimization Research Project, which provides no-cost solid waste assessments to businesses operating in Lucas County. The purpose of the waste assessments was to quantify waste streams by annual weight and composition and to provide economical solutions to reduce, reuse, or recycle components of the waste stream. The companies surveyed included a broad range from manufacturing to hospitals. This section discusses the methodology utilized to conduct the waste assessments, the benefits and drawbacks of this data for this research, and a brief overview of the data. The ECDML solid waste assessment data was used as the baseline for the remainder of the data collection due to its high level of detail and accuracy. The assessment data was compared with data collected using other methods. These solid waste assessments aided in identifying the variables that aided in solid waste prediction via preliminary analyses (Franchetti, 1999).

The solid waste assessment procedure involved several steps including a preliminary questionnaire to collect company information (such as SIC code, days of operation, and number of employees), a facility walk-through to collect annual waste stream composition and weight, and data analysis to estimate annual waste stream composition and weights. The benefits of this data are its high level of accuracy and the ability for the researcher to contact company representatives for more information or clarification. The drawbacks of this data collection source and method are its limited nature (limited by the number of waste assessments that have been conducted by the ECDML) and the high cost and time associated with conducting more assessments. Possible sources of error associated with the data collection method are minimal. This is the most thorough and accurate data collection method. One minor possible source of error is related to the sampling method. The 1-day walk-through at each facility assessment to estimate annual generation amounts may not capture the true amounts or all materials generated. This source of error is minimized by interviews with company personnel and waste records review, which verify the data. The information collected during the assessment is listed below

- Company name
- Business focus
- SIC code
- Total annual waste composition by weigh and volume
- Current recycling levels
- Potential cost and waste reduction improvements
- Number of employees
- Annual waste costs

14.5.2   THE INDUSTRIAL WASTE MINIMIZATION PROJECT (MAHONING COUNTY, OHIO)

Mahoning County developed a project similar to the University of Toledo's ECDML, Waste Minimization Research Project. The Industrial Waste Minimization Project (IWMP), a joint project between Youngstown State University (Ohio) and Mahoning County (Ohio) conducted no-cost solid waste assessments for Mahoning County businesses from 1995 until 1998. The IWMP operated with similar goals to the ECDML. The IWMP conducted on-site audits to estimate annual waste stream weights and compositions of each business surveyed. Environmentally friendly solutions were then developed to reduce, reuse, or recycle components of the waste stream. In From 1995 until 1998 the IWMP conducted 46 industrial solid waste audits. The data the IWMP collected included

- Company name
- Company contact information

- SIC code
- Service provided or product manufactured
- Number of employees
- Volume and weight of the following materials recycled
  - Paper
  - Plastics
  - Styrofoam
  - Cardboard
  - Glass
  - Wood
  - Ferrous metals
  - Nonferrous metals
  - Other materials
  - Additional comments or notes

A benefit of this data source is its level of detail. A drawback is the difficulty associated with collecting data, as in the ECDML assessments, in particular the high cost and time of each assessment. An assumption from this data is that it is accurate and properly collected. This is also a potential source of error. Contacting and verifying information with the Youngtown State University IWMP manager minimized this potential source of error.

14.5.3 STATE AND LOCAL GOVERNMENT WASTE SURVEYS

Many states in the United States have created agencies to monitor and reduce waste generation in their jurisdiction (Arkansas, Indiana, Missouri, New Hampshire, and Ohio). For example, Ohio has created 64 solid waste district throughout the state to monitor and reduce solid waste generation. Several of these solid waste districts conducted industrial surveys to estimate the waste tonnage generated and recycling levels of businesses in their district. The waste tonnage and recycling level reporting are often required by state and federal environmental organizations and aid in measuring progress of waste reduction goals. To begin data collection from the government agencies, a letter was sent to each office requesting information on industrial surveys they have conducted.

Upon receiving responses from many of the districts, follow-up phone calls were made to each responding District. The purposes of the phone calls were to clarify the information desired for the research and provide delivery instructions. Table 14.2 lists the solid waste districts that responded to the letter and the information each district provided.

The benefits of this data source are its abundance (784 businesses surveyed) and the low cost for data collection. Many of the solid waste district directors also pointed out other valuable information sources. One drawback of this data collection method is confidentiality protection of the businesses by the districts. Few districts gave the company names and contact information of the companies they surveyed, making follow-up

TABLE 14.2   DATA PROVIDED FROM U.S. SOLID WASTE DISTRICTS

DISTRICT NAME	DISTRICT LOCATION	INFORMATION PROVIDED
Mahoning County	Youngstown, Ohio	Industrial survey of 58 businesses, including waste compositions and amounts
Erie County	Sandusky, Ohio	Industrial survey of 36 businesses, including waste compositions and amounts
Montgomery County	Dayton, Ohio	Industrial survey of 19 businesses, including waste compositions and amounts
Geauga-Trumbull	Warren, Ohio	Industrial survey of 162 businesses, including waste compositions and amounts
Green County	Xenia, Ohio	Industrial survey of 191 businesses, including waste compositions and amounts
Hancock County	Findley, Ohio	Industrial survey of 70 businesses, including waste compositions and amounts
Darke-Marion County	Marion, Ohio	Industrial survey of 16 businesses, including waste compositions and amounts
Warren County	Lebanon, Ohio	Industrial survey of 102 businesses, including waste compositions and amounts
Fayette-Highland-Pickaway-Ross Counties	Chillicothe, Ohio	Industrial survey of 130 businesses, including waste compositions and amounts
New Hampshire Department of Environmental Service	New Hampshire	Industrial survey of 106 businesses, including waste compositions and amounts
Arkansas Department of Pollution Control and Ecology	Arkansas	Industrial survey of 170 businesses, including waste compositions and amounts
Carroll-Columbiana-Harrison Counties	Carrollton, Ohio	Per capita total industrial waste generation for entire district covering 19 SIC code groupings
Ashland County	Ashland Ohio	Annual industrial waste generation and composition for entire county
Cuyahoga County	Cleveland, Ohio	Annual industrial waste generation and composition for entire county
Hamilton County	Cincinnati, Ohio	Annual industrial waste generation and composition for entire county

questions and data backtracking difficult. A possible source of error is that the data accuracy depends on whether the proper collection method was used.

14.5.4 U.S. GOVERNMENT RESEARCH

Review of government documents and research studies also proved useful for this research. The EPA has conducted no direct research in this field, however state and local agencies of the EPA have conducted studies that are relevant to this research. The research conducted by these agencies provided relevant insights into solid waste generation, but limited direct hard data on individual company waste generation amounts. Each EPA data source is listed in Table 14.3. The table lists the organization

TABLE 14.3 DATA PROVIDED BY EPA AND OTHER GOVERNMENT AGENCIES			
AGENCY NAME	**AGENCY LOCATION**	**INFORMATION PROVIDED**	**RELEVANCY**
Integrated Waste Management Board of California	Sacramento, California	Database that estimates the characteristics of the commercial sector disposed waste stream for any city in California, and determines which business groupings contribute the most material to the waste stream.	Provided compositions of solid waste streams for Californian companies
Indiana Institute on Recycling	Indiana State University	Conducted 80 waste reduction case studies in the U.S., for each a contact is given; summary, actions taken, and payback or benefits	Provided waste reduction information and limited waste generation rates for 80 cases
Wisconsin Department of Natural Resources	Madison, Wisconsin	Conducted 52 waste reduction case studies in the Wisconsin, for each; a contact is given, summary, actions taken, and payback or benefits	Provided waste reduction information and limited waste generation rates for 52 cases
Cornell Waste Management Institute (CWMI)	Cornell University	CWMI researches various aspects of solid waste management including source reduction	Provided composting information for all business types
Florida Department of Environmental Protection	Tallahassee, Florida	Economic study on recycling benefits for Florida companies	Provided limited waste generation poundage for Florida companies

that collected the data, the location of the organization, the types of data collected, and the relevancy of the data to this research. The table provides an overview of each organization and summaries of the data provided by them.

As shown in Table 14.3, U.S. governmental environmental research has provided significant data for this research. The following paragraphs provide a discussion of each data source mentioned in the table, including an overview of the agency, elaboration of the data collected, benefits and drawbacks of the data source, and potential sources of error.

The California Integrated Waste Management Board, referred to as CIWMB (www.ciwmb.ca.gov/WasteChar/), has conducted significant research in characterizing solid waste for individual companies based on SIC codes, in California. The CIWMB has developed a solid waste characterization database that contains waste stream data for different types of businesses. The database combines three types of information, general business data, waste compositions, and waste disposal rates. All the information is based on the 38 business SIC code groupings. Most of the data was collected as part of a 1999 California Statewide Waste Characterization Study. Overall 1200 businesses in California were studied covering all SIC code groupings.

The Indiana Institute on Recycling (www.indstate.edu/recycle/), a state agency, was created in 1989 by the General Assembly of Indiana and is located at Indiana State University. The institute developed concepts, methods, and procedures for assisting Indiana residents and businesses in reducing and recycling solid waste. One aspect of this involved site visits to companies in Indiana to aid them directly in reducing their solid waste stream and increasing recycling. The institute has placed 80 case studies documenting this work on their web site (www.indstate.edu/recycle/caselist.html). Many of these case studies included waste generating and recycling details for each company are described in each case study. Benefits of this data source are the ease of collection and a relatively high amount of data (80 case studies). The case studies also provided contact information, which could be used for clarification or additional information. A drawback of this data source, and a potential source of error, is that it was collected by another agency. Their full methodologies and accuracy level were not completely described in the case studies. In all 80 businesses were studies by the institute, covering 23 SIC code groupings.

The Wisconsin Department of Natural Resources has also conducted case study research in a similar manner to the Indiana Institute of Recycling. The Wisconsin Department of Natural Resources has conducted 52 waste reduction case studies covering 18 SIC code groupings (www.dnr.state.wi.us).

Cornell University has developed a program named the Cornell Waste Management Institute, referred to as CWMI (www.cfe.cornell.edu/wmi). CWMI was established in 1987. CWMI addresses the environmental and social issues associated with waste management by focusing its University resources. Through research, outreach, and teaching activities, CWMI staff and affiliated researchers and educators work to develop technical solutions to waste management problems and to address broader issues of waste generation and composition, waste reduction, risk management, environmental equity, and public decision-making.

The Florida Department of Environmental Protection, referred to as FDEP (www.dep.state.fl.us/), has conducted waste reduction and recycling research. A noteworthy project of FDEP was a recycling economic information study. The goal of the study was to document the size of the recycling and reuse industries in Florida. As part of the study, recycling level estimates were derived via surveys on Florida businesses. The surveys collected the recycling and reuse rates of solid waste materials in Florida.

15

SOLID WASTE CHARACTERIZATION
BY BUSINESS ACTIVITIES

15.1   Introduction

This chapter discusses the characterization of U.S. businesses and government agencies based on solid waste stream composition percentages. The purpose of characterization was to determine and group the business types that generate similar solid waste materials. Figure 15.1 displays the characterization process.

Standard Industrial Classification (SIC) codes were used as the grouping basis to describe types of business functions, as described in Chap. 14. The waste records gathered from each SIC code group were consolidated and the mean and variance waste composition percentages were calculated for each material. SIC code groups have historically been used by U.S. environmental regulators to classify businesses based on solid waste generation. The usage of SIC codes for this purpose was statistically validated and is discussed in Sec. 15.3 of this chapter. After the successful validation of using SIC codes, multivariate cluster analysis was conducted on the SIC code group waste composition data to reduce groups further. Twenty-two of these final clusters, referred to as waste groups, were formed that statistically generate similar compositions of solid waste. In the next chapter, these waste groups are analyzed to determine the significant variables that influence annual solid waste quantities.

15.2   Data Consolidation and Initial Characterization

Several preparations were completed before beginning the solid waste characterization process and waste group determination. Specifically, the 438 waste records (companies) collected from the national survey included 65 SIC code groups. Each waste record was placed in 1 of 65 data matrices; with each matrix representing each of the 65 SIC code

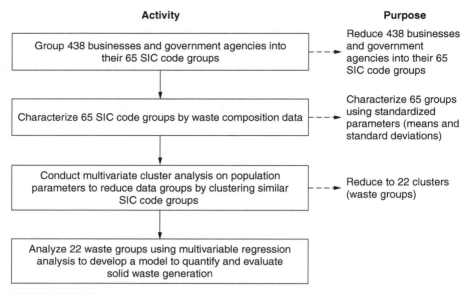

Activity

Purpose

Group 438 businesses and government agencies into their 65 SIC code groups

Reduce 438 businesses and government agencies into their 65 SIC code groups

Characterize 65 SIC code groups by waste composition data

Characterize 65 groups using standardized parameters (means and standard deviations)

Conduct multivariate cluster analysis on population parameters to reduce data groups by clustering similar SIC code groups

Reduce to 22 clusters (waste groups)

Analyze 22 waste groups using multivariable regression analysis to develop a model to quantify and evaluate solid waste generation

Figure 15.1   **Business waste characterization process.**

groups for which data was collected. To initially characterize the solid waste data, mean and variance composition percentages for each material in the 65 SIC code groups (matrices) were calculated using the individual company records collected from the national survey. Materials comprising less than 2 percent of all SIC code group waste streams were not included in the calculations to simplify the analysis. Material composition percentage means and variances of the following materials were calculated.

- Biohazard wastes
- Construction and demolition debris (sand, stone, and concrete)
- Fabric and textiles
- Food waste
- Glass
- Metal
- Old corrugated containers (cardboard)
- Chemicals, sludge, and used oil
- Organic wastes (agricultural)
- Paper (excluding cardboard)
- Plastic
- Rubber
- Wood
- Yard waste

The means and variances of the solid waste composition percentages from the 65 SIC code groups can be seen in Figs. 15.2 to 15.4. The next section provides a further analysis of this data.

Mean Waste Stream Composition Percentage

SIC Group Number	Description	Number of Firms in Group	BioHaz.mean	BioHaz.var	Cand.mean	Cand.var	Fabric.mean	Fabric.var	Food.mean	Food.var	Glass.mean	Glass.var	Metal.mean	Metal.var	OCC.mean	OCC.var	OllChem.mean	OllChem.var	Organic.mean	Organic.var	Paper.mean	Paper.var	Plastic.mean	Plastic.var	Rubber.mean	Rubber.var
01	Agricultural production crops	6	0.0	0.0	0.0	0.0	0.0	0.0	13.0	3.2	1.0	0.3	9.0	2.7	9.0	2.3	0.0	0.0	22.0	5.1	9.0	2.1	7.0	2.1	0.0	0.0
02	Agricultural production livestock and animal specialties	5	1.0	0.5	0.0	0.0	0.0	0.0	13.0	4.4	1.0	0.3	10.0	2.3	9.0	2.2	0.0	0.0	21.0	6.1	8.0	2.6	10.0	2.3	0.0	0.0
07	Agricultural services	2	1.0	0.7	0.0	0.0	0.0	0.0	12.0	3.1	1.0	0.3	9.0	2.4	9.0	2.2	0.0	0.0	20.0	8.2	9.0	2.2	8.0	2.1	0.0	0.0
08	Forestry	3	0.0	0.0	8.0	2.2	0.0	0.0	5.0	1.5	0.0	0.0	9.0	3.2	8.0	1.9	3.0	1.2	6.0	2.1	14.0	4.0	6.0	2.1	0.0	0.0
10	Metal mining	3	0.0	0.0	7.0	2.0	0.0	0.0	5.0	2.0	2.0	0.6	13.0	2.9	8.0	1.9	3.0	1.2	6.0	0.0	16.0	6.0	4.0	0.9	0.0	0.0
12	Coal mining	3	0.0	0.0	9.0	2.7	0.0	0.0	5.0	1.3	2.0	0.6	10.0	2.2	8.0	1.9	3.0	0.8	6.0	0.0	17.0	6.0	4.0	0.9	0.0	0.0
15	Building construction general contractors and operative builders	8	0.0	0.0	21.0	5.9	1.0	0.3	4.0	1.0	4.0	1.1	10.0	3.0	5.0	1.2	0.0	0.0	0.0	0.0	14.0	3.3	13.0	3.9	0.0	0.0
16	Heavy construction other than building construction contractors	7	0.0	0.0	22.0	6.2	1.0	0.3	4.0	1.4	3.0	0.8	12.0	3.9	6.0	1.4	0.0	0.0	0.0	0.0	12.0	3.8	10.0	3.2	0.0	0.0
17	Construction special trade contractors	6	0.0	0.0	20.0	5.6	1.0	0.3	4.0	1.5	4.0	1.1	11.0	3.9	5.0	1.2	0.0	0.0	0.0	0.0	13.0	4.5	9.0	3.2	0.0	0.0
20	Food and kindred products	8	0.0	0.0	0.0	0.0	5.0	1.4	23.0	9.3	1.0	0.3	4.0	1.5	6.0	1.4	5.0	1.4	26.0	1.2	26.0	9.8	17.0	6.4	0.0	0.0
22	Textile mill products	8	0.0	0.0	0.0	0.0	21.0	5.9	5.0	2.2	0.0	0.0	7.0	2.3	10.0	2.4	0.0	0.0	0.0	0.0	13.0	5.3	19.0	6.2	0.0	0.0
23	Apparel and other finished products made from fabrics and similar materials	5	0.0	0.0	0.0	0.0	18.0	5.0	5.0	1.9	0.0	0.0	7.0	1.6	11.0	2.6	0.0	0.0	0.0	0.0	14.0	4.9	19.0	4.3	0.0	0.0
24	Lumber and wood products, except furniture	7	0.0	0.0	0.0	0.0	0.0	0.0	3.0	0.8	2.0	0.6	13.0	3.2	6.0	1.4	0.0	0.0	11.0	3.1	14.0	3.4	5.0	1.2	0.0	0.0
25	Furniture and fixtures	7	0.0	0.0	0.0	0.0	0.0	0.0	3.0	0.8	2.0	0.6	11.0	2.9	7.0	1.7	0.0	0.0	11.0	2.7	15.0	4.0	6.0	1.6	0.0	0.0
26	Paper and allied products	6	0.0	0.0	0.0	0.0	0.0	0.0	3.0	0.9	0.0	0.0	3.0	0.9	11.0	2.6	0.0	0.0	65.0	0.0	18.5	18.5	5.0	1.5	0.0	0.0
27	Printing, publishing, and allied industries	8	0.0	0.0	0.0	0.0	0.0	0.0	4.0	1.3	0.0	0.0	3.0	0.9	13.0	3.1	0.0	0.0	51.0	0.0	16.0	16.0	6.0	1.8	0.0	0.0
28	Chemicals and allied products	5	0.0	0.0	0.0	0.0	0.0	0.0	6.0	2.0	2.0	0.6	7.0	1.7	7.0	1.7	20.0	5.0	0.0	0.0	19.0	6.1	14.0	3.5	0.0	0.0
29	Petroleum refining and related industries	6	0.0	0.0	0.0	0.0	0.0	0.0	5.0	1.4	2.0	0.6	6.0	1.5	8.0	1.9	18.0	6.1	0.0	0.0	18.0	4.8	15.0	3.7	0.0	0.0
30	Rubber and miscellaneous plastics products	5	0.0	0.0	0.0	0.0	0.0	0.0	6.0	1.7	2.0	0.6	7.0	1.6	7.0	1.7	19.0	4.9	0.0	0.0	20.0	5.3	13.0	2.9	0.0	0.0
33	Primary metal industries	4	0.0	0.0	0.0	0.0	0.0	0.0	3.0	0.8	0.0	0.0	61.0	13.8	8.0	1.9	3.0	0.8	0.0	0.0	8.0	1.9	9.0	2.0	0.0	0.0
34	Fabricated metal products, except machinery and transportation equipment	6	0.0	0.0	0.0	0.0	0.0	0.0	3.0	0.8	0.0	0.0	59.0	13.1	9.0	2.2	2.0	0.7	0.0	0.0	9.0	2.2	8.0	1.8	0.0	0.0

Figure 15.2 Means and variances of waste stream composition percentages for the 65 SIC code groups utilized to conduct characterization/cluster analysis. Waste stream composition data (SIC code 01 to 34).

SIC Group Number	Description	Number of Firms in Group	BioHaz.mean	BioHaz.var	CandD.mean	CandD.var	Fabric.mean	Fabric.var	Food.mean	Food.var	Glass.mean	Glass.var	Metal.mean	Metal.var	OCC.mean	OCC.var	OilChem.mean	OilChem.var	Organic.mean	Organic.var	Paper.mean	Paper.var	Plastic.mean	Plastic.var	Rubber.mean	Rubber.var
35	Industrial and commercial machinery and computer equipment	9	0.0	0.0	0.0	0.0	0.0	0.0	5.0	1.3	4.0	1.1	15.0	4.5	7.0	1.7	0.0	0.0	0.0	0.0	29.0	6.9	21.0	6.2	0.0	0.0
36	Electronic and other electrical equipment and components, except computer equipment	8	0.0	0.0	0.0	0.0	0.0	0.0	4.0	1.4	5.0	1.4	13.0	3.8	8.0	1.9	0.0	0.0	0.0	0.0	28.0	8.9	22.0	6.4	0.0	0.0
37	Transportation equipment	14	0.0	0.0	0.0	0.0	2.0	2.0	8.0	2.7	2.0	0.6	11.0	3.1	10.0	2.4	2.0	1.0	0.0	0.0	18.0	5.6	17.0	4.7	1.3	0.4
38	Measuring, analyzing, and controlling instruments; photographic, medical and optical goods; watches and clocks	8	0.0	0.0	0.0	0.0	0.0	0.0	4.0	1.3	5.0	1.4	14.0	3.8	7.0	1.7	0.0	0.0	0.0	0.0	29.0	8.7	21.0	5.7	0.0	0.0
39	Miscellaneous manufacturing industries	7	0.0	0.0	0.0	0.0	0.0	0.0	5.0	1.6	4.0	1.1	15.0	4.9	6.0	1.4	0.0	0.0	0.0	0.0	28.0	8.1	20.0	6.5	0.0	0.0
42	Motor freight transportation and warehousing	5	0.0	0.0	0.0	0.0	1.0	0.3	4.0	1.0	1.0	0.3	8.0	1.0	9.0	1.0	5.0	1.8	0.0	0.0	37.0	12.9	7.0	1.6	3.0	0.8
50	Wholesale trade-durable goods	8	0.0	0.0	0.0	0.0	4.0	1.1	8.0	2.1	2.0	0.6	7.0	1.7	11.0	2.6	0.0	0.0	0.0	0.0	33.0	8.0	11.0	2.6	0.0	0.0
51	Wholesale trade-non-durable goods	7	0.0	0.0	0.0	0.0	3.0	0.8	10.0	2.8	2.0	0.6	7.0	1.3	13.0	3.1	0.0	0.0	0.0	0.0	31.0	8.0	10.0	1.9	0.0	0.0
52	Building materials, hardware, garden supply, and mobile home dealers	8	0.0	0.0	0.0	0.0	3.0	0.8	7.0	1.5	3.0	0.8	7.0	2.5	12.0	2.9	0.0	0.0	0.0	0.0	26.0	5.3	9.0	3.2	0.0	0.0
53	General merchandise stores	7	0.0	0.0	0.0	0.0	2.0	0.6	6.0	2.4	2.0	0.6	6.0	1.8	11.0	2.6	0.0	0.0	0.0	0.0	27.0	10.2	9.0	2.7	0.0	0.0
54	Food stores	9	0.0	0.0	0.0	0.0	0.0	0.0	38.0	4.0	2.0	0.6	5.0	2.0	8.0	1.9	0.0	0.0	0.0	0.0	21.0	6.7	12.0	2.7	0.0	0.0
55	Automotive dealers and gasoline service stations	10	0.0	0.0	0.0	0.0	1.0	0.3	7.0	1.0	2.0	0.6	15.0	1.0	12.0	2.9	9.0	2.7	0.0	0.0	26.0	6.2	12.0	1.0	7.0	2.0
56	Apparel and accessory stores	7	0.0	0.0	0.0	0.0	3.0	0.8	7.0	2.0	3.0	0.8	5.0	1.5	12.0	2.9	0.0	0.0	0.0	0.0	28.0	7.5	8.0	2.4	0.0	0.0
57	Home furniture, furnishings, and equipment stores	8	0.0	0.0	0.0	0.0	4.0	1.1	6.0	2.0	2.0	0.6	7.0	2.0	13.0	3.1	0.0	0.0	0.0	0.0	27.0	8.6	9.0	2.6	0.0	0.0
58	Eating and drinking places	13	0.0	0.0	0.0	0.0	0.0	0.0	61.0	7.0	3.0	1.5	2.0	0.5	10.0	3.0	0.0	0.0	0.0	0.0	12.0	3.8	6.0	1.0	0.0	0.0
59	Miscellaneous retail	8	0.0	0.0	0.0	0.0	2.0	0.6	5.0	1.5	2.0	0.6	6.0	1.5	14.0	3.4	0.0	0.0	0.0	0.0	27.0	7.7	7.0	1.7	0.0	0.0
60	Depository institutions	8	0.0	0.0	0.0	0.0	0.0	0.0	16.0	4.5	2.0	0.6	2.0	0.4	8.0	1.9	0.0	0.0	0.0	0.0	47.0	12.3	3.0	0.6	0.0	0.0
61	Non-depository credit institutions	6	0.0	0.0	0.0	0.0	0.0	0.0	15.0	3.7	2.0	0.6	2.0	1.0	6.0	1.4	0.0	0.0	0.0	0.0	44.0	10.2	3.0	0.6	0.0	0.0
62	Security and commodity brokers, dealers, exchanges, and services	6	0.0	0.0	0.0	0.0	0.0	0.0	14.0	3.0	2.0	0.6	2.0	0.5	8.0	1.9	0.0	0.0	0.0	0.0	46.0	9.3	3.0	0.8	0.0	0.0
63	Insurance carriers	7	0.0	0.0	0.0	0.0	0.0	0.0	16.0	5.0	2.0	0.6	2.0	0.6	7.0	1.7	0.0	0.0	0.0	0.0	48.0	13.9	3.0	0.9	0.0	0.0
64	Insurance agents, brokers, and service	9	0.0	0.0	0.0	0.0	0.0	0.0	17.0	5.8	2.0	0.6	1.0	1.0	7.0	1.7	0.0	0.0	0.0	0.0	44.0	14.0	3.0	0.9	0.0	0.0
65	Real estate	8	0.0	0.0	0.0	0.0	0.0	0.0	16.0	5.7	2.0	0.6	2.0	0.6	8.0	1.9	0.0	0.0	0.0	0.0	46.0	15.3	3.0	0.9	0.0	0.0

Figure 15.3   Waste stream composition data (SIC code 35 to 65).

242

SIC Group Number	Description	Number of Firms in Group	BioHaz.mean	BioHaz.var	CandD.mean	CandD.var	Fabric.mean	Fabric.var	Food.mean	Food.var	Glass.mean	Glass.var	Metal.mean	Metal.var	OCC.mean	OCC.var	OilChem.mean	OilChem.var	Organic.mean	Organic.var	Paper.mean	Paper.var	Plastic.mean	Plastic.var	Rubber.mean	Rubber.var
67	Holding and other investment offices	6	0.0	0.0	0.0	0.0	0.0	0.0	16.0	4.5	2.0	0.6	2.0	0.4	8.0	1.9	0.0	0.0	0.0	0.0	47.0	12.3	3.0	0.6	0.0	0.0
70	Hotels, rooming houses, camps, and other lodging places	7	0.0	0.0	0.0	0.0	2.0	0.6	29.0	0.0	8.0	2.2	2.0	0.5	6.0	1.4	1.0	0.4	0.0	0.0	36.0	0.0	4.0	1.1	0.0	0.0
72	Personal services	8	0.0	0.0	0.0	0.0	0.0	0.0	15.0	4.7	2.0	0.6	2.0	1.0	8.0	1.9	0.0	0.0	0.0	0.0	46.0	13.3	3.0	0.7	0.0	0.0
73	Business services	12	0.0	0.0	0.0	0.0	0.0	0.0	16.0	4.5	2.0	0.6	2.0	0.4	8.0	1.9	0.0	0.0	0.0	0.0	47.0	12.3	3.0	0.6	0.0	0.0
75	Automotive repair, services, and parking	9	0.0	0.0	0.0	0.0	1.0	0.3	7.0	1.0	2.0	0.6	15.0	1.0	12.0	2.9	8.0	1.6	0.0	0.0	26.0	6.0	14.0	1.0	8.0	2.2
76	Miscellaneous repair services	8	0.0	0.0	0.0	0.0	1.0	0.3	6.0	1.0	2.0	0.6	13.0	1.0	12.0	2.9	8.0	1.3	0.0	0.0	27.0	6.0	12.0	1.0	7.0	1.9
78	Motion pictures	6	0.0	0.0	0.0	0.0	1.0	0.0	6.0	1.0	3.0	0.8	4.0	0.1	4.0	1.0	0.0	0.0	0.0	0.0	27.0	6.0	9.0	0.0	0.0	0.0
79	Amusement and recreation services	4	0.0	0.0	0.0	0.0	0.0	0.0	37.0	0.0	3.0	0.8	4.0	0.1	5.0	1.2	0.0	0.0	0.0	0.0	27.0	0.0	9.0	0.0	0.0	0.0
80	Health services	14	22.0	4.3	0.0	0.0	2.0	0.6	14.0	0.0	2.0	0.6	3.0	0.6	4.0	1.0	0.0	0.0	0.0	0.0	36.0	0.0	12.0	2.6	0.0	0.0
81	Legal services	7	0.0	0.0	0.0	0.0	0.0	0.0	15.0	3.7	2.0	0.6	2.0	1.0	8.0	1.9	0.0	0.0	0.0	0.0	45.0	13.0	3.0	1.0	0.0	0.0
82	Educational services	8	0.0	0.0	0.0	0.0	0.0	0.0	33.0	0.0	2.0	0.6	3.0	1.0	6.0	1.4	0.0	0.0	0.0	0.0	32.0	0.0	5.0	1.2	0.0	0.0
83	Social services	6	0.0	0.0	0.0	0.0	0.0	0.0	16.0	4.5	2.0	0.6	1.0	1.0	7.0	1.7	0.0	0.0	0.0	0.0	46.0	12.0	2.0	1.0	0.0	0.0
84	Museums, art galleries, and botanical and zoological gardens	4	0.0	0.0	0.0	0.0	0.0	0.0	38.0	0.0	3.0	0.8	4.0	0.1	4.0	1.0	0.0	0.0	0.0	0.0	26.0	10.0	10.0	0.1	0.0	0.0
86	Membership organizations	6	0.0	0.0	0.0	0.0	0.0	0.0	16.0	5.0	2.0	0.6	2.0	0.5	7.0	1.7	0.0	0.0	0.0	0.0	46.0	13.3	2.0	0.5	0.0	0.0
87	Engineering, accounting, research, management, and related services	5	0.0	0.0	0.0	0.0	0.0	0.0	15.0	4.7	2.0	0.6	4.0	1.1	8.0	1.9	0.0	0.0	0.0	0.0	47.0	13.6	5.0	1.3	0.0	0.0
91	Executive, legislative, and general government, except finance	5	0.0	0.0	0.0	0.0	0.0	0.0	15.0	4.6	1.0	0.3	2.0	0.5	7.0	1.7	0.0	0.0	0.0	0.0	46.0	13.1	3.0	0.8	0.0	0.0
92	Justice, public order, and safety	6	0.0	0.0	0.0	0.0	0.0	0.0	16.0	4.8	2.0	0.6	2.0	0.5	8.0	1.9	0.0	0.0	0.0	0.0	47.0	13.2	2.0	0.5	0.0	0.0
93	Public finance, taxation, and monetary policy	4	0.0	0.0	0.0	0.0	0.0	0.0	16.0	4.4	2.0	0.6	2.0	0.5	8.0	1.9	0.0	0.0	0.0	0.0	48.0	12.2	1.0	0.2	0.0	0.0
94	Administration of human resource programs	5	0.0	0.0	0.0	0.0	0.0	0.0	15.0	3.9	2.0	0.6	1.0	0.2	7.0	1.7	0.0	0.0	0.0	0.0	49.0	11.9	3.0	0.7	0.0	0.0
95	Administration of environmental quality and housing programs	4	0.0	0.0	0.0	0.0	0.0	0.0	15.0	4.0	2.0	0.6	1.0	0.2	8.0	1.9	0.0	0.0	0.0	0.0	48.0	12.0	2.0	0.5	0.0	0.0
96	Administration of economic programs	3	0.0	0.0	0.0	0.0	0.0	0.0	15.0	3.9	2.0	0.6	2.0	0.6	7.0	1.7	0.0	0.0	0.0	0.0	46.0	11.2	3.0	0.9	0.0	0.0
97	National security and international affairs	3	0.0	0.0	0.0	0.0	0.0	0.0	15.0	5.1	2.0	0.6	2.0	2.0	8.0	1.9	0.0	0.0	0.0	0.0	47.0	15.0	2.0	1.0	0.0	0.0

Figure 15.4 Waste stream composition data (SIC code 67 to 97).

15.3   SIC Code Grouping Validation

The initial grouping of business to characterize solid waste generation was based on SIC codes. U.S. environmental regulators typically use SIC codes to report waste generation data of business sectors. The literature review indicated the use of SIC codes for this purpose has not been statistically validated. This section discusses the validation procedure and results to justify this usage.

The following statistical validation procedure was used. Two analyses were conducted, first to examine the variability within SIC code groups and second to examine the variability between SIC code groups:

1 Develop 95 percent confidence intervals on the mean composition percentage for each material in each of the 65 SIC code groups.
2 Conduct one-way ANOVA (analysis of variance) on each waste material across all waste groups at the 95 percent confidence level.
3 Validate the consolidation of businesses by SIC codes to investigate:
   a Similar variation exists within wastes groups (verify each material's confidence interval is equal to or less than plus or minus 5 percentage points)
   b Significant variation exists between waste groups (verify the ANOVA results are statistically different for each waste material)

Ninety-five percent confidence intervals were developed for each waste material in each of the 65 SIC code groups using the following equation:

$$\bar{x} \pm t_{\alpha/2} \frac{s}{\sqrt{n}}$$

where $\bar{x}$ = the average waste material composition percentage for the random sample
   $t_{\alpha/2}$ = the t-value with $v = n - 1$ degrees of freedom
   s = the sample standard deviation
   n = the sample size

All solid waste materials were within the desired 5 percent confidence interval to validate the grouping of SIC codes. The minimum range was 0.5 percent and the maximum range was 4.6 percent. Table 15.1 displays an example of the results for organic wastes in SIC code group 01: agriculture production crops.

The 95 percent confidence interval range of 2.4 percent indicates that 95 percent of the mean percentages of organic waste for SIC code group 01 (agriculture production crops) will be between 19.6 percent and 24.4 percent. The 95 percent confidence interval 2.4 percent is well within the desired range of 5 percent to validate grouping material composition percentages based on SIC codes.

One-way ANOVA was applied to examine difference between all 65 SIC code groups and materials. Table 15.2 is the one-way ANOVA for organic waste. As shown in the table, the test statistic (F = 1354.734) is significantly higher than

TABLE 15.1 NINETY-FIVE PERCENT CONFIDENCE INTERVAL ON MATERIAL COMPOSITION PERCENTAGE EXAMPLE TO VALIDATE SIC CODE GROUPING (ORGANIC WASTE IN THE SIC CODE GROUP 01)

SIC CODE	DESCRIPTION	NUMBER OF COMPANIES SURVEYED	MEAN PERCENTAGE OF ORGANIC WASTE	VARIANCE OF ORGANIC WASTE	95% CONFIDENCE INTERVAL (+/−)
01	Agricultural production crops	6	22%	5%	2.4%

F critical (F crit = 1.344158) at the 95 percent confidence level. This indicates the SIC code groups generate significantly different composition percentages for organic waste. All other waste compositions were also significantly different at the 95 percent confidence level.

The results of the validations tests indicated that the 65 SIC code groups were effective to group businesses based on waste stream composition percentages (means and variances) at the 95 percent confidence level. The companies within each of the 65 SIC code groups generated equal waste stream composition percentages at the 95 percent confidence level. The 65 SIC code groups generated statistically different composition percentages of solid waste as validated by the ANOVA analysis at the 95 percent confidence level. The purpose of the next characterization phase is to cluster the 65 SIC code groups to determine which SIC code groups generate similar solid waste composition profiles as determined by the means and variances for each material.

TABLE 15.2 ANOVA TABLE TO EXAMINE DIFFERENCES BETWEEN SIC CODE GROUPS

SOURCE OF VARIATION	SS	DF	MS	F VALUE	P VALUE	F CRITICAL
Between groups	0.771465	64	0.012054	1354.734	0	1.344158
Within groups	0.003372	379	8.9E-06	–	–	–
Total	0.774837	443	–	–	–	–

SS = sum of squares
DF = degrees of freedom
MS = mean sum of squares

15.4 Multivariate Cluster Analysis and Discussion

This section discusses the final waste grouping process, multivariate cluster analysis, utilized to reduce the 65 SIC code groups further. Cluster analysis is a group of multivariate techniques whose primary purpose is to objectively group objects based on characteristics they possess. Cluster analysis was used to identify SIC code groups that generate similar solid waste material composition percentages. Solid waste stream composition percentage means and variances were calculated for all records gathered from each SIC code group in the previous steps.

Based on previous research, a five-step cluster analysis procedure was applied for this research (Romesburg, 1984):

1 Obtain data matrix
2 Standardize the data matrix (z scores)
3 Compute the resemblance matrix
4 Execute the cluster method
5 Report and evaluate the results (statistical testing)

Each step is discussed in the following sections.

15.4.1 STEP 1: OBTAIN THE DATA MATRIX

A data matrix is a table containing the objects and attributes of each object to be grouped. The columns of the matrix represent each object (t total objects) and the rows represent the attributes or properties of each object (n total attributes). For this research, the objects are the 65 SIC code groups and the attributes are the means and variances of each waste material generated by the respective waste group (μ and σ^2). Figure 15.5 displays the canonical form of the data matrix.

Waste Material	SIC Code Groups					
	1			j		t
	μ	S^2	...	j	...	t
1	X_{11}	X_{12}	...	X_{1j}	...	X_{1t}
2	X_{21}	X_{22}	...	X_{2j}	...	X_{2t}
⋮	⋮	⋮		⋮		⋮
i	X_{i1}	X_{i2}	...	X_{ij}	...	X_{it}
⋮	⋮	⋮		⋮		⋮
n	X_{n1}	X_{n2}	...	X_{nj}	...	X_{nt}

Figure 15.5 Format of cluster analysis data matrix.

For this research, the objects were each SIC code group for which sufficient data was collected (438 company records covering 65 SIC code groups). Details on the attributes (waste material means and variances) for each object (SIC code group) are listed in the following bullet points. As mentioned, if a material comprised less than 2 percent of total waste for all groups, the material was not included in the analysis for simplification and noise reduction (four in total—aerosol cans, rags, lamps, batteries). Material composition percentage means and standard deviations of

- Biohazard wastes
- Construction and demolition debris (sand, stone, and concrete)
- FABRIC and textiles
- Food waste
- Glass
- Metal
- Old corrugated containers (cardboard)
- Chemicals, sludges, and used oil
- Organic wastes (agricultural)
- Paper (excluding cardboard)
- Plastic
- Rubber
- Wood
- Yard waste

15.4.2 STEP 2: STANDARDIZE THE DATA MATRIX

This is an optional step that standardizes the data matrix by converting the original attributes into new unit-less attributes. This is important for two reasons (Romesburg, 1984):

1 The original units for measuring attributes can arbitrarily affect the similarities among objects.
2 Attributes will contribute more equally to the similarities among objects.

To standardize the matrix, a standardizing function is selected and applied to normalize the data matrix. The standardizing function (or standard normal form) that is most commonly used in practice and applied for this research (Romesburg, 1984), was

$$Z_{ij} = \frac{X_{ij} - \overline{X}_i}{S_i}$$

where

$$\bar{X}_i = \frac{\sum_{j=1}^{t} X_{ij}}{t}$$

$$S_i = \left[\frac{\sum_{j=1}^{t} (X_{ij} - X_i)^2}{t-1} \right]^{1/2}$$

Figure 15.6 displays the canonical form of the standardized data matrix.

15.4.3 STEP 3: COMPUTE THE RESEMBLANCE MATRIX

A resemblance coefficient measures the overall degree of similarity between each pair of objects in the standardized data matrix (Romesburg, 1984). Of the many resemblance coefficients available, the Euclidean distance coefficient was chosen. This coefficient is based on the Pythagorean theorem and used in the following calculation:

$$d_{jk} = \left[\frac{\sum_{i=1}^{n} (X_{ij} - X_{ik})^2}{n} \right]^{1/2}$$

The Euclidean resemblance coefficient is considered a dissimilarity coefficient because the smaller the value the more similar two objects are. Resemblance matrices are square and symmetric; each column identifies the first object in the pair and each row identifies the second object. The cell formed by the intersection of a column and

Waste Material	SIC Code Groups					
	1			j		t
	μ	S^2	...	j	...	t
1	Z_{11}	Z_{12}	...	Z_{1j}	...	Z_{1t}
2	Z_{21}	Z_{22}	...	Z_{2j}	...	Z_{2t}
$\vdots$	$\vdots$	$\vdots$		$\vdots$		$\vdots$
i	Z_{i1}	Z_{i2}	...	Z_{ij}	...	Z_{it}
$\vdots$	$\vdots$	$\vdots$		$\vdots$		$\vdots$
n	Z_{n1}	Z_{n2}	...	Z_{nj}	...	Z_{nt}

Figure 15.6 Format of cluster analysis standardized data matrix.

First object in pair (SIC code group)

	1	2	3	...	n
1	–	–	–	–	–
2	d_{21}	–	–	–	–
3	d_{31}	d_{32}	–	–	–
⋮	⋮	⋮	⋮	–	–
n	d_{n1}	d_{n2}	d_{n3}	...	–

(left axis label: Second object in pair (SIC code group))

Figure 15.7 Format of cluster analysis resemblance matrix.

row contains the value of the resemblance coefficient for the given pair of objects (Romesburg, 1984). For these reasons, only the lower half of the resemblance matrix contains values. Figure 15.7 provides an example.

15.4.4 STEP 4: EXECUTE THE CLUSTER METHOD

The k-means method, a nonhierarchical clustering procedure was applied to group the SIC code groups. The k-means method applies an iterative process to find the optimal clustering for a specified number of groups (k). The k-means clustering method splits a set of objects into a selected number of groups by maximizing between-cluster variation (SSA) relative to within-cluster variation (SSE). The following calculations are used to calculate the sums of squares; the nomenclature of the cluster matrix is shown below as well:

	WASTE GROUPS (CLUSTERS)		
Attribute 1	1	2	k
	y_{11}	y_{21}	y_{k1}
	y_{12}	y_{22}	y_{k2}
	y_{1n}	y_{2n}	y_{kn}

$$SST = \sum_{i=1}^{k} \sum_{j=1}^{n} (y_{ij} - \overline{y}..)^2 = \text{ total sum of squares}$$

$$SSA = n \sum_{i=1}^{k} (\overline{y}_i. - \overline{y}..)^2 = \text{ within cluster sum of squares}$$

$$SSE = \sum_{i=1}^{k} \sum_{j=1}^{n} (y_{ij} - \overline{y}_i.)^2 = \text{ between cluster sum of squares}$$

$$SST = SSA + SSE$$

where

k = SIC code groups number

n = number of attributes for the SIC code groups

y_{ij} = the matrix value for attribute i and SIC code group k

$\bar{y}..$ = mean of all standardized attributes

$\bar{y}_i.$ = mean of standardized attributes for SIC code group k

It is similar to doing a one-way analysis of variance where the groups are unknown and the largest F value is sought by reassigning members to each group (Norusis, 1986). The k-means method starts with one cluster and splits it into two clusters by picking the case farthest from the center as a seed for a second cluster and assigning each case to the nearest center. It continues splitting one of the clusters into two (and reassigning cases) until a specified number of clusters are formed. The k-means method reassigns cases until the within-groups sum of squares can no longer be reduced (Norusis, 1986). The k-means method was made possible by the high speed of computer processing available. The k-means method is a rigorous procedure that evaluates all permutations to minimize *SSE* and maximize *SSA*. The software program *SYSTAT*, developed by SPSS, Inc. was used to perform the multivariate cluster analysis.

The drawback of this method is determining the number of clusters to use (k). This was handled by applying a variance analysis technique (Everitt, 1980). Thorndike plotted average within cluster distance (*SSA/k*) against the number of groups (k). With every increase in k, there will be a decrease in this measurement, but Thorndike suggested that a sudden marked flattening of the curve at any point indicated a distinctive, correct value for k (Everitt, 1980). Such a point should occur when the number of groups corresponds to the configuration of points and there is relatively little gain from further increase in k.

Applying the k-means method to all possible optimal grouping for every k (2 through 65) and graphing the results of the Thorndike method, a k = 22 groups was determined as the optimum. Table 15.3 and graphs in Fig. 15.8 display the results.

TABLE 15.3 ANOVA TABLE USED TO DETERMINE OPTIMAL NUMBER OF CLUSTERS

NUMBER OF WASTE GROUPS (CLUSTERS)	SSA (BETWEEN)	SSE (WITHIN)	AVERAGE (*SSE/K*)
2	280	1461	730.50
3	421	1320	440.00
4	570	1171	292.75
5	695	1046	209.20
6	832	909	151.50
7	977	764	109.14

(Continued)

TABLE 15.3   ANOVA TABLE USED TO DETERMINE OPTIMAL NUMBER OF CLUSTERS (*CONTINUED*)

NUMBER OF WASTE GROUPS (CLUSTERS)	SSA (BETWEEN)	SSE (WITHIN)	AVERAGE (*SSE/K*)
8	1084	657	82.13
9	1130	611	67.89
10	1260	481	48.10
11	1321	420	38.18
12	1402	339	28.25
13	1432	309	23.77
14	1488	253	18.07
15	1570	171	11.40
16	1590	151	9.44
17	1617	124	7.29
18	1644	97	5.39
19	1658	83	4.37
20	1676	65	3.25
21	1690	51	2.43
22*	1700	41	1.86
23	1703	38	1.65
24	1707	34	1.42
...	...	...	...
35	1726	15	0.43

* Using the Thorndike method, 22 waste groups ($k = 22$) is the optimal cluster number. This may be observed from the relatively little gain in average variance within groups from increases in k. The Thorndike method involves plotting average within cluster variation (noise) against the number of groups. Thorndike suggested that a marked flattening of the curve at any point indicates the optimal for k (please see the next page for graphs of the curve, with the optimal point identified).

As shown from the statistical analysis in Table 15.4 and multivariate cluster analysis, 22 waste groups is the optimal number of clusters to balance variation between groups and the number of groups. A sum of squares between groups (variation between groups) of 1700.21 and a sum of squares within groups (noise) of 40.96 was achieved using the clustering techniques. The sum of square calculations may be found in Sec. 16.3. The next section discusses the results of the cluster analysis and details of the 22 waste groups.

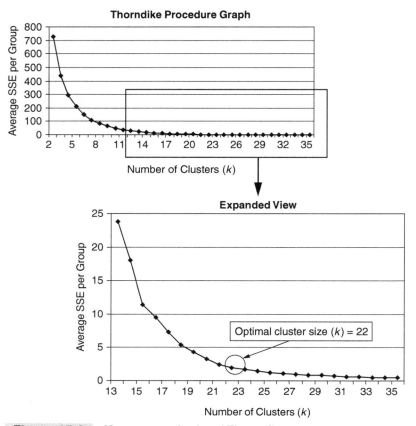

Figure 15.8   *K*-means method and Thorndike procedure graph to determine the optimal number of clusters (*k*).

TABLE 15.4   ANOVA TABLE FOR THE 22 SELECTED WASTE GROUPS (CLUSTERS)

ATTRIBUTE	SSA (BETWEEN 22 CLUSTERS)	DEGREES OF FREEDOM	SSE (WITHIN 22 CLUSTERS)	DEGREES OF FREEDOM	F STATISTIC
BIOMEAN	63.911	21	0.089	43	1463.290
BIOSTD	63.113	21	0.887	43	145.732
CDMEAN	63.817	21	0.183	43	715.906
CDSTD	63.780	21	0.220	43	593.930
FABRICMEAN	63.251	21	0.749	43	173.031
FABRICSTD	63.283	21	0.717	43	180.723
FOODMEAN	63.759	21	0.241	43	540.976
					(Continued)

TABLE 15.4 ANOVA TABLE FOR THE 22 SELECTED WASTE GROUPS (CLUSTERS) (CONTINUED)					
ATTRIBUTE	SSA (BETWEEN 22 CLUSTERS)	DEGREES OF FREEDOM	SSE (WITHIN 22 CLUSTERS)	DEGREES OF FREEDOM	F STATISTIC
FOODSTD	60.935	21	3.065	43	40.703
GLASSMEAN	61.652	21	2.348	43	53.756
GLASSSTD	61.841	21	2.159	43	58.655
METALMEAN	63.730	21	0.270	43	483.753
METALSTD	63.174	21	0.826	43	156.611
OCCMEAN	60.051	21	3.949	43	31.135
OCCSTD	60.355	21	3.645	43	33.906
OILMEAN	63.831	21	0.169	43	772.558
OILSTD	62.584	21	1.416	43	90.516
ORGANICMEAN	63.914	21	0.086	43	1519.041
ORGANICSTD	61.686	21	2.314	43	54.596
PAPERMEAN	63.134	21	0.866	43	149.281
PAPERSTD	61.249	21	2.751	43	45.593
PLASTICMEAN	62.574	21	1.426	43	89.851
PLASTICSTD	62.125	21	1.875	43	67.843
RUBBERMEAN	13.119	21	0.054	43	495.786
RUBBERSTD	63.807	21	0.193	43	676.653
WOODMEAN	63.159	21	0.841	43	153.691
WOODSTD	61.447	21	2.553	43	49.280
YARDMEAN	62.333	21	1.667	43	76.543
YARDSTD	58.596	21	5.404	43	22.203
Total	1700.21	588	40.96		

15.5   Summary of Waste Characterization Findings

Table 15.5 displays the assigned names and number of SIC code groups in each of the 22 waste group clusters. Abbreviations are also given that will be used in future charts and graphs for each waste group.

TABLE 15.5   COMPOSITION OF THE 22 SELECTED WASTE GROUPS (CLUSTERS)

NO.	NAMES OF THE 22 WASTE GROUPS	NUMBER OF SIC CODES INCLUDED	NUMBER OF COMPANIES INCLUDED	ABBREVIATION
1	Agriculture	3	13	ARG
2	Automotive sales, service, and repair	3	27	AUT
3	Chemical and rubber manufacturers	3	16	CHM
4	Commercial and government	20	124	GOV
5	Construction	3	21	CON
6	Education	1	8	EDU
7	Electronic manufacturers	4	32	ELM
8	Food manufacturers	1	8	FDM
9	Food stores	1	9	FDS
10	Forestry	1	3	FOR
11	Hotels	1	7	HTL
12	Medical services	1	14	MED
13	Metal manufacturers	2	10	MLM
14	Mining	2	6	MIN
15	Motor freight and warehousing	1	5	TRN
16	Paper manufacturers and printers	2	14	PPM
17	Recreation and museums	3	14	REC
18	Restaurants	1	13	RST
19	Retail and wholesale trade	7	53	RTL
20	Textile and fabric manufacturers	2	13	FBM
21	Transportation equipment manufacturers	1	14	TRM
22	Wood and lumber manufacturers	2	14	WDM
	Totals	65	438	

Table 15.6 displays the SIC code groups that were clustered from this analysis. As shown in the table, some groups did not cluster with other SIC codes. Based upon mean composition percentages and standard deviations these groups could not be statistically associated with others and were treated as individuals. A review of the Waste Groups indicated a logical grouping. For example, the three agricultural SIC code

TABLE 15.6   SIC CODE GROUPS THAT COMPRISE THE 22 SELECTED WASTE GROUPS (CLUSTERS)

NO.	WASTE GROUP NAME (CLUSTERS)	SIC GROUP NUMBER	DESCRIPTION	NUMBER OF COMPANIES
1	Agriculture (ARG)	01	Agricultural production crops	6
		02	Agricultural production livestock and animal specialties	5
		07	Agricultural services	2
2	Automotive sales, service, and repair (AUT)	55	Automotive dealers and gasoline service stations	10
		75	Automotive repair, services, and parking	9
		76	Miscellaneous repair services	8
3	Chemical and rubber manufacturers (CHM)	28	Chemicals and allied products	5
		29	Petroleum refining and related industries	6
		30	Rubber and miscellaneous plastics products	5
4	Commercial and government (GOV)	60	Depository institutions	8
		61	Non-depository credit institutions	6
		62	Security and commodity brokers, dealers, exchanges, and services	6
		63	Insurance carriers	7
		64	Insurance agents, brokers, and service	9
		65	Real Estate	8
		67	Holding and other investment offices	6
		72	Personal services	8
		73	Business services	12

(Continued)

TABLE 15.6   SIC CODE GROUPS THAT COMPRISE THE 22 SELECTED WASTE GROUPS (CLUSTERS) (CONTINUED)

NO.	WASTE GROUP NAME (CLUSTERS)	SIC GROUP NUMBER	DESCRIPTION	NUMBER OF COMPANIES
		81	Legal services	7
		83	Social services	6
		86	Membership organizations	6
		87	Engineering, accounting, research, management, and related services	5
		91	Executive, legislative, and general government, except finance	5
		92	Justice, public order, and safety	6
		93	Public finance, taxation, and monetary policy	4
		94	Administration of human resource programs	5
		95	Administration of environmental quality and housing programs	4
		96	Administration of economic programs	3
		97	National security and international affairs	3
5	Construction (CON)	15	Building construction general contractors and operative builders	8
		16	Heavy construction other than building construction contractors	7
		17	Construction special trade contractors	6
6	Education (EDU)	82	Educational services	8
7	Electronic manufacturers (ELM)	35	Industrial and commercial machinery and computer equipment	9
		36	Electronic and other electrical equipment and components, except computer equipment	8

(Continued)

TABLE 15.6   SIC CODE GROUPS THAT COMPRISE THE 22 SELECTED WASTE GROUPS (CLUSTERS) (*CONTINUED*)

NO.	WASTE GROUP NAME (CLUSTERS)	SIC GROUP NUMBER	DESCRIPTION	NUMBER OF COMPANIES
		38	Measuring, analyzing, and controlling instruments; photographic, medical and optical goods; watches and clocks	8
		39	Miscellaneous manufacturing industries	7
8	Food manufacturers (FDM)	20	Food and kindred products	8
9	Food stores (FDM)	54	Food Stores	9
10	Forestry (FOR)	08	Forestry	3
11	Hotels (HTL)	70	Hotels, rooming houses, camps, and other lodging places	7
12	Medical services (MED)	80	Health services	14
13	Metal manufacturers (MLM)	33	Primary metal industries	4
		34	Fabricated metal products, except machinery and transportation equipment	6
14	Mining (MIN)	10	Metal mining	3
		12	Coal mining	3
15	Motor freight and warehousing (WAR)	42	Motor freight transportation and warehousing	5
16	Paper manufacturers and printers (PPM)	26	Paper and allied products	6
		27	Printing, publishing, and allied industries	8
17	Recreation and museums (REC)	78	Motion Pictures	6
		79	Amusement and recreation services	4
		84	Museums, art galleries, and botanical and zoological gardens	4

(Continued)

TABLE 15.6 SIC CODE GROUPS THAT COMPRISE THE 22 SELECTED WASTE GROUPS (CLUSTERS) (*CONTINUED*)

NO.	WASTE GROUP NAME (CLUSTERS)	SIC GROUP NUMBER	DESCRIPTION	NUMBER OF COMPANIES
18	Restaurants	58	Eating and drinking places	13
19	Retail and wholesale trade (RTL)	50	Wholesale Trade-durable Goods	8
		51	Wholesale trade-non-durable goods	7
		52	Building materials, hardware, garden supply, and mobile home dealers	8
		53	General merchandise stores	7
		56	Apparel and accessory stores	7
		57	Home furniture, furnishings, and equipment stores	8
		59	Miscellaneous retail	8
20	Textile and fabric manufacturers (FBM)	22	Textile mill products	8
		23	Apparel and other finished products made from fabrics and similar materials	5
21	Transportation equipment manufacturers (TRM)	37	Transportation equipment	14
22	Wood and lumber manufacturers (WDM)	24	Lumber and wood products, except furniture	7
		25	Furniture and fixtures	7

groups were clustered into one group and 20 commercial/government SIC code groups were clustered into one group. These 22 SIC codes primarily generated paper and food wastes.

As shown by the groupings in Table 15.6, logical groupings were made using the multivariate cluster analysis. For example, agricultural SIC codes, similar manufacturing SIC codes, and service oriented businesses were clustered together.

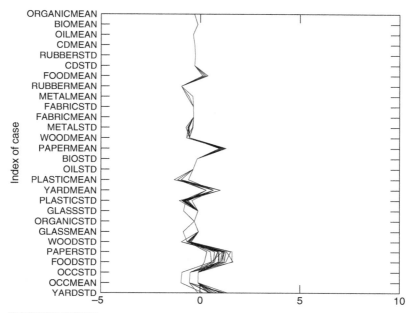

Figure 15.9   Cluster parallel coordinate plot example (for the commercial/government waste group).

Plots were created to visually display the results of the cluster analysis. The two types of graphs developed were cluster parallel coordinate plots and cluster profile plots. The parallel coordinate plots display the z scores for each SIC code population parameter with a line connecting all the scores. The z score is the normalized value for an attribute, as defined by the normal distribution curve. The z score indicated the number of standard deviations from the mean. A value of zero for a z score marks the average for the complete sample. Overlapping lines indicate similarities and gaps indicate discrepancies.

The plot in Fig. 15.9 displays a parallel coordinate plot for the commercial/government waste group. This plot displays the graphical results of the optimal cluster analysis for this group. In the plot, there is one line for each SIC code population in the cluster that connects its z scores for each of the variables. The lines for these 22 SIC code populations all follow a similar pattern: above average values for food, plastic, paper, and so on.

The primary findings from this phase of the research were the determination of companies that generate similar solid waste stream types and compositions, as well as the actual composition percentages. This information is important to begin the next phase of the research, identifying the dominate variables that influence solid waste quantities for companies that generate similar material compositions.

SOLID WASTE ESTIMATION AND PREDICTION

16.1 Introduction

This chapter discusses the process utilized to determine the significant variables that influence annual solid waste quantities for the individual businesses within each of the 22 waste groups determined from the cluster analysis. The cluster analysis provided the SIC code numbers that were grouped into the 22 waste groups. A discussion and analysis of the results are also provided. A three-step process was utilized and it is listed below:

1 Prepare and consolidate the waste records for the 22 selected waste groups for stepwise regression.
2 Perform stepwise regression analysis on the 22 waste groups (using Minitab).
3 Determine the significant variables for the 22 waste groups that influence solid waste quantities using t-tests and F-tests at the 95 percent confidence level.

The remainder of this chapter discusses an overview of the data, multivariable stepwise regression methodology, and the findings from the analysis.

16.2 Overview of Waste Group Data and Consolidation

The first step in the significant variable analysis was to prepare and understand the data. The waste records for each of the 22 waste groups determined from the multivariate cluster analysis were consolidated and initially analyzed to determine basic comparison statistics. Specifically, the following items were calculated for each of the 22 waste groups and are displayed in Table 16.1.

TABLE 16.1   DATA UTILIZED TO DETERMINE ANNUAL SOLID WASTE QUANTITIES FOR INDIVIDUAL COMPANIES FOR THE 20 WASTE GROUPS (SORTED BY AVERAGE WASTE PER COMPANY)

NO.	WASTE GROUP NAME	ABB.	NUMBER OF COMPANIES ANALYZED	AVERAGE SOLID WASTE PER COMPANY PER YEAR (TONS)	AVERAGE NUMBER OF EMPLOYEES PER COMPANY	AVERAGE SOLID WASTE PER EMPLOYEE PER YEAR (TONS)
1	Wood and lumber manufacturers	WDM	14	1528.6	84.3	18.13
2	Metal manufacturers	MLM	10	1313.6	123.8	10.61
3	Food manufacturers	FDM	8	784.8	68.8	11.41
4	Chemical and rubber manufacturers	CHM	16	749.8	84.5	8.87
5	Paper manufacturers and publishers	PPM	14	726.0	84.8	8.56
6	Transportation equipment manufacturers	TRM	14	653.5	120.1	5.44
7	Textile and fabric manufacturers	FBM	13	584.7	97.0	6.03
8	Electronic manufacturers	ELM	32	194.5	59.3	3.28
9	Restaurants	RST	13	158.9	23.5	6.75
10	Mining	MIN	6	105.5	16.5	6.41
11	Food stores	FDS	9	67.7	21.9	3.09
12	Medical services	MED	14	60.7	39.9	1.52
13	Hotels	HTL	7	51.0	23.4	2.18
14	Construction	CON	21	48.4	10.3	4.68
15	Education	EDU	8	46.0	41.4	1.11
16	Recreation and museums	REC	14	44.5	24.0	1.85
17	Retail and wholesale stores	RTL	53	24.3	28.3	0.86
18	Automotive sales, service, and repair	AUT	27	16.3	19.2	0.85
19	Agriculture	ARG	13	15.7	12.4	1.27
20	Commercial and government	GOV	124	2.5	7.9	0.31
21	Transportation	TRN	5	56.9	24.0	2.37
22	Forestry	FOR	3	5.6	11.7	0.48

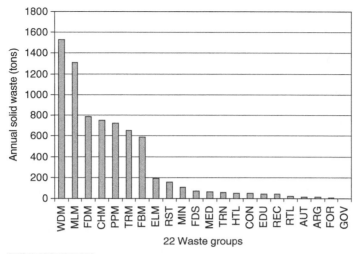

Figure 16.1   **Average annual solid waste per company (tons).**

■ The average annual waste per company in the waste group
■ The average number of employees per company in each waste group
■ The average annual solid waste per company per employee in each waste group

As shown in the table, manufacturing companies generate more average waste per company than service companies. The bar charts in Figs. 16.1 and 16.2 compare the waste groups.

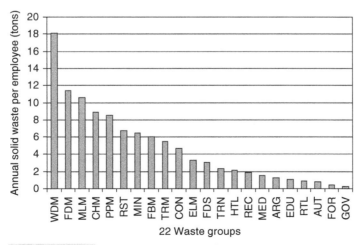

Figure 16.2   **Average annual solid waste per employee per waste group (tons).**

Before conducting the stepwise regression analysis, waste groups with five or less samples were removed from the study. This was done to properly establish relationships among the variables, because sample sizes of less than five could indicate strong relationships when they do not actually exist. Owing to less than five samples, two waste groups were removed from the study: the forestry waste group and the motor freight and transportation waste group. This reduced the number of waste groups from 22 to 20.

The next section discusses the process to determine the significant independent variables that aid in the prediction of annual solid waste for each waste group.

16.3   Stepwise Regression Methodology

Multivariable stepwise regression analysis was conducted on each of the 20 waste group data sets to determine the significant independent variables that influence the annual solid generation. The multivariable stepwise regression analysis of the integrated environmental model functioned to predict and evaluate annual solid waste quantity means and variances of the individual companies in each of the 20 waste groups. Linear, nonlinear, indicator (binary), and interaction variables were considered and statistically tested as potential independent variables. These variables were gathered during the data collection phase. A five-step procedure was utilized to conduct the multivariable stepwise regression analysis and each step is discussed in the following paragraphs. The five-step regression procedure is listed below:

1 Consolidate data for each of the 20 waste groups
2 Determine the multiple regression mathematics and procedure
3 Conduct regression procedure and select independent variables
4 Evaluate results and determine final models
5 Validate regression assumptions

16.3.1   STEP 1: CONSOLIDATE DATA FOR EACH WASTE GROUP

This step prepared the waste records in each of the 20 waste groups for the regression analysis by separating waste records for each waste group and arranging the values in a matrix. The rows of each matrix represented each waste record and the columns represented potential independent variables to be investigated. The value to be predicted was:

y = total annual waste generation (tons/year) for each waste group

The potential independent variables to be investigated were

x_1 = number of employees at the facility

x_2 = ISO 9000 certification duration (months)

x_3 = ISO 14001 certification duration (months)

x_4 = landfill disposal fee per ton in area (dollars)

x_5 = recycling level (percentage of total waste generation)

x_6 = days open per year

Nonlinear variables to be investigated were

x_1^2 = number of employees at the facility squared

x_4^2 = landfill disposal fee per ton in area (dollars) squared

x_5^2 = recycling level (percentage of total waste generation) squared

Several indicator variables to be investigated were

x_7 = indictor variable code 1 non–ISO 9000 certification and code 0 for certification

x_8 = indictor variable code 1 non–ISO 14000 certification and code 0 for certification

Indicator variables were used to investigate the effects of company location. The variables were investigated based on the U.S. state in which the company operated:

x_9 through x_{38} = indictor variable for U.S. state of operation with code 1 for operation in state i and code 0 for operation in a different state

Several interaction variables to be investigated were

x_1x_2 = interaction variable between number of employees and ISO 9000 duration

x_1x_3 = interaction variable between number of employees and ISO 14000 duration

x_1x_4 = interaction variable between number of employees and disposal cost

x_1x_5 = interaction variable between number of employees and recycling level

x_1x_6 = interaction variable between number of employees and days open per year

x_1x_7 = interaction variable between number of employees and ISO 9000 indicator

x_1x_8 = interaction variable between number of employees and ISO 14000 indicator

16.3.2 STEP 2: DETERMINE THE MULTIPLE REGRESSION MATHEMATICS AND PROCEDURE

In fitting a multivariable regression model and when the number of variables exceeds two, as in the case of these models, matrix theory can facilitate the mathematical manipulations considerably (Walpole and Myers, 1993). Minitab was used as the software program to conduct the regression analyses.

For the general model, there are k independent variables, denoted as $x_1, x_2, \ldots, x_k$ and n observations denoted as $y_1, y_2, \ldots, y_n$. Each variable is expressed by the equation

$$y_i = \beta_0 + \beta_1 x_{1i} + \beta_2 x_{2i} + \cdots + \beta_k x_{ki} + \varepsilon_i$$

The model represents n equations describing how the dependent variables (annual solid waste generation per company) are generated. Using matrix notation, the equations can be written as

$$y = X\beta + \varepsilon$$

where

$$
y = \begin{bmatrix} y_1 \\ y_2 \\ \vdots \\ y_n \end{bmatrix}
\qquad
X = \begin{bmatrix}
1 & x_{11} & x_{21} & \cdots & x_{k1} \\
1 & x_{12} & x_{22} & \cdots & x_{k2} \\
\vdots & \vdots & \vdots & & \vdots \\
1 & x_{1n} & x_{2n} & \cdots & x_{kn}
\end{bmatrix}
\qquad
\beta = \begin{bmatrix} \beta_0 \\ \beta_1 \\ \beta_2 \\ \vdots \\ \beta_k \end{bmatrix}
$$

The least squares solution for estimation of β involves finding β for which

$$SSE = (y - X\beta)'(y - X\beta)$$

is minimized. The minimization process involves solving β for the equation

$$\frac{\partial}{\partial b}(SSE) = 0$$

The result reduces to the solution of β in

$$(X'X)\beta = X'y$$

Apart from the initial element, the ith row represents x-values that give rise to the response y_i. Writing

$$
A = X'X = \begin{bmatrix}
n & \sum_{i=1}^{n} x_{1i} & \sum_{i=1}^{n} x_{2i} & \cdots & \sum_{i=1}^{n} x_{ki} \\
\sum_{i=1}^{n} x_{1i} & \sum_{i=1}^{n} x_{1i}^2 & \sum_{i=1}^{n} x_{1i} x_{2i} & \cdots & \sum_{i=1}^{n} x_{1i} \\
\vdots & \vdots & \vdots & & \vdots \\
\sum_{i=1}^{n} x_{ki} & \sum_{i=1}^{n} x_{ki} x_{1i} & \sum_{i=1}^{n} x_{ki} x_{2i} & \cdots & \sum_{i=1}^{n} x_{ki}^2
\end{bmatrix}
$$

and

$$g = X'y = \begin{bmatrix} g_0 = \sum_{i=1}^{n} y_i \\ g_1 = \sum_{i=1}^{n} x_i y_i \\ \vdots \\ g_k = \sum_{i=1}^{n} x_{ki} y_i \end{bmatrix}$$

the normal equations can be put in matrix form

$$A\beta = g$$

If the matrix A is nonsingular, the solution for the regression coefficients is written as

$$\beta = A^{-1}g = (X'X)^{-1}X'y$$

The regression equation is obtained by solving a set of $k + 1$ equations for the like number of unknowns. This involves the inversion of $k + 1$ by $k + 1$ matrix $X'X$.

Step 3 in the following section was used to calculate the regression equation for each waste group. As a visual representation, the scatter diagram for the transportation equipment manufacturing group is displayed in Fig. 16.3 for annual waste versus the number of employees.

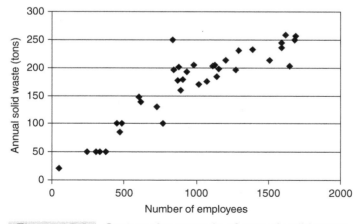

Figure 16.3 Scatter plot example of annual solid waste versus the number of employees for transportation equipment manufacturers' waste group.

16.3.3 STEP 3: CONDUCT REGRESSION PROCEDURE AND SELECT INDEPENDENT VARIABLES (STEPWISE REGRESSION METHOD)

The stepwise regression method was used to select the independent variables for each of the 20 remaining waste groups using the developed matrices. The stepwise method was chosen for several reasons:

- Statistically evaluates the addition or removal of every selected potential independent variable at the specified level of confidence
- Accounts for multicolinearity (correlation between variables)
- Selects the most efficient variables for the model, considering redundancy
- The method is programmable

Stepwise regression is an algorithm that applies an iterative process to determine the optimal independent variables. For the first step, all independent variables are entered into the model and are represented by their partial F statistic. A partial F statistic is the ratio of variance explained by the independent variable divided by total variation between all observations of the sample. An independent variable added at an earlier step may now be redundant because of the relationships between it and the independent variable now in the equations. If the partial F statistic for a variable is less than F_{OUT}, the variable is dropped from the model. For this research, an F_{OUT} corresponding to the 95 percent confidence level was used. Stepwise regression requires two cutoff variables, F_{IN} and F_{OUT}. Some analysts prefer $F_{IN} = F_{OUT}$, although this is not necessary. Frequently $F_{IN} > F_{OUT}$, making it relatively more difficult to add an independent variable than to delete one (Montgomery, 2001). Minitab was applied to conduct this analysis.

The F statistic is calculated based on the sum of squares from the regression results for each variable (the method discussed in the previous section determines the regression results). The mathematics for the F statistic is shown below.

For multiple linear regression, the error and regression sum of squares take the same form as in the simple linear regression case (Walpole and Myers, 1993). The total sum of squares (*SST*) identity is

$$SST = \sum_{i=1}^{n} (y_i - \overline{y})^2 = \sum_{i=1}^{n} (\hat{y}_i - \overline{y})^2 + \sum_{i=1}^{n} (y_i - \hat{y}_i)^2$$

Using a different notation for the *SST* identity:

$$SST = SSR + SSE$$

with

$$SST = \sum_{i=1}^{n} (y_i - \overline{y})^2 = \text{total sum of squares}$$

and

$$SSR = \sum_{i=1}^{n} (\hat{y}_i - \overline{y})^2 = \text{regression sum of squares}$$

There are k degrees of freedom associated with SSR and SST has $n - 1$ degrees of freedom. Therefore, after subtraction, SSE has $n - k - 1$ degrees of freedom. The estimate of σ^2 is given by the error sum of squares divided by the degrees of freedom (Walpole and Myers, 1993). All three of the sums of squares appear on the printout of most multiple regression computer packages, including Minitab, which was used for this research. Minitab and SYSTAT were used to calculate these values. SYSTAT was used to verify results.

The F statistic is calculated using the following equation:

$$f = \frac{SSR/k}{SSE/(n - k - 1)} = \frac{SSR/k}{s^2}$$

The results of the sum of squares and F statistic may be represented in an analysis of variance (ANOVA) table (Table 16.2).

16.3.4 STEP 4: EVALUATE RESULTS AND DETERMINE FINAL MODELS

The goals of this step is to develop the final models and determine the variables that efficiently predict solid waste generation for each waste group. The following procedure is used:

- Apply stepwise regression method to the 20 waste groups (data matrices)
- Remove outliers and recalculate
- Evaluate the results statistically with:
  - ANOVA
  - F-test (strength of entire models)
  - t-test (strength of each independent variable)
  - Coefficient of determination
- Validate regression model assumptions
- Report final results

TABLE 16.2 MULTIVARIABLE ANOVA TABLE FORMAT				
SOURCE OF VARIATION	SUM OF SQUARES	DEGREES OF FREEDOM	MEAN SQUARE	COMPUTED F
Regression	SSR	k	SSR	$(SSR/k)/s^2$
Error	SSE	$n - k - 1$	$SSE/(n - k - 1)$	
Total	SST	$n - 1$		

The above procedure was applied to all business groups. The mathematics, statistical tests, and example calculations for the transportation equipment manufacturer's waste group are discussed in the following paragraphs.

For the multiple regression equation (Walpole and Myers, Theorem 12.1, 1993):

$$y = X\beta + \varepsilon$$

an unbiased estimator of σ^2 is given by the error or residual mean square

$$s^2 = \frac{SSE}{n-k-1}$$

where

$$SSE = \sum_{i=1}^{n} e_i^2 = \sum_{i=1}^{n} (y_i - \hat{y}_i)^2$$

Sum of square calculations for multiple linear regression is similar to the previous simple linear regression equation discussed earlier. One difference is the degrees of freedom discussed previously in this chapter.

One criterion that is commonly used to illustrate the adequacy of a fitted regression line is the coefficient of multiple determination (R^2) (Walpole and Myers, 1993):

$$R^2 = \frac{SSR}{SST} = \frac{\sum_{i=1}^{n} (\hat{y}_i - \bar{y})^2}{\sum_{i=1}^{n} (y_i - \bar{y})^2}$$

The coefficient of determination indicates what proportion of the total variation in the response is explained by the fitted model. The regression sum of squares can be used to give some indication concerning whether or not the model is an adequate explanation of the true situation (Walpole and Myers, 1993). The R^2 value is the percent of variation explained by each independent variable. The higher an R^2 for a dependent and independent variable is, the stronger the relationship among variables. One can test the hypothesis H_0 that the regression is not significant by forming the ratio

$$f = \frac{SSR/k}{SSE/(n-k-1)} = \frac{SSR/k}{s^2}$$

and rejecting H_0 at the α-level of significance when $f > f_\alpha(k, n-k-1)$.

Another test, the t-test is the standard method used to evaluate individual coefficients in a multiple regression model. The addition of any single variable to a regression system will increase the regression sum of squares and thus reduce the error sum of squares (Walpole and Myers, 1993). The decision is whether the increase in the regression is sufficient to warrant using the variable in the model. The use of unimportant variables can reduce the effectiveness of the prediction equation by increasing the variance of the estimated response (Walpole and Myers, 1993). The following mathematics illustrate the method used to evaluate individual coefficients and determine if each variable effectively aids in predicting total annual waste.

A t-test, which is a statistical test on a sample from a normally distributed population, was conducted at the 95 percent confidence level to determine if there was significant correlation between the variables. The null hypothesis (H_0) was defined such that the slope of the population regression line (β_i) is zero, in other words variables are not correlated. This would mean that there is no linear relationship between the independent variables (x_i) and dependent variables (y_i). The alternate hypothesis (H_1) states that the slope of the population regression is not equal to zero, in other words, the tested variables are correlated and do have a relationship. The t-tests were conducted as follows:

Hypothesis test

$$H_0 : \beta_1 = 0$$
$$H_1 : \beta_1 \neq 0$$

Decision rule

$$\text{Reject } H_0 \text{ if } t > t_{\alpha/2,\, n-1} \text{ or } t < t_{-\alpha/2,\, n-1}$$

The t value was calculated at the α significance level and $n - 1$ degrees of freedom. The decision rule is based upon a two-tail test, where $-\alpha/2$ and $+\alpha/2$ define the critical region. Calculated test statistic values (see equation below) with a value less than $-\alpha/2$ or greater than $+\alpha/2$ will indicate a relationship exists between the dependent and independent variable. This rule is based upon the t distribution.

Test statistic

$$t = \frac{b}{s/\sqrt{S_{xx}}}$$

where

$$\beta = \frac{\sum_{i=1}^{n}(x_i - \overline{x})(y_i - \overline{y})}{\sum_{i=1}^{n}(x_i - \overline{x})^2} = \frac{S_{xy}}{S_{xx}}$$

$$\overline{x} = \sum_{i=1}^{n} \frac{x_i}{n}$$

$$\overline{y} = \sum_{i=1}^{n} \frac{y_i}{n}$$

$$n = \text{number of observations}$$

$$S_{xx} = \sum_{i=1}^{n} (x_i - \bar{x})^2$$

$$S_{yy} = \sum_{i=1}^{n} (y_i - \bar{y})^2$$

$$S_{xy} = \sum_{i=1}^{n} (x_i - \bar{x})(y_i - \bar{y})$$

$$s = \sqrt{\frac{S_{yy} - \beta S_{xy}}{n - 2}}$$

In the preceding equations, β is the slope of the regression line and s is an unbiased estimator of σ^2 (the population standard deviation). The variables S_{xx}, S_{yy}, and S_{xy} are the sum of squares for x and y (Walpole and Myers, 1993).

16.3.5 STEP 5: VALIDATE REGRESSION ASSUMPTIONS

There are five major assumptions or ideal conditions for the estimation and inference in multiple regression models (Dielman, 1996). The five assumptions are

1 The relationship is statistically significant.
2 The residuals, e_i, have constant variance σ_e^2.
3 The residuals are independent.
4 The residuals are normally distributed.
5 The explanatory variables are not highly correlated.

The first assumption was tested and validated using the F-test. The F-test was discussed in the previous section at the 95 percent confidence level.

To access the assumptions of constant variance around the regression line, that residuals are randomly distributed, and residuals are normally distributed, residual plot analysis was conducted. A residual describes the error in the fit of the model at the ith data point (Walpole and Myers, 1993) and is described in the following equation:

$$e_i = y_i - \hat{y}_i$$

In a residual plot of $\hat{e}_i$ versus an explanatory variable x, the residuals should appear scattered randomly about the zero line with no difference in the amount of variation in the residuals, regardless of the value of x (Dielman, 1996). If there appears to be a difference in variation (for example, if the residuals are more spread out for larger values of x than for small values), then the assumption of constant variance may be violated.

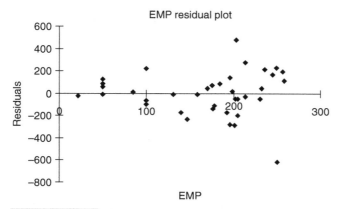

Figure 16.4 **Example residual plot.**

Figure 16.4 is an example of a residual plot for number of employees for the transportation equipment manufacturer's waste group.

As shown by the residual plot, the residuals appear to be random and of equal variance, except for two outliers, which were removed and the model was recalculated. No patterns appear to be present. This validates the constant variance and random error assumptions of the regression model. The same tests were conducted for the other 19 waste groups as well.

The final assumption in a multiple regression model is that explanatory (independent) variables are not correlated with one another. When explanatory variables are correlated with one another, the problem of multicolinearity is said to exist. The presence of a high degree of multicolinearity among the explanatory variables will result in the following problems (Dielman, 1996):

■ The standard deviation of the regression coefficients will be disproportionately large.
■ The regression coefficient estimates will be unstable, and the accuracy will vary significantly for different independent variables.

To detect multicolinearity, several methods have been developed. One method involves computing the pairwise correlations between explanatory variables. One rule of thumb suggested by some researchers is that mutlicolinearity may be a serious problem if any pairwise correlation is greater than 0.5 (Dielman, 1996). Multicolinearity was examined using this method and all pairwise correlation was below 0.5. The correlations between random variables are denoted by r and are calculated as:

$$r = r_{x_1 x_2} = \frac{S_{12}}{\sqrt{S_{11} S_{22}}}$$

S_{11}, S_{12}, and S_{22} may be found in the $(X'X)$ matrix in the off diagonals. All correlations between independent variables were less than 0.001.

16.4   Analysis of Results and Summary of Findings

This section discusses the results and findings from the regression analysis conducted on the 20 waste groups. Included in this discussion are lists of the independent variables that are significant in predicting annual solid waste for each group, the full regression results (F-test, t-test, and R^2-coefficient of determination), ANOVA, and correlation analysis. Table 16.3 and Fig. 16.5 display the significant variables that are

TABLE 16.3   SIGNIFICANT VARIABLES THAT INFLUENCE SOLID WASTE FOR THE 20 WASTE GROUPS

NO.	WASTE GROUP NAME	ABB.	NUMBER OF EMPLOYEES	LANDFILL DISPOSAL COST ($/TON)	ISO 14001 (YES/NO)
1	Wood and lumber manufacturers	WDM	×	×	×
2	Metal manufacturers	MLM	×	×	×
3	Food manufacturers	FDM	×	×	×
4	Chemical and rubber manufacturers	CHM	×	×	×
5	Paper manufacturers and publishers	PPM	×	×	×
6	Transportation equipment manufacturers	TRM	×	×	×
7	Textile and fabric manufacturers	FBM	×	×	×
8	Electronic manufacturers	ELM	×	×	×
9	Restaurants	RST	×	NA	NA
10	Mining	MIN	×	NA	NA
11	Food stores	FDS	×	NA	NA
12	Medical services	MED	×	NA	NA
13	Hotels	HTL	×	NA	NA
14	Construction	CON	×	NA	NA
15	Education	EDU	×	NA	NA
16	Recreation and museums	REC	×	NA	NA
17	Retail and wholesale stores	RTL	×	NA	NA
18	Automotive sales, service, and repair	AUT	×	NA	NA
19	Agriculture	ARG	×	NA	NA
20	Commercial and government	GOV	×	NA	NA

× denotes a significant variable to quantify solid waste for the waste group at the 95 percent confidence level. (Note: 12 of the 20 waste groups were described by the number of employees alone).

Waste Group	Number of Companies	Indicators	Regression Model	t Statistic	t Critical	P Value	Partial R^2	R^2	Adjusted R^2	F Statistic	F Critical	P Value
Agriculture	13	x_1 = # employees	$1.29x_1 - 0.27$					0.79		41.38	4.75	0.000
Automotive sales, service, and repair	27	x_1 = # employees	$0.51x_1 + 6.48$					0.83		122.06	4.23	0.000
Chemical and rubber manufacturers	16	x_1 = # employees, x_2 = disposal \$/ton, x_3 = ISO14001cert.	$7.14x_1 - 3.49x_2 - 140.4x_3 + 371.10$	x_1 = 5.46, x_2 = −3.02, x_3 = −3.47	2.18	x_1 = 0.000, x_2 = 0.011, x_3 = 0.005	x_1 = 0.51, x_2 = 0.14, x_3 = 0.19	0.84	0.80	21.10	3.49	0.000
Commercial and government	124	x_1 = # employees	$0.43x_1 - 0.94$					0.86		749.43	3.92	0.000
Construction	21	x_1 = # employees	$3.54x_1 + 11.81$					0.79		71.48	4.35	0.000
Education	8	x_1 = # employees	$0.25x_1 + 35.68$					0.84		31.50	5.59	0.000
Electronic manufacturers	32	x_1 = # employees, x_2 = disposal \$/ton, x_3 = ISO14001cert.	$2.29x_1 - 0.83x_2 - 56.31x_3 + 126.63$	x_1 =−8.67, x_2 = −2.27, x_3 = −4.51	2.26	x_1 = 0.000, x_2 = 0.033, x_3 = 0.000	x_1 = 0.59, x_2 = 0.04, x_3 = 0.15	0.78	0.76	33.52	3.86	0.000
Food manufacturers	8	x_1 = # employees, x_2 = disposal \$/ton, x_3 = ISO14001cert.	$7.19x_1 - 5.17x_2 - 183.5x_3 + 577.16$	x_1 = 4.53, x_2 = −2.85, x_3 = −2.89	2.57	x_1 = 0.011, x_2 = 0.046, x_3 = 0.044	x_1 = 0.50, x_2 = 0.19, x_3 = 0.21	0.90	0.83	12.35	5.79	0.017
Food stores	9	x_1 = # employees	$2.71x_1 + 8.27$					0.81		29.84	5.32	0.000
Hotels	7	x_1 = # employees	$2.01x_1 + 4.05$					0.82		22.39	5.99	0.000
Medical services	14	x_1 = # employees	$0.98x_1 + 21.64$					0.80		47.11	4.67	0.000
Metal manufacturers	10	x_1 = # employees, x_2 = disposal \$/ton, x_3 = ISO14001cert.	$8.21x_1 - 7.51x_2 - 214.93x_3 + 756.46$	x_1 = 4.81, x_2 = −3.08, x_3 = −2.52	2.48	x_1 = 0.003, x_2 = 0.022, x_3 = 0.045	x_1 = 0.60, x_2 = 0.20, x_3 = 0.10	0.90	0.86	18.86	4.76	0.002
Mining	6	x_1 = # employees	$5.21x_1 + 19.8$					0.85		17.00	7.71	0.001
Paper manufacturers and printers	14	x_1 = # employees, x_2 = disposal \$/ton, x_3 = ISO14001cert.	$6.63x_1 - 3.51x_2 - 114.82x_3 + 340.27$	x_1 = 4.83, x_2 = −2.65, x_3 = −2.48	2.23	x_1 = 0.001, x_2 = 0.024, x_3 = 0.033	x_1 = 0.58, x_2 = 0.15, x_3 = 0.10	0.83	0.78	16.35	3.71	0.000
Recreation and museums	14	x_1 = # employees	$1.87x_1 - 0.41$					0.78		42.55	4.67	0.000
Restaurants	13	x_1 = # employees	$5.91x_1 + 19.87$					0.83		53.71	4.75	0.000
Retail and wholesale	53	x_1 = # employees	$0.78x_1 + 2.27$					0.81		2173.42	4.04	0.000
Textile and fabric manufacturers	13	x_1 = # employees, x_2 = disposal \$/ton, x_3 = ISO14001cert.	$7.29x_1 - 3.78x_2 - 92.36x_3 + 40.5$	x_1 = 5.88, x_2 = −3.38, x_3 = −2.36	2.26	x_1 = 0.000, x_2 = 0.008, x_3 = 0.042	x_1 = 0.56, x_2 = 0.20, x_3 = 0.09	0.85	0.81	17.63	3.86	0.000
Transportation equipment manufacturers	14	x_1 = # employees, x_2 = disposal \$/ton, x_3 = ISO14001cert.	$2.97x_1 - 1.73x_2 - 56.22x_3 + 412.84$	x_1 = 5.06, x_2 = −2.54, x_3 = −2.36	2.23	x_1 = 0.000, x_2 = 0.029, x_3 = 0.040	x_1 = 0.58, x_2 = 0.11, x_3 = 0.10	0.79	0.72	12.29	3.71	0.001
Wood and lumber manufacturers	14	x_1 = # employees, x_2 = disposal \$/ton, x_3 = ISO14001cert.	$17.40x_1 - 7.76x_2 - 217.59x_3 + 473.93$	x_1 = 6.99, x_2 = −3.73, x_3 = −2.99	2.23	x_1 = 0.000, x_2 = 0.004, x_3 = 0.014	x_1 = 0.65, x_2 = 0.15, x_3 = 0.08	0.88	0.85	25.06	3.71	0.000

Figure 16.5 Stepwise regression results for the 20 waste groups (significant variables that influence solid waste quantities).

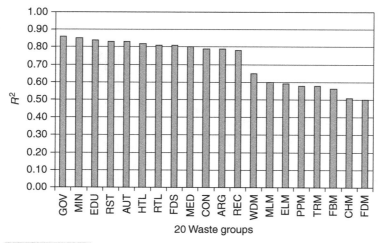

Figure 16.6   Coefficient of determination (R^2) analysis for the variable: number of employees.

influential in predicting annual solid waste at the 95 percent confidence level and the full stepwise regression results for each of the 20 waste groups.

The remainder of this chapter discusses the regression results from the previous table and compares the 20 waste groups to gain insights on waste generation in the United States. As shown in the preceding table, 50 to 85 percent of the total variation (R^2) in annual waste generation is attributed to the number of employees for each waste group. The chart in Fig. 16.6 provides a visual comparison.

The chart in Fig. 16.7 provides a comparison of the total and partial coefficient of determination for each of the 20 waste groups. This chart indicates that 79 percent to 90 percent of the variation in annual solid waste generation was accounted for by the three significant independent variables identified for the 20 waste groups (at the 95 percent confidence level). The magnitudes of the regression coefficients were compared for each of the 20 waste groups. The charts in Figs. 16.8 through 16.10 display this comparison.

The previous procedure determined the significant variables that influence annual solid waste quantities for each group. The next step involved interpreting the results, identifying trends, and creating new knowledge. Two categories of analyses were conducted (from the regression analyses):

■ Analysis of the variables that entered the prediction equations
■ Analysis of the variables that did not enter the prediction equations

First, the variables that did enter the regression equation were examined. Standardized regression equations were established to equally compare the stepwise results. This was accomplished by forcing the constant terms in each regression equation to zero. This was completed using SYSTAT software. Tables 16.4 to 16.6 display the standardized regression coefficients for each waste group.

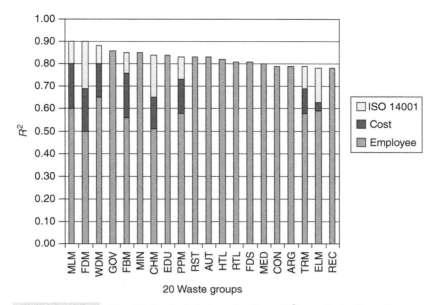

Figure 16.7 Coefficient of determination (R^2) analysis for all influential variables (number of employees, landfill disposal cost, and ISO 14001 certification).

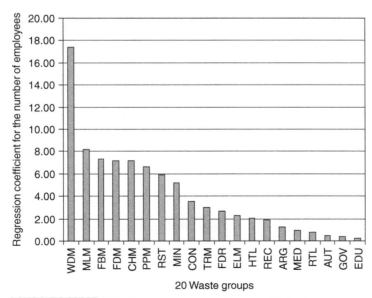

Figure 16.8 Waste group regression coefficient comparison for the number of employees.

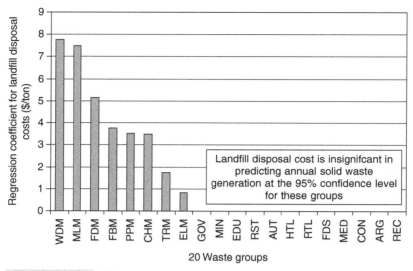

Figure 16.9   Waste group regression coefficient comparison for landfill disposal cost ($/ton).

In all waste groups, the number of employees was the most significant variable to predict annual waste generation. This is logical, because the number of employees should have a high correlation to the on size of a company direct impact on annual waste generation tonnages. The more employees working at a company, the more waste the company generates. The results of the regression

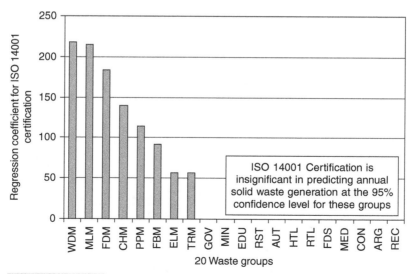

Figure 16.10   Waste group regression coefficient comparison for ISO 14001 certification.

TABLE 16.4   RANKING AND COMPARISON OF SIGNIFICANT VARIABLES TO PREDICT SOLID WASTE BASED ON STANDARDIZED REGRESSION COEFFICIENT MAGNITUDE FOR THE NUMBER OF EMPLOYEES

WASTE GROUP	STANDARDIZED REGRESSION COEFFICIENT FOR THE NUMBER OF EMPLOYEES
FBM	1.32
WDM	1.26
CHM	1.20
MLM	1.16
PPM	1.16
FDM	1.12
TRM	1.03
ELM	0.97
GOV	0.93
EDU	0.92
RST	0.91
AUT	0.91
HTL	0.91
RTL	0.90
FDS	0.90
MED	0.89
ARG	0.89
CON	0.89
REC	0.88

analysis reflected this fact. In order to compare the quantities of solid waste generated by the 20 remaining waste groups, the average solid waste per company in each of the 20 waste groups was calculated. This was completed to gain additional insights into waste generation and to identify trends. Table 16.7 displays the regression coefficients for the number of employees and the average waste generated per company. The graph in Fig. 16.11 displays the relationship between average annual solid waste generation per company for each waste group and the employee regression coefficient.

TABLE 16.5   RANKING AND COMPARISON OF SIGNIFICANT VARIABLES TO PREDICT SOLID WASTE BASED ON STANDARDIZED REGRESSION COEFFICIENT MAGNITUDE FOR LANDFILL DISPOSAL COST

WASTE GROUP	STANDARDIZED REGRESSION COEFFICIENT FOR LANDFILL DISPOSAL COST ($/TON)
MLM	−1.31
FBM	−0.29
WDM	−0.22
CHM	−0.13
PPM	−0.12
ELM	−0.08
FDM	−0.07
TRM	−0.01
GOV	0.00
EDU	0.00
RST	0.00
AUT	0.00
HTL	0.00
RTL	0.00
FDS	0.00
MED	0.00
ARG	0.00
CON	0.00
REC	0.00

For 12 of the 20 waste groups the number of employees alone was the only dominate variable to predict annual solid waste. These waste groups are listed below:

1 Restaurants
2 Mining
3 Food stores
4 Medical services
5 Hotels
6 Construction

TABLE 16.6   RANKING AND COMPARISON OF
SIGNIFICANT VARIABLES TO PREDICT SOLID WASTE
BASED ON STANDARDIZED REGRESSION COEFFICIENT
MAGNITUDE FOR ISO 14001 CERTIFICATION

WASTE GROUP	STANDARDIZED REGRESSION COEFFICIENT FOR ISO 14001 CERT.
CHM	−0.12
ELM	−0.12
FBM	−0.11
WDM	−0.11
FDM	−0.11
PPM	−0.09
MLM	−0.08
TRM	−0.04
GOV	0.00
EDU	0.00
RST	0.00
AUT	0.00
HTL	0.00
RTL	0.00
FDS	0.00
MED	0.00
ARG	0.00
CON	0.00
REC	0.00

7 Education
8 Recreation and museums
9 Retail and wholesale
10 Automotive sales, service, and repair
11 Agriculture
12 Commercial and government

These 12 waste groups were primarily service oriented (nonmanufacturing) and generated less average waste per company than the 8 manufacturing waste groups that required more variables to predict waste. The 12 nonmanufacturing waste groups generated primarily office waste such as papers and food waste. These waste components are

TABLE 16.7   ANNUAL WASTE AVERAGES PER COMPANY FOR EACH WASTE GROUP VERSUS THE REGRESSION COEFFICIENT FOR THE NUMBER OF EMPLOYEES

WASTE GROUP	GRAPH AND CHART CODE	AVERAGE SOLID WASTE PER COMPANY PER YEAR (TONS)	EMPLOYEE REGRESSION COEFFICIENT
Wood and lumber manufacturers	WDM	1707.9	17.40
Metal manufacturers	MLM	1313.6	8.21
Food manufacturers	FDM	784.8	7.19
Chemical and rubber manufacturers	CHM	749.8	7.14
Paper manufacturers and printers	PPM	726.0	6.63
Transportation equipment manufacturers	TRM	653.5	2.97
Textile and fabric manufacturers	FBM	584.7	7.29
Electronic manufacturers	ELM	194.5	2.29
Restaurants	RST	158.9	5.91
Mining	MIN	105.5	5.21
Food stores	FDR	67.7	2.71
Medical services	MED	60.7	0.98
Hotels	HTL	51.0	2.01
Construction	CON	48.4	3.54
Education	EDU	46.0	0.25
Recreation and museums	REC	44.5	1.87
Retail and wholesale	RTL	24.3	0.78
Automotive sales, service, and repair	AUT	16.3	0.51
Agriculture	ARG	15.7	1.29
Commercial and government	GOV	2.5	0.43

generated in direct proportion to the number of employees at each facility. Tables 16.8 and 16.9 and the chart in Fig. 16.12 compare these 12 companies predicted by number of employees alone to the 8 other waste groups.

The chart in Fig. 16.12 displays the average annual waste per company for the 20 waste groups. Highlighted on this chart are the 12 nonmanufacturing waste groups in which the number of employees was the only significant variable to predict annual solid waste generation at the 95 percent confidence level. As shown in the chart, these 12 nonmanufacturing waste groups generate significantly less solid waste than the 8 manufacturing waste groups.

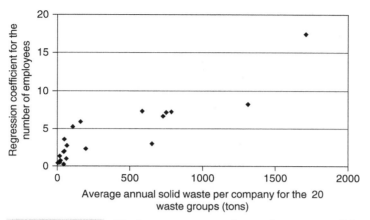

Figure 16.11 Waste group comparison of average solid waste per company in the 20 waste groups versus the regression coefficient for the number of employees.

WASTE GROUP	ABBREVIATION	AVERAGE SOLID WASTE PER COMPANY PER YEAR (TONS)	EMPLOYEE REGRESSION COEFFICIENT
Restaurants	RST	158.9	5.91
Mining	MIN	105.5	5.21
Food stores	FDS	67.7	2.71
Medical services	MED	60.7	0.98
Hotels	HTL	51.0	2.01
Construction	CON	48.4	3.54
Education	EDU	46.0	0.25
Recreation and museums	REC	44.5	1.87
Retail and wholesale	RTL	24.3	0.78
Automotive sales, service, and repair	AUT	16.3	0.51
Agriculture	ARG	15.7	1.29
Commercial and government	GOV	2.5	0.43

TABLE 16.8 NONMANUFACTURING WASTE GROUP COMPARISON OF AVERAGE SOLID WASTE PER COMPANY VERSUS THE REGRESSION COEFFICIENT FOR THE NUMBER OF EMPLOYEES

TABLE 16.9   MANUFACTURING WASTE GROUP COMPARISON OF AVERAGE SOLID WASTE PER COMPANY VERSUS THE REGRESSION COEFFICIENT FOR THE NUMBER OF EMPLOYEES

WASTE GROUP	ABBREVIATION	AVERAGE SOLID WASTE PER COMPANY PER YEAR (TONS)	EMPLOYEE REGRESSION COEFFICIENT
Wood and lumber manufacturers	WDM	1707.9	17.40
Metal manufacturers	MLM	1313.6	8.21
Food manufacturers	FDM	784.8	7.19
Chemical and rubber manufacturers	CHM	749.8	7.14
Paper manufacturers and printers	PPM	726.0	6.63
Transportation equipment manufacturers	TRM	653.5	2.97
Textile and fabric manufacturers	FBM	584.7	7.29
Electronic manufacturers	ELM	194.5	2.29

After investigating the eight solid waste groups that were predicted by variables in addition to the number of employees, several trends were identified. In addition to the number of employees the eight solid waste groups were predicted by solid waste disposal costs per ton and ISO 14001 certification. These two additional variables acted as dampeners in the regression equation by reducing the amounts of waste generated. The results of the regression analysis for the eight manufacturing waste groups provided two insights. First, these waste groups are more sensitive to waste costs and have higher incentives to reduce waste in order to reduce costs over that of the 12 nonmanufacturing

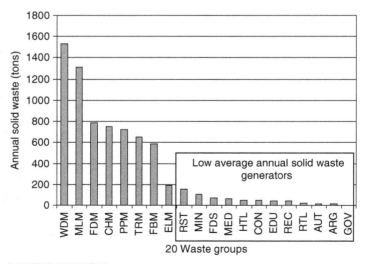

Figure 16.12   Average annual solid waste generation of the waste groups (highlighting nonmanufacturers).

TABLE 16.10 WASTE GROUP COMPARISON OF THE AVERAGE SOLID WASTE PER COMPANY VERSUS THE REGRESSION COEFFICIENT FOR LANDFILL DISPOSAL COSTS

WASTE GROUP	ABBREVIATION	AVERAGE SOLID WASTE PER COMPANY PER YEAR (TONS)	LANDFILL DISPOSAL COST REGRESSION COEFFICIENT	T VALUE
Wood and lumber manufacturers	WDM	1707.9	−7.76	−3.73
Metal manufacturers	MLM	1313.6	−7.51	−3.08
Food manufacturers	FDM	784.8	−5.17	−2.85
Textile and fabric manufacturers	FBM	584.7	−3.78	−3.02
Paper manufacturers and publishers	PPM	726.0	−3.51	−2.65
Chemical and rubber manufacturers	CHM	749.8	−3.49	−2.54
Transportation equipment manufacturers	TRM	653.5	−1.73	−3.38
Electronic manufacturers	ELM	194.5	−0.83	−2.27

waste groups. Second, the regression analysis showed solid waste generation is reduced by ISO 14001 certification of the eight manufacturing groups, indicating the usefulness of the certification to decrease environmental impact. Several trends were also observed. A trend is visible between average waste generation and solid waste disposal costs per ton for these groups as displayed in Table 16.10 and Fig. 16.13.

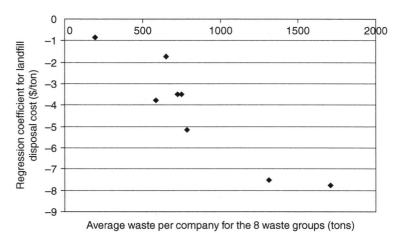

Average waste per company for the 8 waste groups (tons)

Figure 16.13 Waste group comparison of the average solid waste per company versus the regression coefficient for landfill disposal cost.

Examining the t values, allows for a comparison of the landfill disposal cost effects on the eight manufacturing waste groups. Wood and lumber manufacturers indicated the highest absolute value at 3.73 followed by transportation equipment manufacturers at 3.38. These two waste groups achieve greater waste reduction from increased waste costs, indicating they are more sensitive to the economic impact of waste generation. On the contrary, electronic manufacturers and chemical manufacturers were least sensitive.

As shown, the greater the average waste per company a waste group generates, the more costly the regression cost coefficient. This indicates solid waste generation is economically driven. The more waste a company within the groups generates the more sensitive to disposal costs.

Investigating ISO 14001 certification also indicted the more average waste a waste group generated the more sensitive to waste reduction the group. Table 16.11 and Fig. 16.14 display these findings.

As shown by the t value analysis for ISO 14001 certification, transportation equipment manufacturers and chemical manufacturers achieve the largest relative solid waste reductions from certification. This indicates these two waste groups achieve the largest relative waste reduction gains from the attainment of ISO 14001.

An analysis of variance (ANOVA) was conducted to examine the difference in the average waste per company for the eight manufacturing waste groups. This was conducted to examine if significant differences existed in the amount of waste these waste

TABLE 16.11   WASTE GROUP COMPARISON OF THE AVERAGE SOLID WASTE PER COMPANY VERSUS THE REGRESSION COEFFICIENT FOR ISO 14001 CERTIFICATION

WASTE GROUP	ABBREVIATION	AVERAGE SOLID WASTE PER COMPANY PER YEAR (TONS)	ISO 14001 REGRESSION COEFFICIENT	T VALUE
Wood and lumber manufacturers	WDM	1707.9	−217.6	−2.99
Metal manufacturers	MLM	1313.6	−214.9	−2.52
Food manufacturers	FDM	784.8	−183.5	−2.89
Chemical and rubber manufacturers	CHM	749.8	−140.4	−3.47
Paper manufacturers and publishers	PPM	726.0	−114.8	−2.48
Textile and fabric manufacturers	FBM	584.7	−92.3	−2.36
Electronic manufacturers	ELM	194.5	−56.3	−2.36
Transportation equipment manufacturers	TRM	653.5	−56.2	−4.51

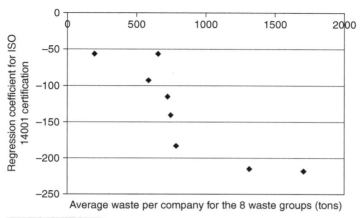

Figure 16.14 **Waste group comparison of the average solid waste per company versus the regression coefficient for ISO 14001 certification.**

groups generate. Bonferroni paired t-tests were conducted to examine if each pair of waste groups generated significantly different amounts of waste at the 95 percent confidence level.

First, a single regression equation for the eight waste groups explained by three significant variables was developed and is displayed below:

$$\text{Annual waste (tons)} = 6.61(\text{number of employees}) - 2.56(\text{landfill disposal cost/ton}) - 116.14(\text{ISO 14001 certification})$$

Figure 16.15 shows the results for the regression equation developed consolidating all eight manufacturing companies into one group. As shown from the ANOVA table in Fig. 16.15, based on the F-statistic a relationship was established ($F = 16.82$). Examining the t statistics for each independent variable indicates only the number of employees was significant at the 95 percent confidence level ($t = 2.345$). When lumping the eight manufacturing groups, detail was lost and landfill disposal costs ($t = -1.465$) and ISO 14001 Certification ($t = -1.907$) were no longer significant at the 95 percent confidence level. Also the coefficient of determination was only 30 percent, indicating the relationship is poor.

An ANOVA analysis on waste groups with three significant variables based on average waste per company using Bonferroni paired t-tests was conducted to examine differences between the eight manufacturing waste groups. Bonferroni paired t-tests is an approach for adjusting the selected alpha level to control the overall Type I error rate (Hair et al., 1998). The Type I error rate is also known as a false positive or the error of rejecting the null hypothesis when it is actually true. This procedure involves computing the adjusted rate as alpha divided by the number of statistical tests to be performed and then using the adjusted rate as the critical value in each separate test. Table 16.12 shows the data used for this analysis.

Summary Output

Regression Statistics	
Multiple R	0.548987304
R Square	0.30138706
Adjusted R Square	0.283473908
Standard Error	335.0315537
Observations	121

ANOVA

	df	SS	MS	F	Significance F
Regression	3	5,665,592	1,888,531	16.82	3.753E-09
Residual	117	13,132,799	112,246		
Total	120	18,798,390			

	Coefficients	Standard Error	t Stat	P Value
Intercept	272.920	116.397	2.345	0.021
Number of Emp.	6.607	1.002	6.596	0.000
Disposal Cost	−2.560	1.748	−1.465	0.146
ISO 14001 Cert.	−116.636	61.166	−1.907	0.059

Figure 16.15 ANOVA output table summary.

TABLE 16.12 DATA UTILIZED FOR THE ANOVA ANALYSIS

WASTE GROUP NAME	ABBREVIATION	NUMBER OF COMPANIES ANALYZED	MEAN SOLID WASTE GENERATION PER COMPANY	STANDARD DEVIATION OF MEAN SOLID WASTE
Wood and lumber manufacturers	WDM	14	1528.6	345.0
Metal manufacturers	MLM	10	1313.6	343.8
Food manufacturers	FDM	8	784.8	217.2
Chemical and rubber manufacturers	CHM	16	749.8	174.4
Paper manufacturers and publishers	PPM	14	726.0	180.6
Transportation equipment manufacturers	TRM	14	653.5	82.7
Textile and fabric manufacturers	FBM	13	584.7	159.2
Electronic manufacturers	ELM	32	194.5	72.0

TABLE 16.13 BONFERRONI PAIRED SAMPLE T-TESTS ON WASTE GROUP'S AVERAGE SOLID WASTE GENERATION PER COMPANY

PAIRS OF WASTE GROUPS TESTED		T-STATISTIC	P VALUE	ADJUSTED BONFERRONI PROBABILITY
FBM	TRM	−4.96	0.000	0.009
FBM	ELM	12.09	0.000	0.000
TRM	ELM	18.76	0.000	0.000
FBM	PPM	−10.25	0.000	0.000
ELM	PPM	−17.31	0.000	0.000
FBM	CHM	−10.74	0.000	0.000
ELM	CHM	−17.93	0.000	0.000
ELM	FDM	−6.58	0.000	0.009
FDM	MLM	−7.38	0.000	0.001
TRM	MLM	−6.38	0.000	0.004
ELM	MLM	−10.63	0.000	0.000
PPM	MLM	−6.38	0.000	0.004
CHM	MLM	−5.22	0.001	0.015
FDM	MLM	−18.51	0.000	0.000
FDM	WDM	−8.94	0.000	0.000
TRM	WDM	−5.77	0.000	0.002
ELM	WDM	−12.10	0.000	0.000
PPM	WDM	−5.55	0.000	0.003
CHM	WDM	−7.71	0.000	0.000

Table 16.13 displays the results of the Bonferroni paired t-tests, indicating which waste groups generated significantly different average annual solid waste quantities at the 95 percent confidence level.

The results of Bonferroni paired t-tests aided in the ranking and comparison of waste groups based on average waste generation per company. Table 16.14 displays these rankings for the highest to lowest average waste generators.

As shown by the table, four of the manufacturing companies generate similar average annual solid waste quantities per company at the 95 percent confidence level. These groups were

1 Transportation equipment manufacturers
2 Paper manufacturers
3 Chemical manufacturers
4 Food manufacturers

TABLE 16.14   RANKING AND COMPARISON OF AVERAGE WASTE GENERATION PER WASTE GROUP PER COMPANY

WASTE GROUP	ABBREVIATION	NUMBER OF COMPANIES ANALYZED	MEAN SOLID WASTE GENERATION PER COMPANY
Wood and lumber manufacturers	WDM	14	1528.6
Metal manufacturers	MLM	10	1313.6
Food manufacturers	FDM	8	784.8
Chemical and rubber manufacturers	CHM	16	749.8
Paper manufacturers and publishers	PPM	14	726.0
Transportation equipment manufacturers	TRM	14	653.5
Textile and fabric manufacturers	FBM	13	584.7
Electronic manufacturers	ELM	32	194.5

Also, wood and metal manufacturers generate similar average annual solid waste quantities per company at the 95 percent confidence level. The results of this analysis are important to identify waste groups that generate high quantities of solid waste. The P values indicate that the mentioned waste groups generate different quantities of waste. Specifically, wood and metal manufacturers generate the largest quantities of waste per company. To most effectively use tax dollars, environmental regulators may target these waste groups to achieve higher waste reduction benefits per company over that of other waste groups. These paired t-tests quantitatively ranked the waste groups based on the quantities generated, but qualitative aspects should also be considered. For example, electronic equipment manufacturers generate higher levels of toxic waste than wood manufacturers (the largest waste generating group). This dimension should also be considered when targeting groups.

After examining the dominant variables that influence solid waste quantities for each waste group, the variables that do not significantly influence solid waste were analyzed. One of the first insignificant variables was the number of days a company is open per year. This was not a significant variable because the companies were all open for approximately the same number of days per year. This variable did little to differentiate the companies.

ISO 9000 certification did not enter the solid waste prediction equations, but this is not to say this variable had no significant effect on solid waste generation. All companies that were ISO 14001 certified were also ISO 9000 certified. This indicates that ISO 9000 may have an effect on waste generation, but ISO 14001 is a

TABLE 16.15 CORRELATION ANALYSIS MATRIX BETWEEN DISPOSAL COST AND RECYCLING LEVEL		
	COST	RECYCLE
Cost	1	
Recycle	0.728647381	1

stronger predictor. This is logical because ISO 14001 is an environmental quality system and ISO 9000 is a general quality system.

Finally, the recycling level of a company also did not significantly influence the quantity of solid waste generated (before subtracting recyclables). Research results indicated the recycling level was positively correlated with disposal cost per ton and positively correlated with ISO 14001 certification. Cost per ton to dispose and ISO 14001 certification had stronger influences on the quantity of solid waste and were included in the equations. The correlation of recycling level forced this variable out in favor of cost per ton to dispose and ISO 14001 certification. Table 16.15 and Fig. 16.16 display the correlation of the recycling level of a company to the cost to dispose for the wood and lumber manufacturing waste group. Table 16.16 and Fig. 16.17 display the correlation of the recycling level of the company to whether or not the company is ISO 14001 certified for the wood and lumber manufacturing waste group. These results were typical for other groups.

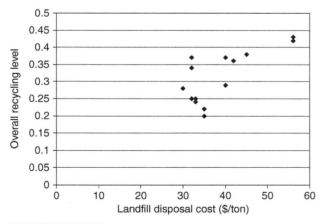

Figure 16.16   Scatter diagram (recycling level and landfill disposal costs).

TABLE 16.16 CORRELATION ANALYSIS MATRIX BETWEEN ISO 14001 CERTIFICATION AND RECYCLING LEVEL		
	ISO	RECYCLE
ISO 14001	1	
Recycle	0.716744711	1

The following equation was used to calculate the correlation coefficients (Walpole and Myers, 1993):

$$r = b\sqrt{\frac{S_{xx}}{S_{yy}}} = b\frac{S_{xy}}{\sqrt{S_{xx}S_{yy}}}$$

An analysis was conducted to examine the effects of company size on waste generation per employee. The purpose of the test was to examine if larger companies were more or less efficient than smaller companies in regards to waste generation. For all groups no significant correlation was found. Table 16.17 and Fig. 16.18 show the results for the wood and lumber manufacturing waste group.

Also notable, nonlinear variables did not aid in the prediction of solid waste. Significant relationships were developed using linear variables at the 95 percent confidence level. The next chapter discusses the development of the performance parameters for the 20 waste groups. The regression equations discussed in this chapter served as the basis of these parameters.

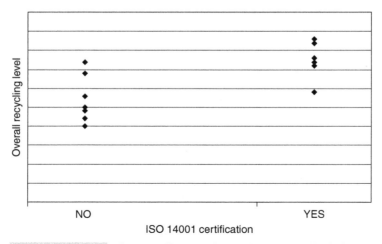

Figure 16.17   Scatter diagram (recycling level and ISO 14001 certification).

TABLE 16.17 CORRELATION ANALYSIS MATRIX BETWEEN WASTE PER EMPLOYEE AND THE NUMBER OF EMPLOYEES		
	EMPLOYEE	WASTE PER EMPLOYEE
Employee	1	
Waste per employee	−0.070714907	1

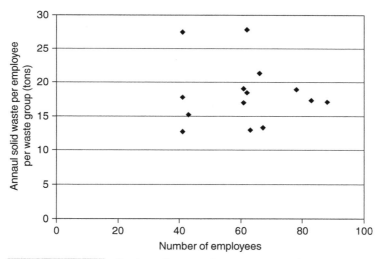

Figure 16.18   Scatter diagram (waste per employee versus the number of employees).

17

BENCHMARKING AND EVALUATION

Based on the previous chapters' waste characterization and significant variable analyses, performance parameters were established for the 20 waste groups. The performance parameters were established integrating the theoretical concepts of statistical quality control into the integrated environmental model. This involved applying control limits to the output of the regression models used to determine the significant variables that influence solid waste. This novel approach to quality control allowed for the monitoring and control of solid waste generation of U.S. businesses and government agencies. In particular, confidence intervals were established for the regression model using a control limit as the level of significance determined by the t value. These established confidence intervals are the performance parameters for U.S. company waste generation.

The following are the single variable upper and lower performance parameter mathematics:

$$\hat{y}_0 - t_{\alpha/2}s\sqrt{\frac{1}{n} + \frac{(x_0 - \overline{x})^2}{S_{xx}}} < \mu_{Y|x_0} < \hat{y}_0 + t_{\alpha/2}s\sqrt{\frac{1}{n} + \frac{(x_0 - \overline{x})^2}{S_{xx}}}$$

where $\hat{y}_0$ = predicted value at x_0

$t_{\alpha/2}$ = value of t-distribution with $n - 2$ degrees of freedom.

s = unbiased estimate of standard deviation

n = sample size

x_0 = value for independent variable

$\overline{x}$ = mean value for independent variable

$S_{xx} = \sum_{i=1}^{n}(x_i - \overline{x})^2$

The following mathematics (in matrix form) were applied to calculate the confidence intervals multivariable regression models:

$$\hat{y}_0 - t_{\alpha/2}s\sqrt{x'_0(X'X)^{-1}x_0} < \mu_{Y|x_{10},x_{20}} < \hat{y}_0 + t_{\alpha/2}s\sqrt{x'_0(X'X)^{-1}x_0}$$

where $\hat{y}_0$ = predicted mean annual solid waste generation

   $t_{\alpha/2}$ = t value at the specified confidence level and $n - k - 1$ degrees of freedom

   s = unbiased estimate of residual mean square

   x_0 = condition vector of independent variable values for prediction

   X = matrix of x_i values that give rise to the response y_i

The performance parameters were established using a $t_{\alpha/2} = 3$. A value of 3 corresponds to approximately 3σ above or below the mean annual waste generation tonnage. A performance parameter of 3σ was chosen to balance the two types of errors associated with statistical quality control, type I and type II. *Type I error* is the risk of a point falling beyond performance parameters, indicating an out of control condition when no assignable cause is present (Montgomery, 1997). *Type II error* is the risk of a point falling between the performance parameters when the process is really out of control. By using a 3σ the probability of type I error is 0.27 percent; that is an incorrect out of control signal will be generated only 27 out of 10,000 data points. A 3σ limit is typically used in statistical quality control and was applied to this research. The performance parameters for the 20 waste groups involved a similar statistical basis used with the $\bar{x}$ (x-bar) chart. The x-bar chart is commonly used in manufacturing environments to track and control the quality of products or services.

The following list displays the three values that were calculated to establish the performance parameters:

1 The expected or mean solid waste quantity waste
2 Upper performance parameter
3 Lower performance parameter

The latter two were calculated from the previous equation. The benefits of this method are: ease of calculations, ease of programming, ability to benchmark, standardization, objectivity, and confidentiality. The confidentiality aspect of the method involves its usage. This method is applicable to Internet systems and may be easily available from a Web site for private use by companies to quickly and privately evaluate their waste generation performance. The performance parameters identify high and low waste generators based on research specific data. The performance parameters developed will aid manufacturing and service companies in evaluating their solid waste generation in comparison with the waste group standards established from this research. The performance parameters determine when a company is out of control in relation to industry standards when they are producing significantly above or below the determined waste generation levels for their industry, represented by their respective waste group.

After establishing the performance parameters and determining if a company is out of control several actions can be taken. First, if the performance parameter indicates the company is below the determined waste generation level, the company should be praised so that best practices can be replicated for similar companies to reduce waste levels. However, if the performance parameter indicates the company is above the determined waste generation level, the company should be investigated so that better waste reduction efforts can be taken up.

Performance parameter analyses were conducted on the 20 waste groups to investigate similarities of highest 5 percent and lowest 5 percent of waste generators for each waste group. This was conducted to identify trends in terms of independent variables and materials generated among the highest and lowest waste generators. Manufacturing waste groups explained by three significant variables (number of employees, landfill disposal cost per ton, and ISO 14001 certification) and non-manufacturing waste groups explained by one significant variable (number of employees) were analyzed separately. Tables 17.1 to 17.4 display the results. The columns labeled higher/lower than average material generation lists the waste material composition percentages that were significantly higher (more/less than 5 percent) than the averages for the waste group discussed. An analysis of the higher/lower than average waste generators is discussed at the conclusion of this chapter. Case studies discussing the application of the performance parameters to two companies are provided later in the book.

As shown by the tables, the manufacturing waste groups (three significant variables) showed some convincing similarities among high waste generators. Specifically, high waste generators were not ISO 14001 certified, had lower than average disposal costs, and generated larger than average composition percentages of wood and cardboard (container waste). Low waste generators were opposite, in that most were ISO 14001 certified, had average disposal costs, and generated lower than average amounts of wood and cardboard. Low waste generators also had higher recycling levels than high waste generators. This is to be expected because recycling levels were correlated with disposal cost and ISO 14001 certification as discussed earlier.

Nonmanufacturing waste groups (one significant variable) also showed similarities between high waste generators, but not as strong as for manufacturing waste groups. Most high waste generators produced higher amounts of mixed office paper (MOP) and some packing materials. High waste generators also had lower recycling levels than low waste generators. Again, as in the manufacturing waste groups, low waste generators displayed opposite trends.

The next chapter discusses the integration of the environmental model, which includes the performance parameters discussed in this section. Chapter 20 discusses two case studies, which apply the performance parameters and provide specific details on waste generation performance.

TABLE 17.1 HIGHEST FIVE PERCENT WASTE GENERATORS IN THE MANUFACTURING WASTE GROUPS (THREE SIGNIFICANT VARIABLES) BASED ON PERFORMANCE PARAMETERS

WASTE GROUP	COMPANIES IN GROUP	NUMBER OF COMPANIES IN THE HIGHEST 5% OF WASTE GENERATORS	ISO 14001 CERTIFICATION	OVERALL RECYCLING LEVEL	LANDFILL DISPOSAL COSTS (DOLLARS PER TON)	HIGHER THAN AVERAGE MATERIAL GENERATION VERSUS WASTE GROUP AVERAGES
Chemical and rubber manufacturers	16	1	No	14%	32	OCC, wood
Electronic manufacturers	32	2	No	13%	33	OCC, wood
Textile and fabric manufacturers	13	1	No	14%	34	OCC, wood
Transportation equipment manufacturers	14	1	No	12%	29	OCC, wood
Wood and lumber manufacturers	14	1	No	18%	36	OCC, wood
Metal manufacturers	10	1	No	34%	32	OCC, wood
Paper manufacturers and printers	14	1	No	25%	33	OCC, wood
Food manufacturers	8	1	No	16%	32	OCC, wood
Totals	121	7				

TABLE 17.2 LOWEST 5 PERCENT WASTE GENERATORS IN THE MANUFACTURING WASTE GROUPS (THREE SIGNIFICANT VARIABLES) BASED ON PERFORMANCE PARAMETERS

WASTE GROUP	COMPANIES IN GROUP	NUMBER OF COMPANIES IN THE LOWEST 5% OF WASTE GENERATORS	ISO 14001 CERTIFICATION	OVERALL RECYCLING LEVEL	LANDFILL DISPOSAL COSTS (DOLLARS PER TON)	LOWER THAN AVERAGE MATERIAL GENERATION VERSUS WASTE GROUP AVERAGES
Chemical and rubber manufacturers	16	1	Yes	29%	36	OCC, wood, plastic
Electronic manufacturers	32	2	Yes	37%	35	OCC, wood, plastic
Textile and fabric manufacturers	13	1	No	38%	41	OCC, wood, plastic
Transportation equipment manufacturers	14	1	Yes	41%	39	OCC, wood, plastic
Wood and lumber manufacturers	14	1	Yes	53%	37	OCC, wood, plastic
Metal manufacturers	10	1	Yes	68%	35	OCC, wood, plastic
Paper manufacturers and printers	14	1	Yes	51%	39	OCC, wood, plastic
Food manufacturers	8	1	Yes	41%	38	OCC, wood, plastic
Totals	121	7				

TABLE 17.3 HIGHEST 5 PERCENT WASTE GENERATORS IN THE NONMANUFACTURING WASTE GROUPS (ONE SIGNIFICANT VARIABLE) BASED ON PERFORMANCE PARAMETERS

WASTE GROUP	COMPANIES IN GROUP	NUMBER OF COMPANIES IN THE HIGHEST 5% OF WASTE GENERATORS	ISO 14001 CERTIFICATION	OVERALL RECYCLING LEVEL	LANDFILL DISPOSAL COSTS (DOLLARS PER TON)	HIGHER THAN AVERAGE MATERIAL GENERATION VERSUS WASTE GROUP AVERAGES
Agriculture	13	1	N/A	8%	42	Organic
Automotive sales, service, and repair	27	1	N/A	11%	29	OCC, metal
Commercial and government	124	6	N/A	9%	33	MOP
Construction	21	1	N/A	13%	38	Wood, OCC
Education	8	1	N/A	8%	41	MOP
Food stores	9	1	N/A	12%	42	OCC
Hotels	7	1	N/A	14%	39	Fabric, MOP
Medical services	14	1	N/A	8%	35	OCC, plastic
Recreation and museums	14	1	N/A	16%	28	MOP
Restaurants	13	1	N/A	7%	31	MOP
Retail and wholesale	53	3	N/A	8%	34	OCC
Totals	303	17				

TABLE 17.4 LOWEST 5 PERCENT WASTE GENERATORS IN THE NONMANUFACTURING WASTE GROUPS (ONE SIGNIFICANT VARIABLE) BASED ON PERFORMANCE PARAMETERS

WASTE GROUP	COMPANIES IN GROUP	NUMBER OF COMPANIES IN THE LOWEST 5% OF WASTE GENERATORS	ISO 14001 CERTIFICATION	OVERALL RECYCLING LEVEL	LANDFILL DISPOSAL COSTS (DOLLARS PER TON)	LOWER THAN AVERAGE MATERIAL GENERATION VERSUS WASTE GROUP AVERAGES
Agriculture	13	1	N/A	15%	42	Organic
Automotive sales, service, and repair	27	1	N/A	21%	33	OCC, metal
Commercial and government	124	6	N/A	14%	40	MOP
Construction	21	1	N/A	22%	38	Wood, OCC
Education	8	1	N/A	18%	31	MOP
Food stores	9	1	N/A	14%	42	OCC
Hotels	7	1	N/A	19%	39	Fabric, MOP
Medical services	14	1	N/A	13%	37	OCC, plastic
Recreation and museums	14	1	N/A	16%	33	MOP
Restaurants	13	1	N/A	9%	31	MOP
Retail and wholesale	53	3	N/A	11%	36	OCC
Totals	303	17				

18

MODEL DEVELOPMENT

AND INTEGRATION

This chapter discusses the integration of the environmental model. A computer program was written based on the characterization analysis, significant variable analysis, and performance parameter development. The program integrated the analysis into an environmental model to characterize, quantify, and evaluate solid waste generation of U.S. businesses and government agencies. The program was written using Excel and Visual Basic.

Inputs to the program are the independent variables for the company to be analyzed

- SIC code
- Number of employees
- Landfill disposal cost per ton
- ISO 14001 certification

Output from the program is

- Annual solid waste expected mean tonnage
- Annual waste stream composition tonnages
- Performance parameters

Figure 18.1 displays a diagram of the model flowchart, including data inputs, model processes, and outputs. Benefits of the program include

- Applicable to Internet systems
- Discrete and confidential for U.S. businesses
- Waste monitoring and control tool
- Serves as an incentive for companies to increase recycling by calculating the potential cost benefits from waste reduction

The integrated environmental model and subsequent programs were utilized to facilitate the validation process and to conduct the case studies discussed in the next chapter.

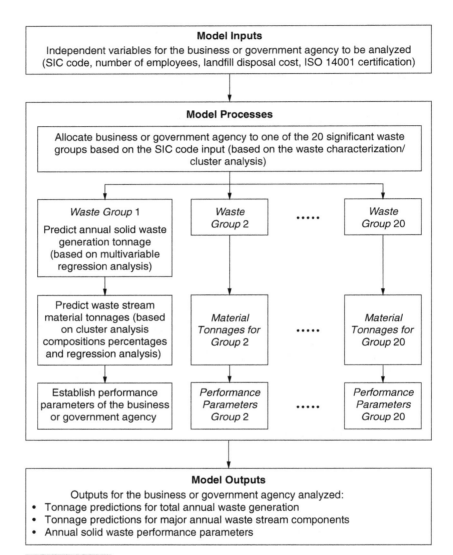

Figure 18.1 Model integration flowchart.

19

MODEL VALIDATION AND CASE STUDY APPLICATION

This chapter discusses the validation and provides a demonstration of the integrated environmental model. Two case studies are discussed that demonstrate the application of the performance parameters. The process used to conduct the validation and demonstration of the model is listed below:

1 Collect additional data from existing sources to validate model outputs (a discussion of the additional data collected from the U.S. government and university solid waste programs is provided).
2 Apply the developed regression model using the additional data for several waste groups and conduct hypothesis tests of the regression coefficients at the 95 percent confidence level to validate the developed model regression coefficients.
3 Predict the annual solid waste generation of a business using the developed model and compare the results to actual data to demonstrate the developed model.
4 Conduct two case studies that establish performance parameters for annual waste generation and discuss findings (one company in control and one out of control).

19.1 Additional Data Collection from Other Existing Sources

A total of 1173 additional solid waste records of U.S. businesses and government agencies were gathered from existing sources other than the research survey data. The waste records are defined as annual solid waste tonnage and compositions as well as the independent variables required for the environmental model. These sources were not collected from the national survey used for this research, but from waste management offices around the United States. The data sources were U.S. government (14 sources) and universities' solid waste assessment programs (6 sources). An overview of the additional data

TABLE 19.1   ADDITIONAL DATA COLLECTED FOR VALIDATION

ADDITIONAL DATA SOURCE	LOCATION	WASTE RECORDS PROVIDED
Arkansas Department of Pollution Control and Ecology	Arkansas	134
Bradley University Industrial Assessment Center	Peoria, IL	8
Carroll-Columbiana-Harrison Counties	Carrollton, Ohio	14
Cornell Waste Management Institute (CWMI)	Cornell University	20
Darke-Marion County Solid Waste District	Marion, Ohio	16
Erie County Solid Waste District	Sandusky, Ohio	36
Fayette-Highland-Pickaway-Ross Counties Solid Waste District	Chillicothe, Ohio	123
Florida Department of Environmental Protection	Tallahassee, Florida	15
Geauga-Trumbull Solid Waste District	Warren, Ohio	141
Green County Solid Waste District	Xenia, Ohio	183
Hancock County Solid Waste District	Findley, Ohio	70
Indiana Institute on Recycling	Indiana State University	43
Mahoning County Solid Waste District	Youngstown, Ohio	58
Montgomery County Solid Waste District	Dayton, Ohio	19
New Hampshire Department of Environmental Service	New Hampshire	86
The University of Toledo Waste Analyses and Minimization Research Project	Toledo, Ohio	24
Warren County Solid Waste District	Lebanon, Ohio	102
West Virginia University Industrial Assessment Center	Morgantown, WV	10
Wisconsin Department of Natural Resources	Madison, Wisconsin	39
Youngstown State University Waste Minimization Research Project	Youngstown, Ohio	32
	TOTAL	1,173

collected is displayed in Table 19.1. The table displays the sources, location, and number of waste records collected.

These data sources were identified from previous research conducted (Franchetti, 1999), professional waste management contacts, information gathered during the literature review, and Internet searches. Much raw data existed on waste generation levels of individual companies, but little company information that quantified waste generation rates. Many state and local environmental agencies collected waste generation data to use for various aggregate reports. These reports examined waste generation for entire geographical regions, not individual companies. The individual data was stored in files and rarely used for other purposes. This research consolidated some of this data to validate the developed model.

The accuracy of this data is considered high because it was collected by the U.S. government at the state and local levels. Different methodologies and survey techniques were utilized to collect the existing data, and may contribute to variation.

19.2   Validation of the Developed Environmental Model

To demonstrate the validation process for the developed environmental model, the waste records were gathered from one existing data source (Greene County, Ohio). Greene County data was used due to its high accuracy and reliability. From these waste records, regression equations were computed for the waste groups with five or more records (this included 16 of the 20 waste groups). Hypothesis tests were conducted to determine if the regression coefficients from the existing data were equal to the developed model's coefficients at the 95 percent confidence level. Table 19.2 displays the results for the 16 waste groups developed from Greene County data.

TABLE 19.2   MODEL VALIDATION TESTS (HYPOTHESIS TESTS* TO EXAMINE IF SIGNIFICANT DIFFERENCES EXIST BETWEEN THE MODELS DEVELOPED FROM SURVEY DATA VERSUS ADDITIONAL DATA COLLECTED[†])

WASTE GROUP	SIGNIFICANT VARIABLE	DEVELOPED MODEL COEFFICIENT USING SURVEY DATA	DEVELOPED MODEL COEFFICIENT USING ADDITIONAL DATA	T VALUE[††]	CRITICAL VALUE	P VALUE
AUT	Employees	0.51	0.62	0.880	2.262	0.408
CHM	Employees	7.14	7.10	0.190	3.182	0.861
	Disposal cost	−3.49	−3.75	−0.880	3.182	0.444
	ISO 14001	−140.40	−132.43	−0.990	3.182	0.395
GOV	Employees	0.43	0.40	0.180	2.052	0.858

(*Continued*)

TABLE 19.2 MODEL VALIDATION TESTS (HYPOTHESIS TESTS* TO EXAMINE IF SIGNIFICANT DIFFERENCES EXIST BETWEEN THE MODELS DEVELOPED FROM SURVEY DATA VERSUS ADDITIONAL DATA COLLECTED†) (CONTINUED)

WASTE GROUP	SIGNIFICANT VARIABLE	DEVELOPED MODEL COEFFICIENT USING SURVEY DATA	DEVELOPED MODEL COEFFICIENT USING ADDITIONAL DATA	T VALUE††	CRITICAL VALUE	P VALUE
EDU	Employees	0.25	0.28	0.270	2.776	0.801
ELM	Employees	2.29	2.58	0.948	2.110	0.349
	Disposal cost	−0.83	−0.78	−0.121	2.110	0.904
	ISO 14001	−53.31	−55.25	−0.136	2.110	0.892
ARG	Employees	1.29	1.38	1.020	2.179	0.326
FDM	Employees	7.19	7.48	0.830	2.228	0.426
	Disposal cost	−5.17	−5.08	−0.410	2.228	0.691
	ISO 14001	−183.50	−173.81	0.730	2.228	0.482
FDS	Employees	2.71	2.58	0.840	2.306	0.425
HTL	Employees	2.01	2.18	0.980	2.447	0.365
MED	Employees	0.98	0.94	1.070	2.365	0.32
MLM	Employees	8.43	8.54	0.270	2.776	0.801
	Disposal cost	−7.51	−7.21	−0.460	2.776	0.669
	ISO 14001	−214.90	−211.05	−0.730	2.776	0.506
REC	Employees	1.87	1.73	0.380	2.447	0.717
RST	Employees	5.91	6.05	0.810	2.306	0.441
RTL	Employees	0.78	0.84	0.180	2.12	0.859
TRM	Employees	2.97	2.81	0.640	3.182	0.568
	Disposal cost	1.73	−1.64	−0.510	3.182	0.645
	ISO 14001	−56.22	−58.92	−0.430	3.182	0.696
WDM	Employees	17.40	15.81	0.790	2.776	0.474
	Disposal cost	−7.76	−7.51	−0.810	2.776	0.463
	ISO 14001	−217.59	−207.73	−1.080	2.773	0.341

*Hypothesis tests conducted at the 95 percent confidence level with H_0: regression coefficients are equal and H_1: regression coefficients are not equal for survey data versus Greene County data for each waste group.

†Waste generation data was available for 16 of the 21 waste groups from the Greene County Solid Waste District (external data source).

††All hypothesis tests resulted in a "do not reject H_0, the null hypothesis."

As shown in the validation table, all the regression coefficients developed from the research data (national survey) and data gathered from other agencies were statistically equal at the 95 percent confidence level. This validated the research data and model findings with an external source.

The integrated model was also validated using an artificial intelligence (AI) program (neural networks) to examine one typical waste group with a relatively large amount of research data. The commercial/government waste group was chosen with 124 company waste records. The results of the AI program yielded similar results to the multivariable regression model developed for this waste group (an average error estimate of 5.8 percent). The benefits of AI are the ability for the program to learn as new data is entered to strengthen the predictions. One drawback of AI for this research is that larger amounts of data are required over that of regression modeling.

19.3 Demonstration of the Prediction of Solid Waste Generation Using the Developed Model

This section discusses a demonstration case study of the integrated environmental model's prediction capability and margin of error. An electronic manufacturing equipment company was randomly selected from the Greene County data set. Table 19.3 displays the actual data that was gathered from the company and the values predicted by the integrated environmental model. The company employed 60 people, paid $60 per ton to dispose of solid waste, and was not ISO 14001 certified.

As shown in the table, the actual data was 13.3 tons or 5.4 percent more than the model prediction. At most, actual material composition tonnages were approximately 10 percent different than predictions with most approximately +/− 5 percent different. This random sample demonstration indicated the effectiveness and accuracy of the integrated environmental model. Figure 19.1 displays the graphical version of the data.

19.4 Performance Parameter Case Study

Two companies were selected from the electronic manufacturer's waste group to demonstrate the solid waste performance parameters. These companies' waste records were provided by the Greene County (Ohio) Solid Waste District. One company

TABLE 19.3 MODEL PREDICTIONS FOR A SINGLE COMPANY IN THE ELECTRONIC MANUFACTURER'S WASTE GROUP VERSUS ACTUAL DATA FROM A RANDOMLY SELECTED COMPANY IN THE WASTE GROUP

VARIABLE NAME	MODEL PREDICTIONS		INDIVIDUAL COMPANY ACTUAL DATA*	DIFFERENCE (TONS/YEAR)	PERCENTAGE DIFFERENCE	COMPOSITION PERCENTAGE DIFFERENCE
Number of employees	Average Comp. %†	$x_1 = 60$	60			
Waste disposal cost per ton ($)		$x_2 = 42$	42			
ISO 14001 certification		$x_3 = $ No	No			
Total annual waste (tons)		230.5	243.8	13.3	5.4%	
MOP (tons/year)	23%	53.0	58.4	5.4	9.2%	−1.0%
Newspaper	3%	6.9	7.3	0.4	5.3%	0.0%
LDPE (tons/year)	7%	16.1	15.3	−0.8	−5.5%	0.7%
PP (tons/year)	4%	9.2	8.7	−0.5	−6.0%	0.4%
PS (tons/year)	4%	9.2	9.1	−0.1	−1.3%	0.3%
HDPE (tons/year)	3%	6.9	7.6	0.7	9.0%	−0.1%
PET (tons/year)	3%	6.9	7.4	0.5	6.5%	0.0%
Ferrous metals (tons/year)	11%	25.4	25.8	0.4	1.7%	0.4%
Nonferrous metals (tons/year)	4%	9.2	10.3	1.1	10.5%	−0.2%
Aluminum cans (tons/year)	2%	4.6	4.5	−0.1	−2.5%	0.2%
OCC (tons/year)	8%	18.4	19.8	1.4	6.9%	−0.1%
Wood (tons/year)	7%	16.1	17.5	1.4	7.8%	−0.2%
Food waste (tons/year)	6%	13.8	14.8	1.0	6.5%	−0.1%
Glass (tons/year)	3%	6.9	7.7	0.8	10.2%	−0.2%
Other (tons/year)	12%	27.7	29.6	1.9	6.5%	−0.1%
Totals	100%	230.5	243.8	13.3	5.4%	

*Randomly selected from surveys provided by Greene County, Ohio.

†From waste characterization (cluster analysis).

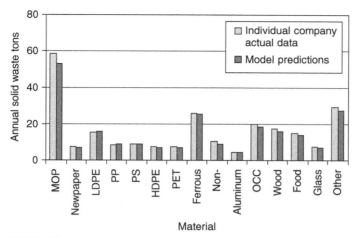

Figure 19.1 Model predictions for a single company in the electronic manufacturer's waste group versus actual data from a randomly selected company in the waste group.

selected was evaluated as in control and one out of control. The contact from Greene County provided insights on each company and additional information. Table 19.4 and Fig. 19.2 provide an overview of the established performance parameter for each company, including data inputs for the integrated model.

TABLE 19.4 COMPARISON OF TWO SELECTED COMPANIES IN THE ELECTRONIC MANUFACTURER'S WASTE GROUP DEMONSTRATING THE PERFORMANCE PARAMETERS

VARIABLES AND PARAMETERS	COMPANY 1	COMPANY 2
Number of employees	83	65
Landfill disposal cost per ton ($)	40	33
ISO 14001 Certification	Yes	No
Actual annual solid waste generation (tons)	221	284
Predicted annual solid waste generation (tons)	229	250
Upper performance parameter*	251	269
Lower performance parameter*	207	231
Performance level	In control (within limits)	Out of control (beyond upper limit)

*Based on 3σ using t values and sample standard deviations.

Performance Parameters—Company 1 (Within Parameters)

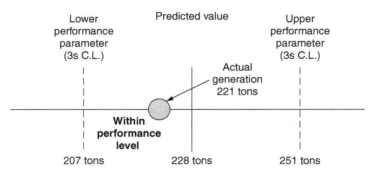

Performance Parameters – Company 2 (Beyond Parameters)

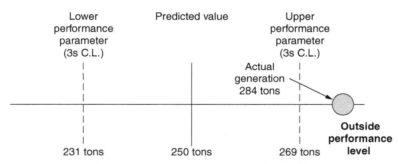

Figure 19.2 Performance parameter case study graphs for company 1 and 2.

Company 1 generated 221 tons of solid waste annually, which was evaluated as within the performance parameters of 207 and 251 tons. Company 1 was ISO 14001 certified and had higher than average disposal costs. Company 1 performed consistently with industry predictions for material composition percentages derived from the integrated model, all within +/− 8 percent of the actual rates.

Company 2 generated 284 tons of solid waste annually, which was evaluated as beyond the performance parameters of 231 and 269 tons. Company 2 was not ISO 14001 certified and had lower than average disposal costs. Company 2 generated higher than average composition percentages (or actual weights) for several materials, specifically cardboard and wood. Further analysis of this company found that management did not use returnable containers (this company used disposable containers of wood and cardboard) and engaged in limited waste reduction activities. No recycling manager or coordinator was employed at the company. Figure 19.2 is visual representation of the performance parameter calculations.

Companies may use the performance parameters to evaluate and improve waste generation performance. The performance parameters are useful to management to efficiently and confidentially identify superior and inferior waste management practices.

19.5   Cost-Benefit Analysis Case Study

A cost-benefit analysis was conducted on the out of control company as determined from the performance parameters discussed in the previous section. This was to demonstrate the potential cost savings and revenues that could be generated from increased recycling. Such an analysis serves as an incentive to company management to increase recycling. The analysis was conducted using the actual annual totals of non recycled materials generated by the company. The national average landfill disposal cost, as of April 2003, of $41 per ton was used for the calculations. National averages for recycling commodities were used to calculate the potential revenues generated from the sale of recyclables. These prices were gathered from Recycler's World as of April 2003 (www.recycle.net/price).

As shown in Table 19.5, the out of control company could annually avoid $7,624 in landfill disposal costs and generate $22,734 from the sale of recyclables. This amounts to a total annual savings of $30,359 from increased recycling. This potential cost benefit serves as an incentive to management to increase recycling, which would both improve the economics for the company and benefit the environment.

TABLE 19.5   POTENTIAL COST BENEFITS FROM INCREASED RECYCLING FOR SPECIFIED OUT OF CONTROL COMPANY

MATERIAL	NON RECYCLED WASTE GENERATED PER YEAR (TONS)	DISPOSAL COST AVOIDANCE PER YEAR (AT $41/TON)	MARKET VALUE PER TON	REVENUE PER YEAR FROM SALE OF RECYCLABLES	TOTAL ANNUAL COST BENEFITS
MOP	65.3	$2,678	$1	$65	$2,743
Newspaper	8.5	$349	$6	$47	$396
LDPE	8.4	$344	$130	$1,092	$1,436
PP	9.3	$381	$55	$512	$893
PS	9.7	$398	$50	$485	$883
HDPE	6.4	$262	$130	$832	$1,094
PET	8.5	$349	$55	$469	$818
Ferrous metals	5.2	$213	$85	$442	$655
Nonferrous metals	7.8	$320	$1,420	$11,076	$11,396
Aluminum cans	5.7	$233	$1,160	$6,589	$6,822
OCC	22.7	$932	$45	$1,022	$1,954
Wood	19.9	$815	$5	$99	$914
Glass	8.5	$349	$1	$4	$354
Totals	186.0	$7,624		$22,734	$30,359

20

MODEL SUMMARY AND RECOMMENDATIONS FOR FUTURE RESEARCH

20.1 Introduction

The goal of this research was to build a model to better understand and evaluate solid waste generation rates of companies and government agencies in the United States. This goal was attained by gathering data, characterizing businesses, and developing mathematical models for companies and government agencies to evaluate individual company waste generation. Figure 20.1 displays the research summary. This chapter discusses the hypothesis testing results, research contributions and limitations, and recommendations for future research.

20.2 Hypothesis Testing Results

Three hypotheses were investigated that were based on the model concept. The investigation of these hypotheses guided the research process. A review of the hypotheses is listed below and discussed in greater detail in the remaining sections:

1 Similar businesses generate similar waste stream compositions and these businesses can be statistically grouped.
2 Significant variables can be identified that aid in the prediction of annual solid waste generation rates.
3 Solid waste generation performance parameters can be established.

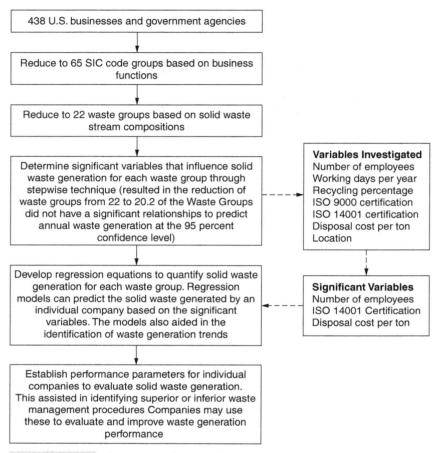

Figure 20.1   Summary of research.

20.2.1   CLASSIFICATION OF U.S. BUSINESSES AND GOVERNMENT AGENCIES

The first hypothesis was successfully validated and proven by research results. U.S. companies and government agencies were rationally grouped by characterizing waste material composition percentages. This was completed by clustering 65 SIC code groups into 20 waste groups by applying multivariate cluster analysis t and associated statistical validation tests. These groups were validated at the 95 percent confidence level, using F-tests and ANOVA.

20.2.2   SIGNIFICANT VARIABLE THAT INFLUENCE SOLID WASTE GENERATION QUANTITIES

The second hypothesis was successfully validated and proven by research results. This was accomplished by applying the stepwise regression method, t tests, and ANOVA. From the 20 solid waste groups established, solid waste quantities were objectively predicted and characterized using statistical techniques to develop a mathematical model. This was completed using multivariable regression analysis for the 20 waste groups and associated statistical validation tests. The number of employees, landfill disposal costs, and ISO 14001 certification were statistically significant at the 95 percent confidence level in predicting annual solid waste generation tonnages.

20.2.3 PERFORMANCE EVALUATION AND PREDICTION

The third hypothesis was successfully validated and proved by research results. This was accomplished by developing confidence intervals based on the regression models and testing the outputs with two case studies. Individual company solid waste generation rates were evaluated in a standardized manner by integrating statistical quality control concepts to monitor and control solid waste generation. This was completed by integrating confidence intervals mathematics with the waste prediction models for the 20 waste groups.

20.3   Research Contributions

The largest contribution of this research was the development of the integrated environmental model to predict and evaluate solid waste generation of individual U.S. businesses and government agencies. The analyses conducted combined with the functionality of the model significantly increased the understanding of solid waste generation of individual U.S. companies and government agencies. This led to the creation of new knowledge. Specific contributions are listed below:

- Objective and rational grouping of businesses that generation similar quantities and compositions of solid waste.
- Ability to predict solid waste generation quantities, material tonnages, and recycling levels of most U.S. businesses and government agencies.
- Standardized evaluation models to monitor and control solid waste generation of U.S. businesses and government agencies.
- The integrated environmental model can be programmed on the Internet for confidential usage by businesses to privately evaluate their solid waste generation.
- Identification of waste generation and recycling trends in the United States and further research opportunities based on findings.

The impact of these contributions will significantly aid businesses and environmental regulators to monitor and control solid waste generation. Businesses may now confidentially evaluate their solid waste generation on an Internet-based system and compare their waste generation to industry-specific benchmarks. This model effectively and efficiently identifies superior or inferior waste management practices based on generation rates. Regulators now have an integrated, standardized model to assist in determining waste generation rates, compositions, and recycling levels of various businesses or groups of businesses. A standardized statistical method was lacking in the past (a nonstandardized system of averages is used by regulators to estimate waste). This research also validated the use of SIC codes to categorize waste rates of businesses.

20.4   Research Limitations

Several limitations of the research were identified. The first limitation was the nature of the data collection instrument and accuracy of the data. The research data was collected using a mailed survey. Nonrespondents may have skewed the research findings, although a validation process using data collected from other sources did not indicate this. Also, since most data was collected from other organizations or respondents, the level of accuracy is dependent on these organizations and people. To combat these issues, the survey was also reviewed by a waste management expert, pretested, and simplified to ensure accurate and consistent responses. A final drawback is the appropriateness of business groupings. More business groupings would allow for a stronger analysis of waste generation, but would have made data collection and analysis very time consuming and costly. More groupings would have extended the data collection beyond budget as well, and statistical relationships may not have been established for smaller group sizes.

20.5   Recommendations for Future Research

This research has led to the identification of several opportunities for additional research. The list below summarizes these opportunities:

■ Collect additional data and recalculate the regression models to strengthen the relationships between the variables. Examine the correlations between variables in greater detail.

■ Consider the further use of artificial intelligence (AI) methods to predict and evaluate solid waste (compare results to regression modeling). Test a variety of AI network types to identify the optimal type to evaluate solid waste generation.

■ Benchmark the waste disposal and recycling practices of low waste generators in greater detail. Perform detailed case studies on these companies to document best practices to aid other companies in improving environmental performance.

■ Conduct detailed case studies applying the performance parameters. Evaluate the effectiveness of the 3σ limits.

■ Improve the model by incorporating an economic benefit module. This would serve as an incentive to industry to increase waste reduction activities for the financial benefits (as demonstrated in Chap. 10).

■ Extend regression models for each waste group by collecting and considering additional independent variables (such as use of returnable containers, employment of recycling coordinator, or level of automation).

- Apply the environmental model to other countries' waste generation and compare the results to this research. This will aid in identifying global leaders in waste reduction and environmental performance.
- Establish a Web site to allow businesses to confidentially evaluate their solid waste generation performance and to collect additional data.

Since this was the first comprehensive analysis and model development of solid waste generation of U.S. companies, improvement upon the performance parameters can be achieved by collecting and analyzing more data. The more data included in the analysis, the more accurate the models and performance parameters will be. The business groupings used may also be revaluated to ensure they best segment businesses based on waste generation.

Another area of potential research involves developing a data collection information system to record and store waste generation and recycling rates of various industries. This information could be easily accessible to government agencies to monitor and control solid waste generation rates. A software program could also be developed to allow company management to easily determine waste generation control limits for their company based on indicator variables. This program could also estimate potential economic benefits from increasing the company's recycling level. This would serve as both a financial and environmental incentive for businesses to reduce waste and promote a green image. This software program would provide the first step in measuring and comparing waste generation to aid in waste reduction. The model developed for this research is applicable to Internet applications. This enhances the confidentiality and ease of use aspects.

One trend observed from the models was that as landfill disposal costs increased, annual solid waste generation decreased for manufacturing waste groups. The methods, recycling procedures, and management tools used by companies that have lower waste generation due to higher disposal costs could be studied to aid other companies in solid waste reduction.

Case studies could be conducted to apply the control limits and economic benefit analysis to various companies to rate the standards. Follow-up studies could be conducted to examine the variance in estimated cost-benefits. Standard procedures could be developed to aid companies in reducing their annual solid waste generation and increase recycling based on the models developed for this research. The case studies could serve as guides and benchmark best practices to reduce solid waste, increase recycling, and economically benefit the company. The cost-benefit analysis discussed in Chap. 18 serves as a starting point to investigate these economical issues from increased recycling.

20.6   Summary

The major objective of this study was to build a model to better understand solid waste generation of U.S. businesses and government agencies. A review of the literature shows that predicting and evaluating solid waste generation is relevant and important to environmental regulators and businesses. However, the majority of the research to date has several critical gaps including a focus on aggregate solid waste analyses for entire regions, not individual companies. In addition, most studies lack empirical support and rigorous statistical support specifically related to waste generation performance.

In this research, a survey was developed to gather data necessary to develop an integrated model to predict and evaluate solid waste generation rates of businesses and government agencies. The conclusions of this research were

- U.S. businesses can be grouped based on waste generation compositions (established 22 waste generating groups).
- Solid waste quantities of individual companies can be predicted and evaluated using statistical techniques and quality control concepts.
- Incentives for businesses to increase recycling levels can be effectively and efficiently determined (in particular cost-benefits).

One important result from this research is the strong evidence that U.S. businesses and government agencies can be statistically grouped based upon business functions and waste stream compositions. Furthermore, the research identified the variables that significantly influence solid waste quantities and should be monitored to predict and control generation rates. Variables that do not aid in the prediction of solid waste were also identified, such as ISO 9000 registration or location.

The findings of this research offer companies and regulators an effective means to improve environmental performance and allow them to predict and evaluate waste generation rates. Using the results of the research, companies can learn more about the strengths and weaknesses in their waste generation performance and assess their performance versus industry specific benchmarks. The results of this study also can form a foundation for further research. Recommendations for further study include conducting a broader study with more data, incorporating an economic evaluation module into the environmental model, and further applying artificial intelligence to improve the model.

Part 4

CASE STUDIES

21

INTRODUCTION TO CASE STUDIES

This part provides an in-depth analysis of the relevant solid waste generation characteristics and minimization tools for 21 industries and business sectors. The data were provided from the environmental model and survey discussed in Part 3. In addition to the data provided, a case study for each sector is also provided. Rather than focusing only on large samples and following a rigid protocol to examine a limited number of variables, the case study method presented in this chapter involves an in-depth, longitudinal examination of single instances. They provide a systematic way of looking at the specific waste generating industries, collecting data, analyzing information, and reporting the results. This will aid the readers in gaining a sharpened understanding of why the instance happened as it did, and what might become important to examine more extensively in terms of waste reduction goals in real-life instances.

The case studies research multiple waste minimization examples in a broad cross section of industries. When selecting the case studies, information-oriented sampling, as opposed to random sampling was utilized. This was done because the typical or average case is often not the richest in information. Also, extreme or atypical cases were not included in this analysis. The industries examined in the case studies were determined based on the North American Industry Classification System (NAICS) and the U.S. Standard Industrial Classification (SIC) system. The NAICS replaced the SIC system and allows for a standardized categorization of businesses and provides an industry breakdown to evaluate the economy. NAICS was developed jointly by the United States, Canada, and Mexico to provide new comparability in statistics about business activity across North America. The U.S. Census Bureau conducts and prepares Economic Census profiles of American business every 5 years, from the national to the local level. At the time of the study, data for 2002 were available. The U.S. Census Bureau sent 2007 Economic Census forms to more than 4 million businesses in December 2007, asking for information about business activity during calendar year 2007. The forms were due back in February 2008. Follow-up is currently underway with businesses from which forms have not been received. Results will be published in late 2009 or early 2010. The data that were collected from this census that is relevant to solid waste generation includes

- The number of companies operating in the United States by state level
- Number of employees in the entire industry
- Total value of shipments and receipts for services

Below is a list of the 21 case studies included in this Part:

Nonmanufacturing applications
   Chapter 22: Agricultural, Forestry, and Fishing Applications
   Chapter 23: Mining Applications
   Chapter 24: Construction Applications

Manufacturing applications
   Chapter 25: Food and Kindred Products
   Chapter 26: Apparel and Other Finished Products Made from Fabrics
   Chapter 27: Lumber and Wood Products
   Chapter 28: Paper and Allied Products
   Chapter 29: Printing, Publishing, and Allied Industries
   Chapter 30: Chemicals, Petroleum, and Allied Products
   Chapter 31: Plastic and Rubber Products
   Chapter 32: Primary and Fabricated Metal Industries
   Chapter 33: Electronics, Semiconductors, and Other Electrical Equipment
   Chapter 34: Industrial and Commercial Machinery Including Automotive

Service applications
   Chapter 35: Transportation, Logistical, and Warehousing Applications
   Chapter 36: Healthcare Systems Applications
   Chapter 37: Commercial, Retail, Financial, and Government Office Applications
   Chapter 38: Accommodations and Food Service Applications
   Chapter 39: Automotive Service Applications
   Chapter 40: Education Applications

International cases
   Chapter 41: Manufacturing Applications
   Chapter 42: Service-Related Applications

Each case study provides an in-depth analysis including

- *Industry overview*
  - The NAICS code
  - Major products and/or services
  - The number of organizations operating in the United States
  - The average number of employees per organization
  - Annual sales for the industry

- The average waste generated per year per employee
- Major waste streams
- *Waste minimization goals and opportunities*—This section lists the major waste streams, typical disposal costs, and list reduction opportunities ranging from low-hanging fruit to options with high initial investments. In addition, a sector overview based on the survey data are provided that includes a breakdown on waste composition percentages and recycling rates. The waste generation equation derived from the model is provided that will allow the reader to quickly estimate waste generation per employee per year for each sector.
- *Economic issues*—This section discusses economic issues specific to each industry such as average revenues, return on investment, and investor interest.
- *Constraints and considerations*—This section provides major industry constraints pitfalls, and issues that must be kept in mind when minimizing waste for each industry.
- *Potential technologies and strategies*—This section highlights the suggested waste minimization alternatives for each industry that have been implanted successfully.
- *Implementation and approach to waste minimization*—This section discusses implementation issues and a recommended structure and process to implement the suggested alternatives.
- *Exemplary performers and industry leaders*—This section lists the organizations that are excelling in terms of waste minimization and reducing their environmental impacts. This section serves as a benchmarking tool that highlights the industry leaders and several of the key waste reduction methods that they have successfully implemented.

22

AGRICULTURAL, FORESTRY, AND FISHING APPLICATIONS

22.1   Industry Overview

NAICS codes: all 11000s

INDUSTRY SNAPSHOT

2.1 million agriculture-related businesses operating in the United States
4.9 million employees
$200.6 billion in annual sales
1.3 tons of solid waste generation per year per employee
Major waste streams: organic wastes (manure and crop residue)

The agriculture and forestry industry covers a wide range of operations in which products are developed from natural resources. Agricultural production covers establishments primarily engaged in the production of crops and plants and the keeping, grazing, or feeding of livestock for the sale of livestock or livestock products. Agricultural services include soil preparation services, crop services, landscape and horticultural services, veterinary and other animal services, and farm labor and management services. Forestry covers establishments primarily engaged in the operation of timber tracts, tree farms, or forest nurseries. Fishing and hunting are also part of the greater agriculture and forestry industry.

As part of the greater food production and distribution operations of the United States, the agriculture sector is subject to considerable regulation. The sector can also be impacted by controls on marketing and advertising placed on the greater food industry. Over 2.1 million farms are operating in the United States with the sector employing over 4.9 million individuals that generate $200.6 billion in annual sales.

The establishments in this sector are often described as farms, ranches, dairies, greenhouses, nurseries, orchards, or hatcheries. A farm may consist of a single tract of land or a number of separate tracts which may be held under different tenures. For example, one tract may be owned by the farm operator and another rented. It may be operated by the operator alone or with the assistance of members of the household or hired employees, or it may be operated by a partnership, corporation, or other type of organization.

The sector distinguishes two basic activities: agricultural-production and agricultural-support activities. Agricultural-production activities includes establishments performing the complete farm or ranch operation, such as farm owner-operators, tenant farm operators, and sharecroppers. Agricultural-support activities include establishments that perform one or more activities associated with farm operation, such as soil preparation, planting, harvesting, and management, on a contract or fee basis.

22.2 Waste Management Goals and Opportunities

The majority of solid waste generated by this sector is organic in nature, such as crop residue, manure, and animal carcasses. Table 22.1 displays the composition breakdown based on survey results.

TABLE 22.1 AGRICULTURAL INDUSTRY SOLID WASTE COMPOSITION (SURVEY RESULTS)

MATERIAL	COMPOSITION (%)	RECYCLING (%)
Organic	20 ± 5.8	74 ± 13.3
Crop residue	8 ± 2.3	67 ± 12.1
Manure	6 ± 1.7	82 ± 14.8
Other organic wastes	6 ± N/A	75 ± 13.5
Paper	13 ± 3.8	5 ± 0.9
Mixed office paper	10 ± 2.9	5 ± 0.9
Newspaper	3 ± 0.9	4 ± 0.7
Food waste	13 ± 3.8	0 ± 0.0
OCC (cardboard)	9 ± 2.6	26 ± 4.7
Metals	9 ± 2.6	61 ± 11.0
Tin/steel cans	4 ± 1.2	62 ± 11.2
Other	5 ± N/A	60 ± 10.8
Wood	9 ± 2.6	3 ± 0.5
Mixed plastics	8 ± 2.3	2 ± 0.4
Yard waste	4 ± 1.2	6 ± 1.2
Other	15 ± N/A	0 ± 0.0
Overall recycling rate		23.9

As shown in the table, the recycling rate for this sector is approximately 24 percent. As derived from the solid waste evaluation model discussed in Chap. 12, the equation that estimates the annual waste generation per year per employee for this sector can be calculated from the following:

Tons of solid waste generated per year $= 1.29 \times$ number of employees $- 0.27$

22.3   Economics

The agriculture sector operates on very tight budgets; this is a double-edged sword for the sector because there is a large incentive to cut costs and operate more efficiently, but at the same time funding is generally limited for capital investments to reduce waste. However, economic waste reduction methods exist with little capital investment as discussed below.

22.4   Constraints and Considerations

The agriculture sector faces unique challenges and constraints for solid waste minimization. Many of these challenges stem from the rural locations of many farms, oftentimes great distances from recycling providers. This can create a cost-prohibitive situation. In addition, many farms are operating on very tight margins with limited budgets, making capital investment difficult to obtain.

Water runoff and the use of pesticides can also poses challenges to composting and recycling waste materials. The strongest negative factor in the use of human and animal wastes for the production of food, feed, or fertilizer is the possibility of disease transmission, which would negate the gains derived from the use of the waste.

22.5   Potential Technologies and Strategies

Biomass is any plant-derived organic matter. Biomass available for energy on a sustainable basis includes herbaceous and woody energy crops, agricultural food and feed crops, agricultural crop wastes and residues, wood wastes and residues, aquatic plants, and other waste materials including some municipal wastes. Biomass is a very heterogeneous and chemically complex renewable resource.

Biomass-processing residues uses by-products from processing all forms of biomass that have significant energy potential. For example, making solid wood products and pulp from logs produces bark, shavings and sawdust, and spent pulping liquors. Because these residues are already collected at the point of processing, they can be convenient and relatively inexpensive sources of biomass for energy.

Most commonly, biomass refers to plant matter grown for use as biofuel, but it also includes plant or animal matter used for production of fibers, chemicals or heat. Biomass may also include biodegradable wastes that can be burned as fuel.

Biomass is grown from several plants, including miscanthus, switch grass, hemp, corn, poplar, willow, sugarcane, and palm oil. The particular plant used is usually not very important to the end products, but it does affect the processing of the raw material. Production of biomass is a growing industry as interest in sustainable fuel sources is growing.

Biomass that is not simply burned as fuel may be processed in other ways such as corn. Low-tech biomass processes are

- Composting (to make soil conditioners and fertilizers)
- Anaerobic digestion (decaying biomass to produce methane gas and sludge as a fertilizer)
- Fermentation and distillation (both produce ethyl alcohol)

More-high-tech processes are

- Pyrolysis (heating organic wastes in the absence of air to produce gas and char; both are combustible).
- Hydro-gasification (produces methane and ethane).
- Hydrogenation (converts biomass to oil using carbon monoxide and steam under high pressures and temperatures).
- Destructive distillation (produces methyl alcohol from high-cellulose organic wastes).
- Acid hydrolysis (treatment of wood wastes to produce sugars, which can be distilled).
- Burning biomass, or the fuel products produced from it, may be used for heat or electricity production.

Other uses of biomass besides fuel and compost include

- Building materials
- Biodegradable plastics and paper (using cellulose fibers)

Most biomass is converted to energy the same way it always has been—by burning it. The heat can be used directly for heating buildings, crop drying, dairy operations, and industrial processes. It can also be used to produce steam and generate electricity. For example, many electric generators and businesses burn biomass by itself or with other fuels in conventional power plants.

Biomass can also be converted into liquids or gases to produce electricity or transportation fuels. Ethanol is typically produced through fermentation and distillation, in a process much like that used to make beer. Soybean and canola oils can be chemically converted into a liquid fuel called biodiesel. These fuels can be used in conventional engines with little, if any, modification.

Biomass can be converted into a gas by heating it under pressure and without oxygen in a gasifier. Manure too can be converted using a digester. The gas can then be burned to produce heat, steam, or electricity.

Other biogas applications are still in development, but show great potential. One promising technology is direct combustion in an advanced gas turbine to run a generator and produce electricity. This process is twice as efficient as simply burning raw biomass to produce electricity from steam. Researchers are also developing small, high-speed generators to run on biogas. These microturbines have no more than three moving parts and generate as little as 30 kW, which could power a medium-sized farm. Several companies are also considering converting gasified biomass into ethanol as a less expensive alternative to fermentation.

Alternatively, biogas can be processed into hydrogen or methanol, which can then be chemically converted to electricity in a highly efficient fuel cell. Fuel cells can be large enough to power an entire farm or small enough to power a car or tractor.

An innovative experiment in Missouri provides one example of the possibilities. Corn is used to produce ethanol, and the waste from the process is fed to cows for dairy production. Cow manure fertilizes the corn and is also run through a digester to produce biogas. A fuel cell efficiently converts the biogas into electricity to run the operation. The end products are ethanol, electricity, and milk. All the waste products are used within the project to lower costs.

22.6   Implementation and Approach

To implement waste minimization strategies several options are available. The federal, state, and local governments offer numerous support programs. In addition, the organization's current waste hauler may have options to reduce solid waste or divert it from landfills, such as recycling and compost programs.

Crops grown for energy could be produced in large quantities, just as food crops are. While corn is currently the most widely used energy crop, native trees and grasses are likely to become the most popular in the future. These perennial crops require less maintenance and fewer inputs than do annual row crops, so they are cheaper and more sustainable to produce. Following is a list of the three commonly used crops to generate energy:

- *Grasses*—Switch grass appears to be the most promising herbaceous energy crop. It produces high yields and can be harvested annually for several years before replanting. Other native varieties that grow quickly, such as big bluestem, reed canarygrass, and wheat grass, could also be profitable.
- *Trees*—Some fast-growing trees make excellent energy crops, since they grow back repeatedly after being cut off close to the ground. These short-rotation woody crops can grow to 40 ft in less than 8 years and can be harvested for 10 to 20 years before

replanting. In cool, wet regions, the best choices are poplar and willow. In warmer areas, sycamore, sweet gum, and cottonwood are best.

- *Oil plants*—Oil from plants such as soybeans and sunflowers can be used to make fuel. Like corn, however, these plants require more intensive management than other energy crops.

Finally, a farm waste management plan is an effective tool to quantify and reduce solid waste. A key element of this plan should address soil runoff and contamination.

22.7   Case Study

In June of 2005, a solid waste minimization analysis was conducted at a dairy farm located in Fort Recovery, Ohio. The farm maintained and milked approximately 700 cows (a mixture of Holsteins, Jerseys, and many crossbreeds) and farmed nearly 1600 acres, of which 1200 were grass pasture. The farm employed 30 people, which included family members. In early 2003, The farm experienced a few years of poor cash flow, labor issues, and other confinement headaches. This encouraged the farm owners to explore cost-reduction and efficiency improvement options. Figure 22.1 displays a typical farm operation.

One such cost-reduction initiative involved conducting a solid waste audit to analyze and minimize the costs associated with solid waste disposal and management. A group consisting of the farm owners and several family members was formed to analyze the farm's solid waste generation. A solid waste audit was conducted with the assistance of industrial engineering students from The University of Toledo. Results

Figure 22.1   Dairy farm operations.

from the audit indicated that the farm generated approximately 31 tons of solid waste per year at a cost of $1700. The majority of the waste was manures and crop residue, of which about 25 percent was recycled in the form of composting.

As a result of the audit, the farm was able to increase its recycling level for crop residue and manure to 100 percent using a combination of improved composting methods and anaerobic digestion to produce fertilizer, which was sold to the neighboring farms. In addition, the farm was able to reduce its solid waste disposal costs by 30 percent to $1200 and generate an additional $1000 per year from the sale of fertilizer resulting in a net annual economic benefit of $1500. The cost of the anaerobic digestion equipment was $600, which was offset by the cost savings.

In addition, the farm implemented computer software to maintain the financial records and monitor income and expense statements on a quarterly basis. The farm also summaries each year and uses those results to track the farm's performance from year to year. The farm has also set stronger financial performance goals including a return on equity and return on capital in the range of 10 to 20 percent each year depending on uncontrollable factors (price of milk, weather, etc).

22.8   Exemplary Performers and Future Research

Oak Ridge National Laboratory (ORNL) is a multiprogram science and technology national laboratory managed for the United States Department of Energy and is located in Oak Ridge, Tennessee, near Knoxville. Scientists and engineers at ORNL conduct basic and applied research and development to create scientific knowledge and technological solutions that build the nation's expertise in key areas of science; increase the availability of clean, abundant energy; restore and protect the environment; and contribute to national security.

An ORNL study found that farmers could grow 188 million dry tons of switch grass on 42 million acres of cropland in the United States at a price of less than $50 per dry ton delivered. This level of production would increase total U.S. net farm income by nearly $6 billion. ORNL also estimates that about 150 million dry tons of corn stover and wheat straw are available annually in the United States at the same price, which could increase farm income by another $2 billion. This assumes about 40 percent of the total residue is collected and the rest is left to maintain soil quality.

22.9   Additional Information

1 "Biosolids Recycling: Beneficial Technology for a Better Environment," Environmental Protection Agency, June 1994 (EPA 832-R-94-009).

2 "AgSTAR Digest," Environmental Protection Agency, February 2003 (EPA 430-F-02-028).

3 "Environmental Fact Sheet: Waste-Derived Fertilizers," Environmental Protection Agency, December 1997 (EPA 530-F-097-053), pp 1–4.

4 U.S. Department of Energy Biopower and Biofuels Programs; www.eere.energy.gov.

5 Institute for Local Self-Reliance, 1313 5th Street SE, Minneapolis, MN 55414-1546; (612) 379-3815; www.carbohydrateeconomy.org.

6 National Renewable Energy Laboratory, 1617 Cole Boulevard, Golden, CO 80401; (303) 384-6979; www.nrel.gov/biomass.

7 U.S. Department of Agriculture 2008 Farm Bill Renewable Energy Incentives; www.rurdev.usda.gov/rbs/coops/csdir.htm.

8 Regional Biomass Energy Program; www.ott.doe.gov/rbep.

9 American Bioenergy Association, 209 Pennsylvania Avenue SE, Washington, DC 20003.

10 Center for the Analysis and Dissemination of Demonstrated Energy Technologies; www.caddet-re.org/technologies.

23

MINING APPLICATIONS

23.1 Industry Overview

NAICS code: all 21000s

INDUSTRY SNAPSHOT

18,491 mining operations in the United States
477,480 employees
$182.9 billion in annual sales
5.2 tons of solid waste generation per employee
Major waste streams: stone/construction and demolition debris

---

The mining sector comprises establishments that extract naturally occurring solid minerals, such as coal and ores; liquid minerals, such as crude petroleum; and gases, such as natural gas. The term *mining* is used in the broad sense to include quarrying, well operations, beneficiating (e.g., crushing, screening, washing, and flotation), and other preparation customarily performed at the mine site, or as a part of mining activity.

The mining industry has been key to the development of civilization, underpinning the iron and bronze ages, the industrial revolution, and the infrastructure of today's information age. In 2001, the mining industry produced over 6 billion tons of raw products valued at several trillion dollars. Downstream beneficiation and minerals processing of these raw materials adds further value as raw materials and products are created to serve all aspects of industry and commerce worldwide. The last decade of the 20th century saw the creation of megacommodity corporations that increasingly moved downstream into the beneficiation area, leaving exploration for new mineral deposits increasingly to small junior mining companies. Application of new technology has led to productivity gains across the value chain.

Apart from Antarctica (which has a treaty in place preventing short- to medium-term exploitation and exploration of minerals), mining takes place in all of the world's continents. Traditional mining countries such as the United States, Canada, Australia, South Africa, and Chile dominate the global mining scene. These countries have become the traditional leaders in mining and exploration methods and technology. Exploration and development funding has changed over the past few years with emphasis shifting to areas that have been poorly explored or have had poor access for reasons of politics, infrastructure, or legislation. Gold, base metals, diamonds, and platinum group elements are the more important commodities explored for and developed globally.

The mining sector distinguishes two basic activities: mine operation and mining support activities. Mine operation includes establishments operating mines, quarries, or oil and gas wells on their own account or for others on a contract or fee basis. Mining support activities include establishments that perform exploration (except geophysical surveying) and/or other mining services on a contract or fee basis (except mine site preparation and construction of oil/gas pipelines).

Establishments in the mining sector are grouped and classified according to the natural resource mined or to be mined. Industries include establishments that develop the mine site, extract the natural resources, and/or those that beneficiate (i.e., prepare) the mineral mined. Beneficiation is the process whereby the extracted material is reduced to particles that can be separated into mineral and waste, the former suitable for further processing or direct use. The operations that take place in beneficiation are primarily mechanical, such as grinding, washing, magnetic separation, and centrifugal separation. In contrast, manufacturing operations primarily use chemical and electrochemical processes, such as electrolysis and distillation. However, some treatments, such as heat treatments, take place in both the beneficiation and the manufacturing (i.e., smelting/refining) stages. The range of preparation activities varies by mineral and the purity of any given ore deposit. While some minerals, such as petroleum and natural gas, require little or no preparation, others are washed and screened, while yet others, such as gold and silver, can be transformed into bullion before leaving the mine site.

Mining, beneficiating, and manufacturing activities often occur at a single location. Separate receipts will be collected for these activities whenever possible. When receipts cannot be broken out between mining and manufacturing, establishments that mine or quarry nonmetallic minerals, beneficiating the nonmetallic minerals into more finished manufactured products is classified according to the primary activity of the establishment. A mine that manufactures a small amount of finished products will be classified in sector 21, mining. An establishment that mines and its primary output is a more finished manufactured product will be classified under sector 31 to 33, Manufacturing.

23.2   Waste Management Goals and Opportunities

The majority of solid waste generated by this sector is stone and construction and demolition debris. Table 23.1 displays the composition breakdown based on survey results.

TABLE 23.1 MINING INDUSTRY SOLID WASTE COMPOSITION (SURVEY RESULTS)

MATERIAL	COMPOSITION (%)	RECYCLING (%)
Stone/C & D	17 ± 5.3	0 ± 0.0
Paper	16 ± 5.0	7 ± 1.5
Mixed office paper	13 ± 4.0	8 ± 1.7
Newspaper	3 ± 0.9	3 ± 0.6
Wood	12 ± 3.7	81 ± 17.0
Metals	11 ± 3.4	97 ± 4.2
Ferrous metals	6 ± 1.8	99 ± 4.0
Nonferrous metals	4 ± 1.2	99 ± 4.0
Aluminum cans	1 ± 0.3	78 ± 16.4
Mixed plastics	9 ± 2.8	8 ± 1.7
OCC (cardboard)	8 ± 2.5	6 ± 1.3
Food waste	5 ± 1.6	0 ± 0.0
Yard waste	4 ± 1.2	0 ± 0.0
Chemical/oils	3 ± 0.9	100 ± 15.4
Glass	1 ± 0.3	9 ± 1.9
Other	14 ± 4.3	0 ± 0.0
Overall recycling level		26.0

Mining wastes include waste generated during the extraction, beneficiation, and processing of minerals. Most extraction and beneficiation wastes from hard rock mining (the mining of metallic ores and phosphate rock) and 20 specific mineral processing wastes are categorized by EPA as special wastes and have been exempted by the mining waste exclusion from federal hazardous waste regulations under Subtitle C of the Resource Conservation and Recovery Act (RCRA).

Mining wastes are generally by-products of two types: (a) mining-and-quarrying extraction wastes, which are barren soils removed from mining and quarrying sites during the preparation for mining and quarrying and do not enter into the dressing and beneficiating processes; and (b) mining-and-quarrying dressing and beneficiating wastes, which are obtained during the process of separating minerals from ores and other materials extracted during mining-and-quarrying activities.

These wastes occupy valuable land and cause harm to stream life when they are deposited near the drainage area of a stream. As shown in Table 23.1, the recycling rate for this sector is approximately 26 percent. As derived from the solid waste evaluation model discussed in Chap. 12, the equation that estimates the annual waste generation per year per employee for this sector can be calculated from the following:

Tons of solid waste generated per year = $5.21 \times$ number of employees + 19.8

23.3   Economics

Generally speaking, the mining sector generates relatively high revenue per employee versus other sectors. This increased revenue and cash flow allow this sector to select waste minimization opportunities with high initial investments.

23.4   Constraints and Considerations

There are several constraints to recycling in the mining sector, such as

- Limits to market access
- Limits on the importation of recyclable materials
- Added transportation and administrative costs
- Added government regulations

Extraction is the first phase of hard rock mining which consists of the initial removal of ore from the earth. Beneficiation follows and is the initial attempt at liberating and concentrating the valuable mineral from the extracted ore. After the beneficiation step, the remaining material is often physically and chemically similar to the material (ore or mineral) that entered the operation, except that particle size has been reduced. Beneficiation operations include crushing; grinding; washing; dissolution; crystallization; filtration; sorting; sizing; drying; sintering; pelletizing; briquetting; calcining; roasting in preparation for leaching; gravity concentration; magnetic separation; electrostatic separation; flotation; ion exchange; solvent extraction; electrowinning; precipitation; amalgamation; and heap, dump, vat, tank, and in situ leaching. The extraction and beneficiation of minerals generates large quantities of waste.

Mineral processing typically generates waste streams that generally bear little or no resemblance to the materials that entered the operation. These operations most often destroy the physical structure of the mineral, producing product and waste streams that are not earthen in character. Mineral-processing operations generally follow beneficiation and include techniques that often change the chemical composition of the ore or mineral, such as smelting, electrolytic refining, and acid attack or digestion. Regulation affecting mineral processing wastes was developed through a long process covering the period from 1980 to 1991. It involved numerous proposed and final rule makings and federal litigation.

23.5   Implementation and Approach

To implement waste minimization strategies several options are available. The federal, state, and local governments offer numerous support programs. In addition, the organization's current waste hauler may have options to reduce solid waste or divert it from landfills, such as recycling and compost programs. The pace of growth in the global economy and demand for resources suggests that raw materials will remain the primary source of mineral and metal commodities. Recycled materials, however, have become an increasingly important source. Minerals and metals provide a unique environmental advantage. Because of their value, consistent performance characteristics, durability, chemical properties, and wide-range of uses, many minerals and metals can be recycled again and again. As a result, recycled materials have become a critical source of supply, and are now traded on national and global markets.

A progressive metal recycling policy and strategy can make a tangible contribution to an organization's innovation strategy by

- Positioning the organization as a leader in sustainable natural resource development
- Positioning the organization as a global recycling leader
- Fostering the transition from commodity suppliers to suppliers of innovative products and services that support recycling
- Encouraging life cycle product design
- Fostering public understanding and awareness of the benefits of recycling

23.6   Case Study

In response to concerns about abandoned nonmetallic mining sites, the state of Wisconsin passed legislation in 2000. Through this law, the legislature directed the Wisconsin Department of Natural Resources to write a reclamation rule that could be used to implement uniform statewide mine reclamation standards. The purpose of legislation is to establish county and municipal reclamation programs through the enactment of an applicable ordinance as a means of ensuring that uniform reclamation standards are applied consistently throughout the state. In this way, it provides assurance that a stable and productive postmining condition will be achieved at all active nonmetallic mines in the state of Wisconsin.

This new rule made it mandatory for counties to enact ordinances by June 1, 2001 for the purpose of establishing and administering programs to address the reclamation of nonmetallic mining sites. Figures 23.1 and 23.2 show before and after photographs of a reclaimed metallic mine, respectively. Although mandatory for counties, the rule allows the option of enacting an ordinance establishing a reclamation program for cities, villages, and towns. In general, a new or reopened mining site will follow these steps in developing a reclamation plan:

- The applicant will need to obtain the information necessary to complete the application, addressing the specific requirements that may arise while the mine reclamation plan is being drafted.
- In order to develop a mine reclamation plan an applicant must first decide upon and propose a target postmining land use(s). It may be beneficial to discuss the proposed land use(s) with the state.

Figure 23.1   Metallic mining operations.

- Once all maps and background information have been assembled and developed into the reclamation plan, the applicant shall submit the plan to the regulatory authority for review. The state provides an opportunity for affected members of the public to request an informational hearing on the reclamation plan.
- The state's review of the reclamation plan will focus on whether or not the plan provides adequate detail on how reclamation will be conducted. The state will evaluate

Figure 23.2   Reclamation of a former metallic mine.

the plan to determine how well it may meet the uniform statewide reclamation standards and whether or not the target post-mining land use(s) can be achieved.

■ When all requirements are met the state will inform the applicant of its decision to approve the plan, deny the plan, or approve it with conditions. If approved, the state will issue a reclamation permit or other approval decision to the applicant.

■ Once the operator pays the annual fees (for unreclaimed acreage) and provides for financial assurance, they may commence mining.

23.7 Exemplary Performers and Industry Leaders

The Safe Management of Mining Waste and Waste Facilities (SAFEMANMIN) project is an excellent example of mining waste minimization efforts. The research project is a joint effort between several European countries. The project focuses on the

■ Review methods for the characterization of mining waste
■ Review risk-assessment methodologies for the classification of mining waste facilities
■ Review techniques for the prevention and abatement of pollution generated by mining wastes
■ Develop a decision-support tool for minimizing the impact of the mining industry on the environment

Key deliverables of the project include a series of reports on characterization, handling, and safe recycling or disposal of mining wastes as well as a research review in Europe on risk-assessment strategies for mining waste.

24

CONSTRUCTION APPLICATIONS

24.1   Industry Overview

NAICS code: all 23000s

INDUSTRY SNAPSHOT

710,307 construction operations in the United States
7,193,069 employees
$1,196.6 billion in annual sales
3.6 tons of solid waste generation per employee
Major waste streams: stones, construction and demolition debris, plastics

---

The construction sector comprises establishments primarily engaged in the construction of buildings or engineering projects (e.g., highways and utility systems). Establishments primarily engaged in the preparation of sites for new construction and establishments primarily engaged in subdividing land for sale as building sites are also included in this sector.

Construction work done may include new work, additions, alterations, or maintenance and repairs. Activities of these establishments are generally managed at a fixed place of business, but they usually perform construction activities at multiple project sites. Production responsibilities for establishments in this sector are usually specified in (a) contracts with the owners of construction projects (prime contracts) or (b) contracts with other construction establishments (subcontracts).

Establishments primarily engaged in contracts that include responsibility for all aspects of individual construction projects are commonly known as general contractors, but also may be known as design-builders, construction managers, turnkey contractors, or (in cases where two or more establishments jointly secure a general contract) joint-venture contractors. Construction managers who provide oversight and scheduling only (i.e., agency) as well as construction managers who are responsible for the entire project (i.e., at risk) are included as general contractor-type establishments. Establishments of the general contractor type frequently arrange construction of separate parts of their projects through subcontracts with other construction establishments.

Establishments primarily engaged in activities to produce a specific component (e.g., masonry, painting, and electrical work) of a construction project are commonly known as specialty trade contractors. Activities of specialty trade contractors are usually subcontracted from other construction establishments, but especially in remodeling and repair construction, the work may be done directly for the owner of the property. Construction offers more opportunities than most other industries for individuals who want to own and run their own business. The 1.9 million self-employed and unpaid family workers in 2006 performed work directly for property owners or acted as contractors on small jobs, such as additions, remodeling, and maintenance projects. The rate of self-employment varies greatly by individual occupation in the construction trades, partially dependent on the cost of equipment or structure of the work. The construction industry provides services in the following areas:

- Heavy construction
- Plumbing, heating, and air-conditioning
- Painting, paper hanging, and decorating
- Mobile home construction
- Prefabricated wood buildings and components
- Masonry, stonework, tile work, and plastering
- Carpentering and floor work
- Concrete work
- Roofing and sheet metal work
- Glass and glazing work
- Wrecking and demolition

Establishments primarily engaged in activities to construct buildings to be sold on sites that they own are known as operative builders, but also may be known as speculative builders or merchant builders. Operative builders produce buildings in a manner similar to general contractors, but their production processes also include site acquisition and securing of financial backing. Operative builders are most often associated with the construction of residential buildings. Like general contractors, they may subcontract all or part of the actual construction work on their buildings.

There are substantial differences in the types of equipment, work force skills, and other inputs required by establishments in this sector. To highlight these differences and variations in the underlying production functions, this sector is divided into three subsectors.

Subsector 236, construction of buildings, comprises establishments of the general contractor-type and operative builders involved in the construction of buildings.

Subsector 237, heavy and civil engineering construction, comprises establishments involved in the construction of engineering projects. Subsector 238, specialty trade contractors, comprises establishments engaged in specialty trade activities generally needed in the construction of all types of buildings.

Force account construction is construction work performed by an enterprise primarily engaged in some business other than construction for its own account and use, using employees of the enterprise. This activity is not included in the construction sector unless the construction work performed is the primary activity of a separate establishment of the enterprise. The installation and the ongoing repair and maintenance of telecommunications and utility networks are excluded from construction when the establishments performing the work are not independent contractors. Although a growing proportion of this work is subcontracted to independent contractors in the construction sector, the operating units of telecommunications and utility companies performing this work are included with the telecommunications or utility activities.

24.2   Waste Management Goals and Opportunities

The majority of solid waste generated by this sector is stone, wood, and construction and demolition debris. Table 24.1 displays the composition breakdown based on survey results.

As shown in the table, the recycling rate for this sector is approximately 21 percent. As derived from the solid waste evaluation model discussed in Chap. 12 the equation that estimates the annual waste generation per year per employee for this sector can be calculated from the following:

Tons of solid waste generated per year = 3.54 × number of employees + 11.8

Most construction and demolition (C&D) recycling sites accept the following materials:

- Untreated and treated wood
- Shingles
- Sheetrock (gypsum/drywall)
- Plant-mixed asphalt
- PVC piping
- Plastic buckets (open)
- Brick
- Concrete

TABLE 24.1   CONSTRUCTION INDUSTRY SOLID WASTE COMPOSITION (SURVEY RESULTS)

MATERIAL	COMPOSITION (%)	RECYCLING (%)
Stone/C & D composite	21 ± 5.5	0 ± 0.0
Wood	18 ± 4.7	34 ± 11.6
Paper	13 ± 3.4	4 ± 1.4
Mixed office paper	10 ± 2.7	4 ± 1.4
Newspaper	3 ± 0.8	4 ± 1.4
Plastics	12 ± 3.1	25 ± 8.5
PVC	5 ± 1.4	24 ± 8.2
HDPE	4 ± 1.1	25 ± 8.5
LDPE	1 ± 0.3	24 ± 8.2
Other	2 ± 0.5	25 ± 8.5
Metals	11 ± 2.9	89 ± 30.3
Ferrous metals	8 ± 2.2	89 ± 30.3
Nonferrous metals	3 ± 0.8	90 ± 30.6
OCC (cardboard)	6 ± 1.6	14 ± 4.8
Yard waste	6 ± 1.6	0 ± 0.0
Food waste	4 ± 1.0	0 ± 0.0
Glass	3 ± 0.8	24 ± 8.2
Fabrics/textiles	1 ± 0.3	0 ± 0.0
Other	5 ± NA	0 ± 0.0
Overall recycling level		21.0

24.3   Economics

In the past, when landfill capacity was readily available and disposal fees were low, recycling or reuse of construction material was not economically feasible. Construction materials were inexpensive compared to the cost of labor; thus construction site job managers were focused on worker productivity rather than material reuse or conservation. In addition, recycling infrastructure and a recycled materials marketplace that processes and resells construction debris did not exist. In recent years, with the increase in international competition for both recycled and raw materials, the economics of recycled materials have improved. During this same time period disposal costs have increased. Recyclable materials have differing market values depending on the presence of local recycling facilities, reprocessing costs, and the availability of virgin materials on the market. In general, it is economically feasible for construction sites to recycle metals, concrete, asphalt, and cardboard. If revenue can not be generated, cost benefits can still be achieved with reduced disposal or hauling costs.

24.4   Constraints and Considerations

Construction and demolition activities generate very large quantities of solid waste. The EPA estimates that 136 million tons of construction and demolition waste was generated in 1996. Recycling opportunities are expanding in many communities. Metal, vegetation, concrete, and asphalt recycling have long been available and economically justified in most communities. Paper, cardboard, plastics, and clean wood markets vary by regional and local recycling infrastructure. Some materials, such as gypsum wallboard, have recycling opportunities only in communities where reprocessing plants exist or where soil can handle the material as a stabilizing agent.

The recyclability of demolished materials is often dependant on the amount of contamination attached to it. Demolished wood, for instance, is often not reusable or recyclable unless it is deconstructed and the nails are removed.

Another consideration is that recycling of construction and demolition debris reduces demand for virgin resources, and in turn, reduced the environmental impacts associated with resource extraction, processing, and in many cases transportation. Through effective construction waste management, it is possible to extend the lifetime of existing landfills, avoiding the need for expansion or new landfills sites. The benefits of a construction and demolition recycling program include

- *Avoid trash collection and disposal fees*—To avoid the high cost of disposal, a construction or demolition company can reduce the amount of waste produced during a project by reusing and recycling waste materials.
- *Save resources and money through deconstruction*—Deconstruction is an expression describing the process of selective dismantling or removal of materials from buildings before or instead of demolition. Reuse and recycling examples include electrical and plumbing fixtures that are reused; steel, copper, and lumber that are reused or recycled; wood flooring that is remilled; and doors and windows that are refinished for use in new construction. Also, by donating reusable excess construction and demolition debris, a business not only helps to keep reusable material out of landfills and incinerators, but can also help to reduce costs for future projects.
- *Improve your organization's public image*—By using fewer resources and reducing the amount of waste sent to landfills and incinerators, a company can enhance its image in the community and with its customers.
- *Make new products from old materials*—Six major constituents of construction and demolition debris, including concrete, asphalt, metals, and wood, and to a much lesser degree, gypsum wallboard and asphalt shingles, have all been recovered and processed into recycled content products and successfully marketed in the United States.
- *Improve the market for recycled content products*—Because of the effort being exerted to develop markets for recovered materials, the numbers of construction and demolition facilities are continuing to grow. It was recently estimated by *C&D Debris Recycling* magazine that there are now more than 3500 construction and demolition debris recycling facilities in operation throughout the United States.

■ *Help your community meet local and state waste reduction goals*—Many communities have established waste reduction goals, since construction and demolition projects generate a large amount of debris. Finding new uses or recycling these materials can significantly help in these efforts.

24.5   Potential Technologies and Strategies

Tools available include jobsite waste guidelines, a waste management plan template, sample waste recycling specifications, directory of local construction waste recyclers, and more. Available assistance includes presentations to jobsite workers on building material reuse, salvage, and recycling, site visits to assess diversion options, and research on recycling options for hard-to-recycle commodities.

There are several key tips to build a successful construction and demolition diversion program. Preventing waste and practicing recycling pays off by lowering supply and disposal costs. But even when a sound waste management plan has been implemented, difficulties may still arise. Here are solutions to some common challenges:

Manage your program—Designate a leader, either an individual or a team, who will be responsible for educating the crew and subcontractors, setting up the site and coordinating and supervising recycling efforts to prevent the contamination of recycling loads. The development of a waste management plan will also help keep the solid waste minimized and diverted from landfills. The plan should consist of identifying the types of debris that will be generated by the project and identifying how all waste streams will be handled. To finalize the plan, all receipts from reuse, recycling, and disposal activities must be submitted.

Involve subcontractors—Require subcontractors to use the on-site recycling and disposal bins or require them to recycle their own waste and provide documentation.

Find appropriate space—Recycling and reuse efforts require space. Set aside an area of the jobsite to store salvaged building materials and house recycling bins for either commingled or source-separated loads.

Promote and educate—Communicate your plan to the crew and subcontractors on site. They will need to know:
■ How materials should be separated?
■ Where materials should go?
■ How often the materials will be collected and delivered to the appropriate facilities?
■ Include waste-handling requirements and expectations in all project documents.

Prevent contamination—Adopt strategies to prevent contamination, such as
■ Clearly label the recycling bins and waste containers on site.
■ Post lists of recyclable and nonrecyclable materials.

- Conduct regular site visits to verify that bins are not contaminated.
- Provide feedback to the crew and subcontractors on the results of their efforts.

The choice of recovery methods will depend on many factors. These include the quantity and type of construction and demolition debris, availability of space for on-site recovery, existence of waste haulers and/or end users for off-site recovery, and program costs. Construction and demolition debris recovery methods include reducing waste at the source, reuse of scrap materials, recycling materials, and use of recycled content construction materials.

Source reduction is also a feasible strategy to reduce solid waste. A business can save money by reducing the amount of waste it creates. Source reduction decreases disposal costs, lowers labor costs due to a reduction in handling and cutting materials, and reduces expenditures for materials because less is wasted. Ways to reduce waste are

- *Design*—Ask your architect for building designs that use standard material sizes.
- *Plan*—Plan ahead so that fewer supply runs need to be made to local suppliers.
- *Reduce packaging*—Ask suppliers to remove packaging before shipping materials to your site, wrap materials in reusable blankets or padding, or take back the packaging after the materials have been delivered.
- *Include waste disposal costs in bids*—Require subcontractors to include the cost of removing their waste in their bids to give them an incentive to produce less waste.
- *Deconstruction*—Require the process of selective dismantling or removal of materials from buildings before or instead of demolition. Reuse and recycling examples include electrical and plumbing fixtures that are reused; steel, copper, and lumber that are reused or recycled; wood flooring that is remilled; and doors and windows that are refinished for use in new construction.

In addition, the construction team can reuse scrap materials on-site. Many building materials may be reusable during renovation projects and also in projects where a new building is built following the demolition of another. Planners can increase reuse potential by making efforts to use the same size and types of materials as in the old construction. Inadequate storage space for materials during the interim from removal to reinstallation may limit reuse as a materials-recovery option. Typical materials suitable for reuse include plumbing fixtures, doors, cabinets, windows, carpeting, bricks, light fixtures, ceiling and floor tiles, wood, HVAC equipment, and decorative items (including fireplaces and stonework). Below are several basic construction and demolition reuse tips

- Leftover masonry material can be crushed on site and used for fill or as bedding material for driveways.
- Joist off-cuts can be cut up and used as stakes for forming or for headers around openings in the floor assembly.
- Leftover rigid insulation can be used as ventilation baffles in attics or installed into house envelopes at joist header assemblies.
- Pallets can be reused or returned to vendors.
- Salvageable materials can be given to businesses that collect and resell used construction materials.

The construction team may also consider donating or reselling reusable materials. Many materials can be salvaged from demolition and renovation sites and sold or donated. By selling or donating unwanted reusable materials, contractors can avoid disposal costs.

Many construction and demolition wastes can be recycled into new materials. Keep in mind that local recycling options vary across United States. You can obtain information about recycling opportunities in your project area from municipal solid waste managers, regional offices of state solid waste management agencies, and waste haulers. In addition, the construction team can make an effort to buy recycled-content construction materials. To help expand markets for recyclable materials, it is important to buy building supplies that contain recycled materials. Some of these materials have been used for years by the construction industry, but they have not been advertised as recycled. There are also many new recycled-content building materials of which you may not be aware.

24.6   Implementation and Approach

In implementing solid waste minimization strategies, the first step is to identify construction haulers and recyclers to handle the designated materials. They often serve as valuable partners in the process. Make sure jobsite personnel understand and participate in the program, with updates throughout the construction process. Obtain and retain verification records (waste hauler receipts and waste management reports) to confirm that diverted materials have been recycled or diverted as intended. Note that diversions may include donations to charitable organizations such as Habitat for Humanity.

The following are options for potentially reducing the amount of waste generated by manufacturing concrete slabs. Many of these options will require further investigation to determine if they are feasible:

Use as rip rap—Broken concrete is used to stabilize banks and shores of creeks and small rivers. The U.S. Department of Transportation or local Park Services should be contacted to determine if there is a need for this material. One major drawback to this option is the need for the concrete to be ground into baseball- or softball-sized rocks to be useful as riprap.

Landscape—Local landscape companies may be able to use this material in their landscaping jobs. It is doubtful that there would be enough demand for the broken concrete to be able to use all that is currently being generated.

Other concrete recyclers—Other concrete recycling companies were contacted but none could offer a better price than the $90 per ton currently being charged for landfill disposal.

Sand from the sand blasting operation is generated on the order of over 1600 tons per year. Although this material is relatively inexpensive (about $0.04 per pound) and is recycled for free, there are other alternatives that may be more cost-effective.

Use of alternative abrasives—The use of alternative abrasive media can be used in many of these types of applications such as switching to a steel shot or grit because, overall, these abrasives are more cost-effective. The steel shot or grit can be used over 3000 times before having to be replaced. To reuse the abrasive, a capture and separation system consisting of an enclosure, bar grading, screw conveyors, and a wash system would need to be installed. The steel shot or grit abrasive is more expensive ($0.30 per pound f.o.b. [freight on board]), but can be reused many times to make it more cost-beneficial. Additional investigation into the cost of the equipment and the feasibility is needed to determine if this option is viable.

Collect sand and reuse—Another option to reduce sand blast waste is to vacuum up used sand back into the feed hopper and reuse for next job. This option is not very feasible because of the contamination present in the used sand and the inability to separate the particles from the concrete blast and the sand used for sand blasting. The sand also pulverizes when shot at concrete at a high pressure, and therefore would not be reusable.

Workers use shop towels throughout the facility to clean various items and to wipe their hands. These towels become soiled and greasy and are disposed of with the municipal solid waste. Approximately 5000 lb of shop towel waste is generated each year.

Launder on site—The rags could be laundered on site using a standard washer and dryer for reuse. By purchasing a washer and dryer (about $1200), one person from the facility could be assigned the duty to wash and dry the towels. More durable rags should be purchased to be able to withstand repeated washings. Drawbacks for this option include the need for additional labor to wash and dry the towels, and the possibility of discharging oil, which is an environmental and regulatory concern.

Use laundry service—Laundry services exist that will wash and deliver fresh towels to industrial customers. The average cost for laundering industrial shop towels is $0.75 per pound. Assuming the amount of shop towels that would need to be purchased is 1500 lb, below is a simplified cost analysis:

$$\text{Laundry service } 1500 \text{ lb} \times \$0.75 \text{ per pound} = \$1,125 \text{ per year}$$

Potential cost savings
Assuming a purchase cost of $0.50 per pound, the savings in purchase of shop towels would be

$$\$0.50 \text{ per pound} \times (5000 - 1500) \text{ lb} = \$1750$$

Disposal of the current shop towels is assumed to be one 20 yd^3 dumpster at $224 per haul. The overall estimated cost savings for this option is

$$\$1750 + \$224 - \$1125 = \$849 \text{ per year}$$

24.7   Case Study

Clearview Elementary School in Hanover, Pennsylvania is an excellent example of an organization that diverted construction waste during a reconstruction phase. The school was the first elementary school in Pennsylvania to achieve the leadership in energy and environmental design (LEED) certification. Located in a mixed-use neighborhood as part of the Hanover School District, the project diverted 90 percent of their construction waste from the landfill by recycling materials, such as concrete from the project—which was removed from the site and reused as clean backfill. The construction administration supervised the contractor's performance in managing construction waste. In addition, the school teaches its students the construction aspects of the building and its green features in their curriculum. Figures 24.1 and 24.2 show the classroom renovations and exterior renovations for the LEED project.

Some project highlights include building commissioning, a building integrated sundial, a 30 percent reduction in water use, a 40 percent reduction in energy use, detailed attention to materials selection, superior indoor air quality, controllable building systems, and construction waste management. The school district has further committed to augmenting its curriculum to include the lessons that the building teaches. The district will also conduct a student performance study in partnership with a local university to compare test results in this daylight facility with test results in its other elementary schools. Building materials were selected based on life-cycle analysis (LCA) comparisons considering their recycled content, recyclability, renewability, and the

Figure 24.1   Classroom renovations.

Figure 24.2   Exterior building renovations.

environmental or energy-consumption impact of their production processes. Basic comparisons were produced on BEES software and detailed analyses utilized Athena Environmental Impact Estimator software. Over 50 percent of the building materials, by cost, contain in aggregate, a minimum weighted average of 20 percent postconsumer or 40 percent postindustrial recycled content. Over 40 percent of all selected building materials are manufactured within 500 miles of the project site. The contractor was given specifications to divert over 50 percent of construction waste from the landfill. As executed, the construction waste management plan diverted over 75 percent of the construction waste.

The "green strategies" applied at the school included

- *Plan for materials longevity*—Use materials and systems with low-maintenance requirements
- *Jobsite recycling*—Require a waste management plan from the contractor
- *Design for adaptability*—Use an access floor to facilitate reconfiguring of spaces and cabling systems
- *Toxic upstream or downstream burdens*—Choose naturally rot-resistant wood species for exposed applications
- *Greenhouse gas emissions from manufacture*—Replace up to 30 percent of the cement in concrete with fly ash
- *Postconsumer recycled materials*—Use plastic toilet partitions made from recycled plastic and specify heavy steel framing with highest recycled content
- *Preconsumer recycled materials*—Use concrete masonry units with recycled or industrial-waste aggregates, use recycled-content rubber flooring, and use agricultural-waste-fiber panels for millwork and interior finish

24.8   Additional Information

1 www.usgbc.org.
2 www.metrokc.gov/dnrp/swd/greenbuilding/construction-recycling/index.asp.
3 www.cdrecycling.org.
4 www.ciwmb.ca.gov/condemo.
5 www.charmeck.org/Departments/LUESA/Solid+Waste/Construction+Recycling/Home.htm.
6 www.cdrecycler.com.
7 www.recycleworks.org/con_dem/index.html.
8 www.wastecap.org/wastecap/commodities/construction/construction.htm.

25

FOOD AND KINDRED PRODUCT
MANUFACTURING APPLICATIONS

25.1   Industry Overview

NAICS code: all 31100s

INDUSTRY SNAPSHOT

27,915 food product manufacturers' operations in the United States
1,506,932 employees
$458.8 billion in annual sales
2.7 tons of solid waste generation per employee
Major waste streams: paper, food by-products, and plastics

Workers in the food-manufacturing industry link farmers and other agricultural producers with consumers. They do this by processing raw fruits, vegetables, grains, meats, and dairy products into finished goods ready for the grocer or wholesaler to sell to households, restaurants, or institutional food services.

Food-manufacturing workers perform tasks as varied as the many foods we eat. For example, they slaughter, dress, and cut meat or poultry; process milk, cheese, and other dairy products; can and preserve fruits, vegetables, and frozen specialties; manufacture flour, cereal, pet foods, and other grain mill products; bake bread, cookies, cakes, and other bakery products; manufacture sugar, candy, and other confectionery products; process shortening, margarine, and other fats and oils; and prepare packaged seafood, coffee, potato and corn chips, and peanut butter. Although this list is long, it is not exhaustive. Food-manufacturing workers also play a part in delivering numerous other food products to our tables.

Quality control and quality assurance are vital to this industry. The U.S. Department of Agriculture's (USDA's) Food Safety and Inspection Service branch oversees all aspects of food manufacturing. In addition, other food safety programs have been adopted recently as issues of chemical and bacterial contamination and new foodborne pathogens remain a public health concern. For example, by applying science-based controls from raw materials to finished products, a food safety program called Hazard Analysis and Critical Control Point focuses on identifying hazards and preventing them from contaminating food in early stages of meat processing. The program relies on individual plants developing and implementing safety measures along with a system to intercept potential contamination points, which is then subject to USDA inspections.

About 34 percent of all food-manufacturing workers are employed in plants that slaughter and process animals and another 19 percent work in establishments that make bakery goods. Seafood product preparation and packaging, the smallest sector of the food-manufacturing industry, accounts for only 3 percent of all jobs.

Industries in the food-manufacturing subsector transform livestock and agricultural products into products for intermediate or final consumption. The industry groups are distinguished by the raw materials (generally of animal or vegetable origin) processed into food products.

The food products manufactured in these establishments are typically sold to wholesalers or retailers for distribution to consumers, but establishments primarily engaged in retailing bakery and candy products made on the premises not for immediate consumption are included.

Establishments primarily engaged in manufacturing beverages are classified in subsector 312, beverage and tobacco product manufacturing.

25.2 Waste Management Goals and Opportunities

The majority of solid waste generated by this sector is mixed paper and food wastes. Table 25.1 displays the composition breakdown based on survey results.

As shown in the table, the recycling rate for this sector is approximately 31 percent. Food residuals are defined as source separated residuals produced from food preparation and consumption activities at homes, restaurants, commercial businesses, and institutions, which consist of fruits, vegetables, grains, fish and animal by-products, and soiled paper unsuitable for recycling. Examples of food residuals that can be composted are leftovers, bread products, outdated/expired foods, produce and vegetables, meat and fish scraps, plate scrapings, coffee grounds, soups, paper products such as napkins and wax corrugated cardboard, and wood chips and shavings. Major generators of food and organic materials include restaurants, supermarkets, hotels, produce centers, food processors, school and business cafeterias, hospitals, prisons, farmers, and community events. As derived from the solid waste evaluation model

TABLE 25.1   FOOD-MANUFACTURING INDUSTRY SOLID WASTE COMPOSITION (SURVEY RESULTS)

MATERIAL	COMPOSITION (%)	RECYCLING (%)
Paper	30 ± 8.4	13 ± 5.1
Mixed office paper	24 ± 7.0	12 ± 4.7
Newspaper	2 ± 0.6	15 ± 5.9
Paper (other)	4 ± 1.2	11 ± 4.3
Food waste	30 ± 8.4	41 ± 14.2
Plastics	18 ± 5.0	18 ± 7.0
HDPE	5 ± 1.4	25 ± 9.8
LDPE	4 ± 1.1	25 ± 9.8
PP	3 ± 0.8	11 ± 4.3
PET	2 ± 0.6	11 ± 4.3
Other	4 ± N/A	14 ± 5.5
Metals	6 ± 1.7	90 ± 35.1
Ferrous metals	2 ± 0.7	89 ± 34.7
Nonferrous metals	2 ± 0.7	94 ± 36.7
Aluminum cans	2 ± 0.7	81 ± 31.6
OCC (cardboard)	6 ± 1.7	82 ± 12.4
Wood	3 ± 0.8	24 ± 9.4
Glass	1 ± 0.3	28 ± 10.9
Yard waste	1 ± 0.3	0 ± 0.0
Other	5 ± N/A	0 ± 0.0
Overall recycling level		30.8

discussed in Chap. 12, the equation that estimates the annual waste generation per year per employee for this sector can be calculated from the following:

Tons of solid waste generated per year = 2.71 × number of employees + 8.27

25.3   Economics

Food waste recycling often entails specialized equipment or hauling agreements which translates into additional cost. Specialized equipment, such as in-vessels for composting can cost more than $50,000. If in-house composting is not used, specialized hauling agreements and composting sites must be contracted. Other wastes, such as paper and plastics may require separation or cleaning before recycling, also translating into additional cost.

25.4 Constraints and Considerations

The key constraints and considerations for food processors can be separated into five areas:

1 *Contamination*—Items in the waste stream that are recyclable may be contaminated with food waste or organic by-products. Most vendors require a minimum level of contamination (such as food debris on plastics or papers) in order to accept materials for recycling. If a high level of contamination is present, the vendors may refuse service, increases charges, or terminate contracts.

2 *Separation and disease control*—Tying in with contamination, the separation of recyclables in the waste stream becomes an issue to minimize solid waste. Food processors are under increased scrutiny from the FDA in regards to hygiene, cleanliness, and disease control. Creating separations and containers for recyclables may increase these risks if not handled properly. A clearly defined separation system and process must be clearly identified and executed in this environment.

3 *Cost of equipment*—The cost of equipment can be prohibitive, in-vessel composting units can cost thousands of dollars to purchase and install.

4 *Invest the extra time, space, and employee training and retraining to sort waste and recyclable materials*—Training and employee involvement are key elements to the success of the recycling program. Creating a strong, well planned system at the beginning will save many headaches and additional costs for the organization at a later point in time.

5 *Specialized haulers for food waste*—As mentioned in Sec. 24.3, specialized vendors and composting sites are required for food wastes to minimize disease and bacteria infestations. At the present time, most regular waste haulers do not offer food waste composting options; a separate hauler must be contracted.

25.5 Potential Technologies and Strategies

The choice of recovery methods will depend on many factors. These include the quantity and type of food discards, availability of space for on-site recovery, existence of haulers and/or end users for off-site recovery, and program costs. Food discard–recovery methods include making donations, processing this waste into animal feed, rendering, and composting. Off-site methods involve food discard generators, haulers, and end users.

The following is a list of diversion options for their food waste in preferred order of implementation:

1 *Food donations*—Nonperishable and unspoiled perishable food can be donated to local food banks, soup kitchens, and shelters. Local and national programs frequently offer free pickup and provide reusable containers to donors. Because these donations recycle food and help feed people in need of assistance, this option should be considered before looking at other alternatives. Smaller food collection organizations are also appropriate. For a list of contact information and needs of small food collection organizations check the yellow pages under "food pantries" or "shelters."

2 *Source reduction*—Source reduction, including reuse, can help reduce waste disposal and handling costs, because it avoids the costs of recycling, municipal composting, landfilling, and combustion. Source reduction also conserves resources and reduces pollution, including greenhouse gases that contribute to global warming. By doing a careful audit of the waste stream a business can determine the percentage of food and organic wastes that are present in their trash. Once the potential for waste reduction is established a business can reduce the quantity of food they buy, purchase precut foods, or explore the possibilities of portion control at restaurants.

3 *Animal feed*—Recovering food discards as animal feed is not new. In many areas hog farmers have traditionally relied on food discards to feed their livestock. Farmers may provide storage containers and free or low-cost pickup service. Coffee grounds and foods with high salt content are not usually accepted, because they can be harmful to livestock. At least one company is using technology to convert food discards into a high-quality, dry, pelletized animal feed.

4 *Rendering*—Liquid fats and solid meat products can be used as raw materials in the rendering industry, which converts them into animal food, cosmetics, soap, and other products. Many companies will provide storage barrels and free pickup service. Check the yellow pages for "rendering" or "grease trap."

5 *Composting*—Composting can be done both on- and off-site. The availability of land space, haulers, and/or end users in your area will help you decide which option is best for you. For on-site composting, companies should consider feed stocks, siting, and operational issues. Composting can take many forms:

- *Un-aerated static pile composting*—Organic discards are piled and mixed with a bulking material. This method is best suited for small operations; it cannot accommodate meat or grease.

- *Aerated windrow/pile composting*—Organics are formed into rows or long piles and aerated either passively or mechanically. This method can accommodate large quantities of organics. It cannot accommodate large amounts of meat or grease.

- *In-vessel composting*—Composting that occurs in a vessel or enclosed in a building that has temperature- and moisture-controlled systems. They come in a variety of sizes and have some type of mechanical mixing or aerating system. In-vessel composting can process larger quantities in a relatively small area more quickly than windrow composting and can accommodate animal products.

- *Vermi composting*—Worms (usually red worms) break down organic materials into a high-value compost (worm castings). This method is faster than windrow or in-vessel composting and produces high-quality compost. Animal products or grease cannot be composted using this method.

25.6   Implementation and Approach

Tips for implementing solid waste reducing in the food-processing industry

- Consult with local and state recycling coordinators. These solid waste planners may help locate a market for food discards or provide technical advice. Some agencies award grant money for innovative projects.
- Ask the solid waste planners to provide you with contacts and information about businesses with successful food-recovery programs. By networking with other businesses you will be able to learn from their experiences. These organizations can also provide assistance in finding haulers and end users in your area.
- Anticipate barriers to a successful program and how you will overcome them. Learn from others. Ask employees what potential problems they see. They, after all, will be responsible for running the program.
- Train food service workers well, and well ahead of program implementation.
- Monitor and periodically reevaluate your program.
- Use composting diversion to reduce your waste hauling and tipping costs.
- Be creative.

25.7   Case Study

A major food-processing company was experiencing difficulties with their waste management program both in cost overruns and inaccurate goal setting for their units. The programs were running over budget due to poor cost estimates and lack of employee involvement. This resulted in a $12,000 over budget expense for the fiscal year 2005. In addition, they were concerned about not minimizing all possible waste streams, especially food by-products. The food-processing company concluded that it might be in their best interest to explore outsourcing their waste management activities. After an internal ROI (return on investment) analysis, the food-processing company decided to look at the possibility of outsourcing this business to a third party.

The food-processing company worked with The University of Toledo Waste Minimization group to audit their solid waste streams and identify minimization opportunities. After an in-depth analysis, the group was able to demonstrate that by focusing on food by-product composting, the food-processing company could save close to $15,000 per year. The group's analysis outlined the potential savings that could be gained through purchasing a composting vessel and working with supplies to reduce packing materials.

As a result of implementing the recommendations, the food-processing company was able to reduce their annual operating cost by $13,000 and generate $3000 from the sale of the compost. This year's program came in under budget and below the cost of past programs.

Figure 25.1   Composter unit.

The composting machine used applied high heat and aerobic action to turn food waste into compost in a few days. A blower pushes the exhaust gases through a biofilter to remove odors. Liquids collected can be recycled back into the composter. The pilot program showed some promising results; the composter could pay for itself in lower trash hauling fees in about 3 years. Figure 25.1 shows a composter unit currently in use in Ohio.

25.8   Additional Information

1 www.wastecap.org/wastecap/commodities/organics/organics.htm.
2 www.cetonline.org/FarmBusiness/farm_composting.php.
3 www.asiaisgreen.com/2008/01/15/recycling-of-food-waste-in-singapore.
4 www.ur.umich.edu/9798/Apr22_98/cafe.htm.
5 www.recycle.net/recycle/Organic/index.html.
6 www.sppscafe.org/Food_Waste_Recycling.html.
7 www.uvm.edu/~recycle/?Page=Composting/Composting.html.
8 www.wmnorthwest.com/guidelines/foodwastea.htm.
9 www.recyclenow.org/b_restaurant.html.

26

APPAREL- AND FABRIC-
MANUFACTURING APPLICATIONS

26.1   Industry Overview

NAICS code: all 31500s

INDUSTRY SNAPSHOT

13,038 textile, apparel, and fabric operations in the United States
343,450 total employees
$44.5 billion in annual sales
7.2 tons of solid waste generation per employee
Major waste streams: fabric, paper, and plastics

Textile, apparel, and furnishings workers produce fibers, cloth, and upholstery, and fashion them into a wide range of products that we use in our daily lives. Textiles are the basis of towels, bed linens, hosiery and socks, and nearly all clothing, but they also are a key ingredient in products ranging from roofing to tires. Jobs range from those that involve programming computers to those in which the worker operates large industrial machinery and to those that require substantial handwork.

Textile machine setters, operators, and tenders run machines that make textile products from fibers. The first step in manufacturing textiles is preparing the natural or synthetic fibers. Extruding and forming machine operators, synthetic and glass fibers; set up and operate machines that extrude or force liquid synthetic material such as rayon, fiberglass, or liquid polymers through small holes and draw out filaments. Other operators put natural fibers such as cotton, wool, flax, or hemp through carding and combing machines that clean and align them into short lengths collectively called "sliver." In making sliver, operators may combine different types of natural fibers and synthetic filaments to give the product a desired texture, durability, or other characteristic. Textile winding, twisting, and drawing-out machine operators take the sliver and draw out, twist, and wind it to produce yarn, taking care to repair any breaks.

Industries in the apparel-manufacturing subsector group include establishments with two distinct manufacturing processes

1 Cut and sew (i.e., purchasing fabric and cutting and sewing to make a garment).
2 The manufacture of garments in establishments that first knit fabric and then cut and sew the fabric into a garment.

The apparel-manufacturing subsector includes a diverse range of establishments manufacturing full lines of ready-to-wear apparel and custom apparel: apparel contractors, performing cutting or sewing operations on materials owned by others; jobbers performing entrepreneurial functions involved in apparel manufacture; and tailors, manufacturing custom garments for individual clients are all included. Knitting, when done alone, is classified in the textile mills subsector, but when knitting is combined with the production of complete garments, the activity is classified in apparel manufacturing.

In order to make textiles, the first requirement is a source of fiber from which a yarn can be made, primarily by spinning. The yarn is processed by knitting or weaving, which turns the yarn into cloth. The machine used for weaving is the loom. For decoration, the process of coloring yarn or the finished material is dyeing. Typical textile processing includes four stages: yarn formation, fabric formation, wet processing, and fabrication.

The four main types of fibers include natural vegetable fibers (such as cotton, linen, jute and hemp), man-made fibers (those made artificially, but from natural raw materials such as rayon, acetate, Modal, cupro, and the more recently developed Lyocell), synthetic fibers (a subset of man-made fibers, which are based on synthetic chemicals rather than arising from natural chemicals by a purely physical process), and protein-based fibers (such as wool, silk, and angora).

26.2   Waste Management Goals and Opportunities

The majority of solid waste generated by this sector is fabrics, mixed paper, plastics, and wood. Table 26.1 displays the composition breakdown based on survey results.

As shown in the table, the recycling rate for this sector is approximately 34 percent. Textile waste can be classified as either preconsumer or postconsumer. Preconsumer textile waste consists of by-product materials from the textile, fiber, and cotton industries. Each year 750,000 tons of this waste is recycled into new raw materials for the automotive, furniture, mattress, coarse yarn, home furnishings, paper, and other industries.

As derived from the solid waste evaluation model discussed in Chap. 12, the equation that estimates the annual waste generation per year per employee for this sector can be calculated from the following:

Tons of solid waste generated per year $= 7.29 \times$ number of employees $- 3.78$
$\times$ solid waste disposal cost per ton $+ 40.5$

TABLE 26.1   APPAREL- AND FABRIC-MANUFACTURING INDUSTRY SOLID WASTE COMPOSITION (SURVEY RESULTS)

MATERIAL	COMPOSITION (%)	RECYCLING (%)
Fabric	19 ± 4.6	81 ± 25.1
Paper	18 ± 4.3	15 ± 4.7
Mixed office paper	16 ± 3.8	16 ± 5.0
Newspaper	2 ± 0.5	14 ± 4.3
Plastics	16 ± 4.2	8 ± 2.5
LDPE	6 ± 1.6	9 ± 2.8
HDPE	5 ± 1.3	8 ± 2.5
PP	3 ± 0.8	7 ± 2.2
Other	2 ± 0.5	6 ± 1.9
Wood	10 ± 2.6	32 ± 9.9
OCC (cardboard)	9 ± 2.3	48 ± 14.9
Metals	8 ± 2.1	83 ± 8.1
Food waste	5 ± 1.3	0 ± 0.0
Yard waste	1 ± 0.3	0 ± 0.0
Glass	1 ± 0.3	3 ± 0.9
Other	13 ± N/A	0 ± 0.0
Overall recycling level		33.6

26.3   Constraints and Considerations

A key constraint in the textile industry is the use of dyes and bleaches to prepare the raw materials for processing. This contamination can render the potentially recyclable material unusable as it contains high amounts of dangerous chemicals. Additionally, establishing relationships with potential recyclers or organizations that may reuse the wastes as a raw material may also present a challenge. Establishing transportation and estimating periodic generation rates may present obstacles. Another organization that may be able to use a by-product as a raw material (for example, a pet bed manufacturer using scrap fibers for fillings) generally will require an estimate of the daily or weekly generation amounts. This will allow the production planners to plan and control internal processes and establish the required transportation routes. Finally, investing the extra time, space, and employee training and retraining to sort waste and recyclable materials will add additional costs.

Clothing and fabric generally consist of composites of cotton (biodegradable material) and synthetic plastics. The textile's composition will affect its durability and method of recycling. A composition analysis of the material will be necessary to evaluate recycling alternatives. Fiber reclamation mills can aid in this process and perform sorting if required, but usually at an additional cost.

26.4   Potential Technologies and Strategies

Fiber reclamation mills grade incoming material into type and color. The color sorting means no re-dying has to take place, saving energy and pollutants. The textiles are shredded into "shoddy" fibers and blended with other selected fibers, depending on the intended end use of the recycled yarn. The blended mixture is carded to clean and mix the fibers and spun ready for weaving or knitting. The fibers can also be compressed for mattress production. Textiles sent to the flocking industry are shredded to make filling material for car insulation, roofing felts, loudspeaker cones, panel linings, and furniture padding.

For specialized polyester-based materials the recycling process is significantly different. The first step is to remove the buttons and zippers and then to cut the garments into small pieces. The shredded fabric is then granulated and formed into small pellets. The pellets are broken down, polymerized, and turned into polyester chips. The chips are melted and spun into new filament fiber used to make new polyester fabrics.

Some companies create new pieces of clothing from scraps of old clothes. By combining and making new additions, the eclectic garments are marketed as a type of style. Thrift stores and secondhand stores may be an outlet for these types of fabrics.

Many companies have been able to cost-effectively reuse fiber scraps or "cutouts" as a raw material. Carpet padding, mattress filling, firewalls, and pet beds are several examples in which these materials may be reused. The organizations that will reuse these materials will require the material safety data sheets (MSDS) for each material to ensure no toxins or dangers exist for workers or future customers. Numerous material exchanges exist where organizations may list material available or needed. The EPA provides a Web site with several popular material exchanges at http://www.epa.gov/jtr/comm/exchstat.htm.

26.5   Implementation and Approach

The hierarchy of solid waste minimization plays a very large role in this sector. As the majority of the wastes generated are fibers, emphasis should first be placed on eliminating the waste via more efficient material usage. If a cutout is made, steps could be taken to reduce the leftover material by adjusting patterns. The organization could also work with the supplier for additional insights.

Next, with the remaining waste generated, a feasibility analysis should be conducted to analyze in-process recycling. In other words, can the scrap material be processed in the same facility to be used as a raw material input into the same process? This may require specialized equipment and training and a full economic evaluation should be performed.

Next, material exchanges should be explored to search for another organization that may be able to use the scrap as a raw material. Several national and state-specific exchanges are available from the EPA. Finally, any remaining waste should be sent to a third-party processor for recycling, such as a broker or fiber reclamation mill.

26.6 Case Study

In the spring of 2002, The University of Toledo Waste Minimization Team conducted a solid waste minimization audit for a leather product manufacturer located in the Midwest that specialized in gloves. The company employed 230 people and had annual revenue of $1.1 million. The company generated 1500 tons of solid waste per year, primarily fabric and leather hide by-products for an annual solid waste management cost of $65,000 (see Fig. 26.1 for photos of the raw material).

From the results of the solid waste audit, the waste minimization team identified opportunities to reduce the amount of solid waste disposed of at landfills by 25 percent or 375 tons per year. This was accomplished by establishing a strategic partnership with a pet-bedding manufacturer to collect the fiber scraps. These fiber scraps were processed and used as filler in the pet beds. The pet-bedding company provided containers and transportation for the materials.

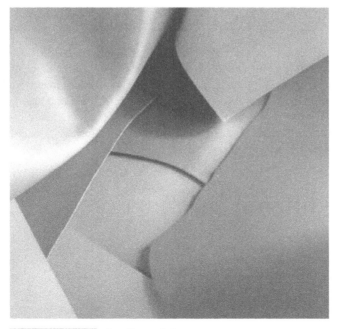

Figure 26.1   Leather picture.

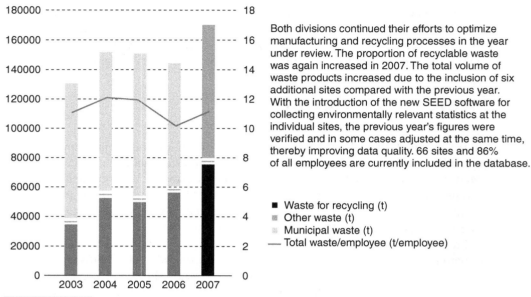

Both divisions continued their efforts to optimize manufacturing and recycling processes in the year under review. The proportion of recyclable waste was again increased in 2007. The total volume of waste products increased due to the inclusion of six additional sites compared with the previous year. With the introduction of the new SEED software for collecting environmentally relevant statistics at the individual sites, the previous year's figures were verified and in some cases adjusted at the same time, thereby improving data quality. 66 sites and 86% of all employees are currently included in the database.

■ Waste for recycling (t)
▦ Other waste (t)
▨ Municipal waste (t)
── Total waste/employee (t/employee)

Figure 26.2 Rieter waste and recycling annual trends.

26.7 Exemplary Performer—Textile By-Product Reuse

Rieter Automotive is an excellent example of a textile and automotive supply company excelling in waste minimization. Rieter is an industrial group based in Winterthur, Switzerland, and operating on a global scale. Formed in 1795, the company is a leading supplier to the textile and automotive industries. Rieter has a presence in 20 countries with some 70 manufacturing facilities and has a total worldwide workforce of approximately 15,500 employees. Figure 26.2 provides a snapshot of Rieter's waste and recycling trends from 2003 through 2007.

26.8 Additional Information

1 www.textilerecycle.org.
2 www.epa.gov/garbage/textile.htm.
3 www.en.wikipedia.org/wiki/Textile_manufacturing.
4 www.reiter.com.

27

WOOD- AND LUMBER-
MANUFACTURING APPLICATIONS

27.1 Industry Overview

NAICS code: all 32100s

INDUSTRY SNAPSHOT

17,202 wood and lumber operations in the United States
540,565 employees
$89.1 billion in annual sales
2.7 tons of solid waste generation per employee
Major waste streams: wood, paper, and metals

Despite the abundance of plastics and other materials, wood products continue to be useful and popular. Woodworkers help to meet the demand for wood products by creating finished products from lumber. Many of these products are mass produced, such as many types of furniture, kitchen cabinets, and musical instruments. Other products are crafted in small shops that make architectural woodwork, handmade furniture, and other specialty items.

Industries in the wood-product-manufacturing subsector manufacture wood products, such as lumber, plywood, veneers, wood containers, wood flooring, wood trusses, manufactured homes (i.e., mobile homes), and prefabricated wood buildings. The production processes of the wood-product-manufacturing subsector include sawing, planing, shaping, laminating, and assembling of wood products starting from logs that are cut into bolts, or lumber that may then be further cut or shaped by lathes or other shaping tools. The lumber or other transformed wood shapes may also be subsequently planed or smoothed, and assembled into finished products, such as wood containers.

The wood-product-manufacturing subsector includes establishments that make wood products from logs and bolts that are sawed and shaped, and establishments that purchase sawed lumber and make wood products. With the exception of sawmills and wood preservation establishments, the establishments are grouped into industries mainly based on the specific products manufactured.

Although the term *woodworker* often evokes images of a craftsman who builds ornate furniture using hand tools, the modern wood industry is highly technical. Some woodworkers still build by hand, but more often hand tools have been replaced by power tools, and much of the work has been automated. Work is usually done on an assembly line, meaning that most individuals learn to perform a single part of a complex process. Different types of woodworkers are employed in every stage of the building process, from sawmill to finished product. Their activities vary greatly.

Many woodworkers use computerized numerical control (CNC) machines to operate factory tools. Using these machines, woodworkers can create complex designs with fewer human steps. This technology has raised worker productivity by allowing one operator to simultaneously tend a greater number of machines. The integration of computers with equipment has improved production speed and capability, simplified setup and maintenance requirements, and increased the demand for workers with computer skills.

Production woodworkers set up, operate, and tend all types of woodworking machines. In sawmills, sawing machine operators and tenders set up, operate, or tend wood-sawing machines that cut logs into planks, timbers, or boards. In manufacturing plants, woodworkers first determine the best method of shaping and assembling parts, working from blueprints, supervisors' instructions, or shop drawings that woodworkers themselves produce. Before cutting, they often must measure and mark the materials. They verify dimensions and may trim parts using hand tools such as planes, chisels, wood files, or sanders to ensure a tight fit.

Woodworking machine operators and tenders set up, operate, or tend specific woodworking machines, such as drill presses, lathes, shapers, routers, sanders, planers, and wood-nailing machines. New operators may simply press a switch on a woodworking machine and monitor the automatic operation, but more highly skilled operators set up the equipment, cut and shape wooden parts, and verify dimensions using a template, caliper, or rule.

After wood parts are made, woodworkers add fasteners and adhesives and connect the pieces to form a complete unit. The product is then finish-sanded; stained, and, if necessary, coated with a sealer, such as lacquer or varnish. Woodworkers may perform this work in teams or be assisted by helpers.

Precision or custom woodworkers, such as cabinetmakers and bench carpenters, model makers and pattern makers, and furniture finishers, often build one-of-a-kind items. These highly skilled precision woodworkers usually perform a complete cycle of tasks—cutting, shaping, and preparing surfaces and assembling complex wood components into a finished wood product. Precision workers normally need substantial training and an ability to work from detailed instructions and specifications. In addition, they often are required to exercise independent judgment when undertaking an assignment. They may still use heavy machinery and power tools in their everyday work. As CNC machines have become less expensive, many smaller firms have started using them.

27.2   Waste Management Goals and Opportunities

The majority of solid waste generated by this sector is wood. Table 27.1 displays the composition breakdown based on survey results.

As shown in the table, the recycling rate for this sector is approximately 52 percent. As derived from the solid waste evaluation model discussed in Chap. 12, the equation that estimates the annual waste generation per year per employee for this sector can be calculated from the following:

$$\text{Tons of solid waste generated per year} = 17.40 \times \text{number of employees}$$
$$- 7.76 \times \text{solid waste disposal cost per ton}$$
$$- 217.59 \text{ if the company is ISO 140001 certified}$$
$$+ 473.93$$

TABLE 27.1   WOOD- AND LUMBER-MANUFACTURING INDUSTRY SOLID WASTE COMPOSITION (SURVEY RESULTS)

MATERIAL	COMPOSITION (%)	RECYCLING (%)
Wood	35 ± 8.8	84 ± 11.3
Paper	15 ± 3.8	14 ± 4.3
Mixed office paper	14 ± 3.5	14 ± 4.3
Newspaper	1 ± 0.3	2 ± 0.6
Metals	13 ± 3.3	84 ± 6.1
Ferrous metals	9 ± 2.3	85 ± 6.4
Nonferrous metals	4 ± 1.0	81 ± 6.0
Composite organic	11 ± 2.8	61 ± 18.9
OCC (cardboard)	6 ± 1.5	36 ± 11.2
Mixed plastics	5 ± 1.3	9 ± 2.8
Food waste	3 ± 0.8	0 ± 0.0
Yard waste	1 ± 0.3	0 ± 0.0
Glass	1 ± 0.3	6 ± 1.9
Other	10 ± N/A	0 ± 0.0
Overall recycling level		51.8

27.3   Constraints and Considerations

Sawmilling has been compared to juggling. There is a need to constantly balance a log inventory with the demands of the market. A sawmill has little control over its log resource. This lack of control extends to the market. Orders can't be taken for products without a customer base. A sawmill caught between these two variables (resource and market), over which it exercises little control, has only the flexibility anticipated, planned for, and designed into the mill to rely upon for economic survival.

Following is a list of considerations that are designed around a series of questions. These questions are arranged in a particular sequence:

- *Timber resource characteristics*—Many potential reuse options for wood by-products exist, as described later in this chapter. A key consideration is the type, quality, and size of the wood. Some outlets may be interested in a specific type of wood or size (sawdust to large fragments). The condition of the by-product will play a large role in determining the potential market.
- *Market characteristics*—Understanding and predicting the potential market for wood processing by-products is also necessary. Estimating the market size and marketing approaches will aid in establishing a sustainable system.
- *Regulatory considerations*—Investigating whether federal, state, and local laws and regulations relating to such concerns as worker's compensation, fees, taxes, registration, zoning safety, and the environment have all been identified and found to be compatible with the proposed enterprise and whether past uses present any regulation issues.
- *Logistical and facility issues*—Transportation of the by-product is also a concern and constraint. Adequate storage and dock space is necessary to store the by-product prior to shipment. In a facility limited in size, this may adversely impact potential options.
- *Contamination*—Most recyclers will set a maximum allowable level of contamination for all materials collected, including wood debris. If the wood is contaminated with fluids such as oils or cleaning agents, recyclers may reject loads.
- *Cost of equipment*—The cost of equipment can be prohibitive.
- *Invest the extra time, space, and employee training and retraining to sort waste and recyclable materials.* Training and employee involvement are key elements to the success of the recycling program. Creating a strong, well planned system at the beginning will save many headaches and additional costs for the organization at a later point in time.

27.4   Potential Technologies and Strategies

Technology has changed sawmill and wood-processing operations significantly in recent years, emphasizing increasing profits through waste minimization and increased energy efficiency as well as improving operator safety. The once ubiquitous, rusty, steel conical sawdust burners have for the most part vanished, as the sawdust and other mill waste is now processed into particleboard and related products, or used to heat wood-drying kilns. Cogeneration facilities will produce power for the operation and may also feed superfluous energy onto the grid. While the bark may be ground for landscaping bark mulch, it may also be burned for heat. Sawdust may make particleboard or be pressed into wood pellets for pellet stoves. The larger pieces of wood that won't make lumber are chipped into wood chips and provide a source of supply for paper mills. Wood by-products of the mills will also make oriented strand board paneling for building construction, a cheaper alternative to plywood for paneling. In the United States annual cut of timber for lumber products is equivalent to approximately 240 million trees. However, if the sawmills operated at a 70 percent recovery efficiency, the same annual harvest of lumber could be derived from 171 million trees. The saving would be the equivalent of 69 million trees annually if recovery efficiency improved from 50 to 70 percent in the primary processing industry. In addition, these same 69 million trees, if permitted to grow in the forest, would continue to absorb about 900,000 tons of carbon dioxide and produce about 650,000 tons of oxygen each year.

Following is a brief case study describing "thin kerf" sawing as related to improving process efficiency. Today, a typical circular sawmill converts 50 percent of the log into primary product with band mill conversion at about 57 percent. Saw kerf averages 21 percent for the circular sawmill and as low as 12 percent for high production band mills. Obviously, sawmill efficiencies have increased since the 1920s but there is still much room for improvement. In addition, trends such as environmental constraints on timber harvesting, smaller logs from the forest, and an increased demand for wood products makes it imperative that we improve sawmill efficiencies.

The good news is that technologies currently exist that can enable 70 percent or more conversion efficiencies at sawmills. However, the adoption of new technologies such as thin kerf sawing have not yet come state-of-the-art in many mills. A saw kerf is the width of the path cut by the sawteeth as the saw blade moves through the log. *Thin kerf* is a relative term, however, because it only has meaning when compared to something else. If circular saws are compared to band saws, then band saws would be considered thin kerf since they are generally 50 percent thinner than circular saws. If today's saw kerfs are compared with those of the past, we can generally say that today we have thinner kerfs.

The saw kerf has a significant impact on conversion efficiency, often referred to as lumber recovery. A crude but effective way of calculating the amount of sawdust that develops during sawing is to determine the total wood usage per pass, the action of logs moving or passing back and forth through the saw blade. Wood usage per pass includes the average thickness of the piece being sawn plus the saw kerf. For example, in cutting a board that is 1.125 in thick with a saw kerf of 0.300 in, the total wood usage per pass is 1.425 in. Calculating the saw kerf as a percentage of the total wood usage per pass results in 21 percent of the wood removed as sawdust or about one-fifth of the log resource. A band saw with a kerf of 0.140 in would result in an increase in lumber recovery of about 10 percent.

A recent study by the U.S. Forest Service, State & Private Forestry (S&PF), demonstrated the potential of thin kerf sawing when used in combination with other lumber-recovery practices. Twenty red oak logs with lengths of 4 to 6 ft and diameters between 12 and 20 in were sawn with a saw kerf band saw (0.062 in saw kerf). Boards sawn were 1-in thick and ranged in lengths from 2 to 6 ft. Lumber recovery or conversion efficiency for the 20-log sample averaged 82 percent, which is 30 percent greater than the typical circular sawmill.

A similar study of small-diameter (4–8 in) red and white pine logs was conducted by S&PF. Using the same thin kerf band saw as noted above the lumber recovery for the 47-log sample was 67 percent. Sawdust averaged 12 percent for the study.

A third study conducted at a Missouri pallet mill in 1993 produced results very similar to the pine study. The Missouri mill reduced short hardwood logs (length of 45 in) to 5/8-in-thick pallet parts. The band saws utilized in the pallet operation had saw kerfs of 0.050 in. The average conversion efficiency at this mill was calculated at 69 percent which is remarkable considering 5/8-in-thick lumber, rather than 1-in lumber, was produced. Thin lumber such as 5/8 in means more cuts and, consequently, more kerf removed, resulting in reduced conversion efficiency. However, 69 percent is still very impressive.

As well as the obvious benefits described above, on a more local level, thin kerf sawing will enable the use of lower quality and/or smaller diameter logs, which otherwise may have little or no economic value. Consequently forest management could be stimulated to improve and expand the resource base and lead to more successful rural development efforts in retaining, expanding, and attracting wood-using industries.

In increasing lumber recovery and simultaneously reducing waste, thin kerf sawing has the added potential benefit of keeping some sawmills profitable and in-business. In effect, the adoption of thin kerf technologies can save jobs by enabling mills to continue to operate.

Following is a list of diversion options for their wood waste in preferred order of implementation:

1 *Source reduction*—Source reduction, including reuse, can help reduce waste disposal and handling costs, because it avoids the costs of recycling, municipal composting, landfilling, and combustion. Source reduction also conserves resources and reduces pollution, including greenhouse gases that contribute to global warming.

2 *In-process and off-site recycling*—Wood shavings and sawdust may be reused internally or as a raw material for other operations. Sawdust is a primary material in particleboard.

3 *Energy recovery*—Energy recovery includes any technique or method of minimizing the input of energy to an overall system by the exchange of energy from one subsystem of the overall system with another. For example, wood scrap may be burned to heat a building or water tanks. Biofuels potential is another option. Energy production from unused wood fiber has been identified as a potential opportunity for the forest industry. Higher energy costs may make this opportunity economically viable.

27.5   Implementation and Approach

Tips for implementing solid waste reducing in the wood-processing industry:

- Work with suppliers to provide raw materials that generate less waste, including lumber that better meets production requirements, so less scrap is generated.
- Establish relationships with other organizations that may use the wood by-products as a raw material.

27.6   Case Study

In 2001, The University of Toledo waste minimization team conducted a solid waste audit for a manufacturer of ready-to-assemble furniture. The company employed 230 people and generated $49 million in annual sales. A review of waste records and an on-site audit indicated that the company generated 4000 tons of solid waste per year with an annual disposal cost of around $86,000. Figure 27.1 displays a portion of the workroom floor for this facility.

Its key challenges were to reduce solid waste generation, increase material reuse, and increase recycling levels. The company set additional goals to increase visibility on execution processes, improve manufacturing flow and efficiency by following a

Figure 27.1   Woodworking application.

strategic direction for lean manufacturing, and simplify the work-order reporting process. The waste minimization team determined that wood by-products, office paper, and cardboard were the major waste items and comprised 43 percent of the entire waste stream. To address the wood by-products, a collection system was developed to collect and reuse appropriately sized pieces of wood that were in good condition. Wood pieces that could not be reused were shredded and sold to a local fishery. To accomplish this, a chipping machine was purchased and installed near the waste removal dock and a strategic partnership was established with a local company. As a result of the process changes, 80 percent of the wood by-products that were previously disposed of at the landfill were reused or recycled resulting in an annual waste hauling savings of over $25,000. To address the office paper and cardboard, a baler was purchased and installed. This material is now collected, baled, and sold generating net revenue of over $9,000 per year. In addition, the solid waste hauler contract was reviewed and the number of container pulls per week was reduced by 35 percent.

27.7   Exemplary Performers— Environmental Management Systems in Woodworking

Because of pollution control and monitoring equipment difficulties experienced at Louisiana Pacific's (LP) Olathe, Colorado facility, the company implemented an environmental management system (EMS) to further ensure that their business practices protect the environment and community. Acknowledging the problem, LP invested substantial financial and human resources into analyzing the causes and potential solutions. Through an evaluation of the cause of environmental problems at Olathe and throughout the company, LP decided to strengthen its environmental management at the plant level. The Olathe mill was selected for the initial pilot of LP's EMS because of its immediate need for change. If this particular facility could be successful in changing, it was likely that the EMS would enhance environmental performance company-wide. Olathe's transformation is a notable success. Soon after the EMS was implemented, a comprehensive 2-day, unannounced inspection by EPA was conducted. EPA considered permit compliance to be well-managed, contingency plans appropriately linked to the community, and training thorough. EPA's lead inspector, who had conducted numerous inspections of wood product–manufacturing operations across the country, likened the LP Olathe plant to a bakery because it was so clean and well run. The company now has a cutting edge EMS and can transfer this knowledge to employees everywhere. Louisiana Pacific has realized the following benefits from the EMS program:

■ Mill employees have an enhanced sense of environmental awareness. Mill employees report that changes implemented through the EMS have greatly improved individual performance and consequently that of the whole facility. Knowledge of and compliance with environmental regulations is at an all-time high.

■ As an outgrowth of the EMS program, LP's laminated veneer lumber and I-joist plant in Hines, Oregon, began recycling planer shavings in early 1998. The wood waste material, originally disposed of at a cost to the company, is now being used in the production of other wood products such as medium density fiberboard. In February and March 1998, an estimated $12,900 was generated as a result of recycling the material rather than sending it to the local landfill. Total revenue from planner shavings is estimated at $90,000 for 1998.

■ Soon after becoming a part of LP, Associated Chemists, Inc. (ACI) in Portland, Oregon, implemented the EMS program and discovered how to save thousands of dollars by retaining and filtering their wastewater. After a few months with the filtration system, cost savings were estimated in the thousands of dollars. In 1997, ACI received the Best Practice Award for Water Conservation from the City of Portland.

■ Significant savings in waste management are being reported company-wide. At LP's particleboard plant in Missoula, employees have reduced annual landfill disposal costs by $31,000 by reducing waste, reusing, and recycling. Much of this savings is the direct result of their pallet rebuild and reuse program. Rather than dispose of pallets, old or broken ones either are repaired or sold to other companies. Recycling paper, cardboard, and metal has also resulted in additional plant profits and/or savings. Overall "scrap sales" led to a $5500 profit in 1998.

■ During the development of a Standard Operating Procedure for wood-burner slag at one of LP's facilities, the EMS team discovered that each employee performed disposal tasks in different, sometimes overly time-consuming, ways. The implementation of an SOP eliminated the confusion and unnecessary steps by clearly stating the standard disposal requirements and procedures. Consistent handling of wood-burner slag reduces environmental risk for local communities and increases efficiency in company operations.

27.8   Additional Information

1 Associated Builders & Contractors Inc. (ABC).
2 Building Materials Reuse Association (BMRA).
3 Pennsylvania Resources Council.
4 Southeast Recycling Development Council, Inc. (SERDC).
5 The American Wood Council (www.awc.org/helpoutreach/faq/FAQfiles/Recycled WoodReclaimLumber.html).

PAPER MANUFACTURING

APPLICATIONS

28.1   Industry Overview

NAICS code: all 32200s

INDUSTRY SNAPSHOT

5520 paper and allied product operations in the United States
491,436 employees
$153.8 billion in annual sales
7.0 tons of solid waste generation per employee
Major waste streams: pulp and paper by products

Paper is an indispensable component in the production, packaging, and delivery of a wide variety of products used daily by most Americans. For many it begins with breakfast. Most of us would be lost without our newspaper, coffee filters, napkins, cereal boxes, and milk and juice cartons. In some cases, our entire breakfast is baked, sold, reheated, and served in its original paperboard box.

Americans consume more paper than do the citizens of most countries. Compared with the 1994 world average of 97 lb, the U.S. per capita consumption of paper is more than 700 lb, approximately 2 lb per person per day. Per capita consumption of paper products in the United States has grown 43 percent since 1980.

Steady increases in paper consumption have had divergent impacts on employment in manufacturing production, in wholesale trade distribution, and in recycling collection.

Employment in paper manufacturing, which historically has been less volatile, has fluctuated less in recent years. Employment gains have been elusive, however, and 16,000 jobs have been lost, on net, since 1990. At the same time, employment in recycling, collection, and paper distribution reflects a steady and increasing rate of growth. Increases in these jobs in recent history have far outweighed employment declines in paper manufacturing.

Industries in the paper manufacturing subsector make pulp, paper, or converted paper products. The manufacturing of these products is grouped together because they constitute a series of vertically connected processes. More than one is often carried out in a single establishment. There are essentially three activities. The manufacturing of pulp involves separating the cellulose fibers from other impurities in wood or used paper. The manufacturing of paper involves matting these fibers into a sheet. Converted paper products are made from paper and other materials by various cutting and shaping techniques and include coating and laminating activities.

The paper manufacturing subsector is subdivided into two industry groups, the first for the manufacturing of pulp and paper and the second for the manufacturing of converted paper products. Paper making is treated as the core activity of the subsector. Therefore, any establishment that makes paper (including paperboard), either alone or in combination with pulp manufacturing or paper converting, is classified as a paper or paperboard mill. Establishments that make pulp without making paper are classified as pulp mills. Pulp mills, paper mills, and paperboard mills comprise the first industry group.

Establishments that make products from purchased paper and other materials make up the second industry group, converted paper product manufacturing. This general activity is then subdivided based, for the most part, on process distinctions. Paperboard container manufacturing uses corrugating, cutting, and shaping machinery to form paperboard into containers. Paper bag and coated and treated paper manufacturing establishments cut and coat paper and foil. Stationery product–manufacturing establishments make a variety of paper products used for writing, filing, and similar applications. Other converted paper product manufacturing includes, in particular, the conversion of sanitary paper stock into such things as tissue paper and disposable diapers.

An important process used in the paper-bag and coated and treated paper manufacturing industry is lamination, often combined with coating. Lamination and coating makes a composite material with improved properties of strength, impermeability, and so on. The laminated materials may be paper, metal foil, or plastics film. While paper is often one of the components, it is not always. Lamination of plastics film to plastics film is classified in the North American Industry Classification System (NAICS) subsector 326, plastics and rubber products manufacturing, because establishments that do this often first make the film. The same situation holds with respect to bags. The manufacturing of bags from plastics only, whether or not laminated, is classified in subsector 326, plastics and rubber products manufacturing, but all other bag manufacturing is classified in this subsector.

28.2 Waste Management Goals and Opportunities

The majority of solid waste generated by this sector is paper pulp and mixed paper wastes. Table 28.1 displays the composition breakdown based on survey results.

As shown in the table, the recycling rate for this sector is approximately 37 percent. Paper pulp constitutes the majority of the solid waste stream. As derived from the solid waste evaluation model discussed in Chap. 12, the equation that estimates the annual waste generation per year per employee for this sector can be calculated from the following:

Tons of solid waste generated per year = 6.63 × number of employees
$$- 3.51 \times \text{solid waste disposal cost per ton}$$
$$- 114.82 \text{ if the company is ISO 140001 certified}$$
$$+ 340.27$$

TABLE 28.1 PAPER MANUFACTURING INDUSTRY SOLID WASTE COMPOSITION (SURVEY RESULTS)		
MATERIAL	**COMPOSITION (%)**	**RECYCLING (%)**
Pulp	43 ± 8.9	28 ± 6.4
Paper	14 ± 8.1	81 ± 9.2
Composite paper	6 ± 4.2	80 ± 8.1
Mixed office paper	6 ± 5.1	79 ± 9.4
Newspaper	1 ± 6.2	80 ± 10.4
OCC (cardboard)	11 ± 3.1	80 ± 28.8
Wood	6 ± 1.7	51 ± 18.4
Mixed plastics	5 ± 1.4	8 ± 2.9
Food waste	4 ± 1.1	0 ± 0.0
Metals	3 ± 0.8	51 ± 14.3
Chemical	2 ± 0.6	8 ± 2.9
Yard waste	2 ± 0.6	0 ± 0.0
Glass	1 ± 0.3	8 ± 2.9
Other	9 ± N/A	0 ± 0.0
Overall recycling level		37.4

28.3   Constraints and Considerations

The number of trees and other vegetation cut down in order to make paper can be enormous. Paper companies insist that they plant as many new trees as they cut down. Environmentalists contend that the new growth trees, so much younger and smaller than what was removed, cannot replace the value of older trees. Efforts to recycle used paper (especially newspapers) have been effective in at least partially mitigating the need for destruction of woodlands, and recycled paper is now an important ingredient in many types of paper production. The key constraints for paper production can be separated into five areas:

- *Pulp fiber length*—Paper fibers can be recycled only a limited number of times before they become too short or weak to make high-quality paper. This means the broken, low-quality fibers are separated out to become waste sludge. These paper mill sludges consume a large percentage of our local landfill space each year. Worse yet, some of the wastes are land spread on cropland as a disposal technique, raising concerns about trace contaminants building up in soil or running off into area lakes and streams. Some companies burn their sludge in incinerators, contributing to our serious air pollution problems. Finding alternative uses for this sludge versus landfill disposal will divert a large amount of waste for this sector.
- *Chemical contamination*—All the inks, dyes, coatings, pigments, staples, and sticky items (tape, plastic films, etc.) are also washed off with the recycled fibers to join the waste solids. The shiny finish on glossy magazine–type paper is produced using a fine kaolin clay coating, which also becomes solid waste during recycling. The chemicals used in paper manufacture, including dyes, inks, bleach, and sizing, can also be harmful to the environment when they are released into water supplies and nearby land after use. The industry has, sometimes with government prompting, cleared up a large amount of pollution, and federal requirements now demand pollution-free paper production. The cost of such clean-up efforts is passed on to the consumer.
- *Governmental regulations*—These regulations are intended primarily to control conventional pollutants from pulp and paper mills in order to protect fish and their habitat. Pollutants of concern are suspended solids, biochemical oxygen-demanding matter, and effluent that are acutely lethal to fish. The regulations often require that the industry establish an environmental effects monitoring program. The program is intended to provide information to evaluate the need for further control measures, by evaluating the effectiveness of existing control measures and by assessing changes in the receiving environment.
- *Cost of equipment*—The cost of equipment can be prohibitive and can cost thousands of dollars to purchase and install.
- *Invest the extra time, space, and employee training and re-training to sort waste and recyclable materials*—Training and employee involvement are key elements to the success of the recycling program. Creating a strong, well planned system at the beginning will save many headaches and additional costs for the organization at a later point in time.

28.4   Potential Technologies and Strategies

The following is a list of diversion options for waste in preferred order of implementation:

1 *Source reduction*—Source reduction, including reuse, can help reduce waste disposal and handling costs, because it avoids the costs of recycling, municipal composting, landfilling, and combustion. Source reduction also conserves resources and reduces pollution, including greenhouse gases that contribute to global warming.

2 *In-process and off-site recycling*—Using recycled paper as a raw material versus virgin trees offers double environmental benefits. Fewer trees will be harvested for paper manufacturing and this waste will be diverted from landfills. Most estimates indicate that recycling 1 ton of paper typically saves approximately 17 trees and about 6.7 yd^3 of landfill space (2500 lb). In addition, paper pulp waste has been tested as a new source of raw material for the synthesis of a porous ceramic composite. A synthetic porous ceramic composite material consisting of the mullite, cordierite, and cristobalite phases is produced from a mixture of paper pulp waste and clay by reaction sintering at 1400°C. For more information visit the Web site www.springerlink.com/content/h7j112mv54112648.

3 *Land spread on cropland or daily landfill cover*—If the pulp is contaminate free, the wastes may be used as land spread on cropland or as daily cover for local landfills. Using the material as daily landfill cover may also reduce disposal fees and related taxation.

4 *Energy recovery*—Energy recovery includes any technique or method of minimizing the input of energy to an overall system by the exchange of energy from one subsystem of the overall system with another. For example, pulp scrap may be burned to heat a building or water tanks. Biofuels potential is another option. Energy production from unused wood fiber has been identified as a potential opportunity for the forest industry. Higher energy costs may make this opportunity economically viable.

28.5   Implementation and Approach

Tips for implementing solid waste reduction in the paper processing industry:

■ Consult with local and state recycling coordinators. These solid waste planners may help locate a market for paper-product discards or provide technical advice. Some agencies award grant money for innovative projects.

■ Ask the solid waste planners to provide you with contacts and information about businesses with successful paper-product recovery programs. By networking with other businesses you will be able to learn from their experiences. These organizations can also provide assistance in finding haulers and end users in your area.

■ Anticipate barriers to a successful program and how you will overcome them. Learn from others. Ask employees what potential problems they see. They, after all, will be responsible for running the program.

■ Train paper manufacturing workers well, and well ahead of program implementation.

■ Monitor and periodically reevaluate your program.

■ Use composting diversion to reduce your waste hauling and tipping costs.

■ Be creative.

28.6 Case Study

This case study examines a 300 ton per day paper mill in Ohio. For the purposes of this case study, the facility will be referred to as "the mill." Figure 28.1 is a picture of the factory floor of the mill. The mill consists of a de-inking facility (which recycles waste paper into reusable fiber) and a paper mill. Approximately 85 to 90 percent of the fiber used as raw material at the paper mill is generated at the de-inking facility (approximately 30 percent is post-consumer fiber). The de-inking facility produces approximately 225 tons of fiber daily. The waste fiber is combined with approximately 25 tons of virgin fiber, 40 tons of calcium carbonate, and 20 tons of starch per day to produce the mill's paper products. The facility employs approximately 340 people. The largest waste stream generated by the facility is waste fiber, which is disposed of at the facility's landfill. The mill spends approximately $1.2 to $1.5 million per year to operate and maintain this landfill. The projected capacity for the landfill was about

Figure 28.1   Paper manufacturing operations.

8 years in 1996. The mill has the added difficulty of not being able to landfill the waste in a municipal solid waste landfill due to the waste's high water content. Reducing generation of waste fiber and solid waste at the facility will extend the life of the landfill and is an important goal for the company.

Developing a formal pollution-prevention program and performing the associated assessments can often help companies identify Pollution Prevention (P2) options, even if they are already doing P2. Performing initial and detailed assessments of pollution-prevention opportunities is an integral part of developing a P2 plan. The mill owners were a little skeptical of the program at first, because they were already aware of pollution-prevention concepts and had implemented some projects. However, the company used the P2 process (as required in the Supplement Environmental Project [SEP]) to define previously unidentified projects, better track waste streams and their associated costs, and to reinvestigate previously identified projects that had never been implemented.

The mill developed a P2 team that included members of all levels, from staff to upper management. Empowering the team to make decisions on behalf of the company resulted in more timely development and implementation of the P2 program. Pollution prevention was a success at the mill, in part, because management was willing to commit the necessary resources to develop the program. While the implementation of P2 projects identified in the assessment did require capital investments from the company, the initial investments for the projects have been recovered through annual cost savings.

Twenty-two potential P2 projects were identified in the assessment process, and at publication of this case study, 19 of these projects had been implemented by the company. Implementation and results of the projects are briefly described below. The projects are sorted into various groups by type of project. Cost savings are indicated for each project when available.

The cost savings presented below will vary according to market prices. In the paper-milling business, profits and expenditures are heavily dependent on current market prices of paper fiber. The mill purchases predominantly waste fiber, which ranges from $200 to $400 per ton, and at time of publication, was around $250 per ton.

The paper machine improvement projects represent the largest savings for the company from P2 projects. These projects alone have improved plant yield by over 1 percent. The first project involved diverting approximately 3.6 tons per day of fiber rejected by the paper machines to the de-ink plant to be used as raw material. Formerly the fiber was discharged to the wastewater treatment plant (WWTP). The rejected fiber generated from the four paper machine cleaners is now used as raw material instead of lost as waste. This project saves 1300 tons per year of purchased waste fiber, for a total savings of $325,000 annually at current fiber rates. Last year the mill saved over $500,000.

The second project involved diverting one paper machine's electrifier reject stream (bleed) to a Johnson screen, instead of to the WWTP. This project had a payback period of only 4 months. The project cost approximately $30,000 to install, and saves the company 1.08 tons per day of lost fiber (395 tons annually), for an annual savings of $100,000 at current market value.

The company is replacing the forward cleaners in the reject loop and improving reject sorter operations to reduce fiber losses. The forward cleaners are being replaced in increments of 10 cleaners at a time. The new cleaners operate more efficiently, reducing fiber losses. Since this project is incremental, it is difficult to measure results, but the company is confident that they will be saving a significant amount of fiber and money.

To reduce fiber losses in the reject sorter at the de-ink plant; improvements included replacing the screen, increasing the dilution flow, and installing a strainer, costing approximately $10,000. This improved plastic scrap going to the landfill and saves the mill $5300 annually, along with the elimination of the disposable totes.

The mill also reduced its fresh water usage in the de-ink plant by returning vacuum pump seal water to the reservoir, reducing fresh water usage significantly and reducing final effluent volume.

The company instituted a beverage container–recycling program for glass bottles and aluminum cans. The company also extended its paper-recycling efforts to include mixed paper (magazines and newspapers) and corrugated cardboard. These programs were relatively inexpensive to implement, and reduce disposal of solid waste significantly. The mill recycles approximately 1.5 tons of waste beverage containers, more than 4 tons of mixed paper waste, and approximately 100 tons of corrugated paper (cardboard) annually.

The mill also cuts the metal end caps off their paper roll cores and returns them to the supplier, who recycles them. This project costs approximately $12,000 a year in labor, but reduces solid waste by approximately 55 tons per year. The company continues this project because the reduction in solid waste is substantial.

In addition, the mill attempted to reduce the generation of waste paper in its office procedures. Prior to the distribution of many in-house reports, the management information services (MIS) department now sends a questionnaire to the distribution list asking whether or not it is necessary to print the report, and who wants copies.

While there is not a large financial incentive for the recycling and MIS projects, the company benefits by reducing their landfilled solid waste, and the employees feel that the company is actively working to reduce solid waste.

28.7 Exemplary Performers— Recycling Paper Waste and Paper By-Products

EcoCover is a mulch mat designed to replace plastic on the ground, that is historically used in landscaping, horticulture, agriculture, and forestry. EcoCover is an organically certified paper mulch mat primarily manufactured from waste paper removed from the landfill waste stream. The mat is uniquely manufactured in the form of a continuous laminate (sheet) that will suppress weeds; promote plant growth; conserve water; is fully biodegradable; will moderate (extreme) soil temperatures; eliminate or reduce the use of herbicide; reduce plant mortality; carry beneficial additives within the laminate to deliver to both the soil and plant root system; allows for flexibility in the manufacturing process in terms of roll and square sizes/and useful life as a result of the laminate thickness, and much more. EcoCover biodegradable paper mulch mats are manufactured from up to 87 percent (by weight) waste paper. The company's manufacturing processes create zero waste. The products are organically certified and are used to improve plant growth, conserve soil moisture and suppress weed growth until they biodegrade into soil humus and food for earthworms and soil microorganisms that help to create natural soil fertility.

Zero waste is a goal that is both pragmatic and visionary, to guide people to emulate sustainable natural cycles, where all discarded materials are resources for others to use. Zero waste means designing and managing products and processes to reduce the volume and toxicity of waste and materials, conserve and recover all resources, and not burn or bury them. Implementing Zero waste will eliminate all discharges to land, water, or air that may be a threat to planetary, human, animal or plant health.

The Zero Waste International Alliance adopted the following Zero Waste Business Principles in 2005 to evaluate current and future zero waste policies and programs established by businesses and as the basis for evaluating the commitment of companies to achieve zero waste as well as to enable workers, investors, customers, suppliers, policymakers, and the public in general to better evaluate the resource efficiency of companies.

1 *Commitment to the triple bottom line*—The invention of EcoCover is based on the idea that social, environmental, and economic performance standards are met together. Read EcoCover's Principles of Sustainability.

2 *Use precautionary principle*—EcoCover's products and processes are organically certified (or in the case of several new products, certifiable). Their products have been subjected to extensive independent university research as well as the International Federation of Organic Agricultural Movements (IFOAM) international certification standards. The claims that they make about product performance are based on independently conducted research and trials.

3 *Zero waste to landfill or incineration*—This is a very lofty and worthy goal for manufacturers. In the case of EcoCover, they take waste paper from the waste stream going to landfill and manufacture 100 percent biodegradable products that benefit agriculture, forestry, horticulture, landscaping, and other plant-growing endeavors. Read about the impact of unsustainable plastic film mulch by comparison.

4 *Responsibility: take back products and packaging*—Most EcoCover products are shipped unpackaged, tied or bundled with natural fiber twine. Cardboard cartons used to package are biodegradable. EcoPins are recyclable.

5 *Buy reused, recycled or composted*—EcoCover uses recycled content up to 87 percent by weight in its mulch mats.

6 *Prevent pollution and reduce waste*—EcoCover's supply, production, and distribution systems are destined to reduce the use of natural resources and eliminate waste. The EcoCover manufacturing process emit no toxins or volatile organic compounds (VOCs). Their processes emit a little water vapor and modest amounts of CO_2 from their drying ovens.

7 *Highest and best use*—EcoCover biodegradable mulch mats are self composting, adding organic material to sustain soils and help eliminate the use of chemical fertilizers and herbicides. Their products can be used to reduce erosion and have been shown through research to do a better job of retaining soil moisture than unsustainable plastic sheeting or other forms of sheet mulch.

8 *Use of economic incentives*—EcoCover encourages its customers, workers, and suppliers to eliminate waste and maximize the reuse, recycling and composting of discarded materials.

9 *Products or services sold are not toxic or wasteful*—EcoCover does not use or produce toxins. Their products turn into earthworm castings and soil humus.

10 *Use nontoxic production, reuse, and recycling processes.*

Manufacturing EcoCover is a worldwide business opportunity to purchase the capital equipment to make EcoCover patented products under license. Waste in manufacturing processes is an indication of inefficiency which typically exchanges environmentally damaging pollution for profits. True profitability comes from zero waste manufacturing. EcoCover is an amazingly profitable business opportunity.

EcoCover is also a sustainable business idea that takes into account the huge market opportunities that are here today and coming in the future trends of agriculture. For more information on EcoCover, please visit www.ecocover-america.com.

28.8   Additional Information

1 www.p2pays.org/ref/10/09395.htm.

2 www1.eere.energy.gov/industry/forest/pdfs/erving.pdf.

PRINTING AND PUBLISHING

APPLICATIONS

29.1   Industry Overview

NAICS code: all 32300s

INDUSTRY SNAPSHOT

37,532 publishing operations in the United States
715,777 employees
$95.7 billion in annual sales
3.6 tons of solid waste generation per employee
Major waste streams: paper

The printing and related support activities subsector produces print products, such as newspapers, books, labels, business cards, stationery, business forms, and other materials, and performs support activities, such as data imaging, plate-making services, and bookbinding. The support activities included here are an integral part of the printing industry, and a product (a printing plate, a bound book, or a computer disk or file) that is an integral part of the printing industry is almost always provided by these operations.

Processes used in printing include a variety of methods used to transfer an image from a plate, screen, film, or computer file to some medium, such as paper, plastics, metal, textile articles, or wood. The most prominent of these methods is to transfer the image from a plate or screen to the medium (lithographic, gravure, screen, and flexographic printing). A rapidly growing new technology uses a computer file to directly "drive" the printing mechanism to create the image and new electrostatic and other types of equipment (digital or nonimpact printing).

Printing machine operators, also known as press operators, prepare, operate, and maintain printing presses. Duties of printing machine operators vary according to the type of press they operate. Traditional printing methods, such as offset lithography, gravure, flexography, and letterpress, use a plate or roller that carries the final image that is to be printed and copies the image to paper. In addition to the traditional printing processes, plate-less or nonimpact processes are coming into general use. Plate-less processes—including digital, electrostatic, and ink-jet printing—are used for copying, duplicating, and document and specialty printing. Plate-less processes usually are done by quick printing shops and smaller in-house printing shops, but increasingly are being used by commercial printers for short-run or customized printing jobs.

Machine operators' jobs differ from one shop to another because of differences in the types and sizes of presses. Small commercial shops can be operated by one person and tend to have relatively small presses, which print only one or two colors at a time. Large newspaper, magazine, and book printers use giant "in-line web" presses that require a crew of several press operators and press assistants.

After working with prepress technicians to identify and resolve any potential problems with a job, printing machine operators prepare machines for printing. To prepare presses, operators install the printing plate with the images to be printed and adjust the pressure at which the machine prints. Then they ink the presses, load paper, and adjust the press to the paper size. Operators ensure that paper and ink meet specifications, and adjust the flow of ink to the inking rollers accordingly. They then feed paper through the press cylinders and adjust feed and tension controls. New digital technology, in contrast, is able to automate much of this work.

While printing presses are running, printing machine operators monitor their operation and keep the paper feeders well stocked. They make adjustments to manage ink distribution, speed, and temperature in the drying chamber, if the press has one. If paper tears or jams and the press stops, which can happen with some offset presses, operators quickly correct the problem to minimize downtime. Similarly, operators working with other high-speed presses constantly look for problems, and when necessary make quick corrections to avoid expensive losses of paper and ink. Throughout the run, operators must regularly pull sheets to check for any printing imperfections. Most printers have, or will soon have, presses with computers and sophisticated instruments to control press operations, making it possible to complete printing jobs in less time. With this equipment, printing machine operators set up, monitor, and adjust the printing process on a control panel or computer monitor, which allows them to control the press electronically. In most shops, machine operators also perform preventive maintenance; they oil and clean the presses and make minor repairs.

In contrast to many other classification systems that locate publishing of printed materials in manufacturing, the North American Industry Classification System (NAICS) classifies the publishing of printed products in subsector 511, publishing industries (except Internet). Though printing and publishing are often carried out by the same enterprise (a newspaper, for example), it is less and less the case that these distinct activities are carried out in the same establishment. When publishing and printing are done in the same establishment, the establishment is classified in sector 51, information, in the appropriate NAICS industry even if the receipts for printing exceed those for publishing.

29.2   Waste Management Goals and Opportunities

The majority of solid waste generated by this sector is mixed papers. Table 29.1 displays the composition breakdown based on survey results.

As shown in the table, the recycling rate for this sector is approximately 60 percent. Wastes and emissions from printing facilities can cause environmental compliance headaches and unnecessary drain on the operation's profits. Eliminating the sources of solid waste can often reduce costs and simplify environmental compliance. Prevention of waste, rather than after the fact treatment of wastes, attacks the problem at the source. The primary nonhazardous solid wastes generated by this sector include

- Waste substrate—Paper, plastic, foil, textiles and metals from trimmings, rejects, and excess quantities
- Nonhazardous waste inks—Inks and empty containers
- Rags and wipes—Without solvents or hazardous residues
- Empty packaging—Including cartons, wrappers and roll cores

As derived from the solid waste evaluation model discussed in Chap. 12, the equation that estimates the annual waste generation per year per employee for this sector can be calculated from the following:

Tons of solid waste generated per year = 9.37 × number of employees + 105.32

TABLE 29.1   PRINTING AND PUBLISHING INDUSTRY SOLID WASTE COMPOSITION (SURVEY RESULTS)

MATERIAL	COMPOSITION (%)	RECYCLING (%)
Paper	55 ± 8.1	81 ± 9.2
Composite paper	27 ± 4.2	80 ± 8.1
Mixed office paper	25 ± 5.1	79 ± 9.4
Newspaper	3 ± 6.2	80 ± 10.4
OCC (cardboard)	13 ± 3.6	80 ± 28.8
Wood	7 ± 2.0	51 ± 18.4
Mixed plastics	4 ± 1.1	8 ± 2.9
Food waste	4 ± 1.1	0 ± 0.0
Metals	2 ± 0.6	51 ± 14.3
Yard waste	2 ± 0.6	0 ± 0.0
Glass	2 ± 0.6	8 ± 2.9
Other	11 ± N/A	0 ± 0.0
Overall recycling level		60.2

29.3   Constraints and Considerations

The constraints and considerations that the publishing and printing sector face involve paper waste, its primary waste stream. The following is a list of key constraints:

- *Pulp fiber length*—Paper fibers can be recycled only a limited number of times before they become too short or weak to make high-quality paper. This means the broken, low-quality fibers are separated out to become waste sludge. These paper mill sludges consume a large percentage of our local landfill space each year. Worse yet, some of the wastes are land spread on cropland as a disposal technique, raising concerns about trace contaminants building up in soil or running off into area lakes and streams. Some companies burn their sludge in incinerators, contributing to our serious air pollution problems. Finding alternative uses for this sludge versus land-fill disposal will divert a large amount of waste for this sector.
- *Paper color and quality*—Changes to input materials may alter the quality or color of the final product. This is especially important for printers. If switching to a recy-cling paper will create a grey appearance to the article it may diminish the quality in the customers' eyes.
- *Chemical contamination*—All the inks, dyes, coatings, pigments, staples, and sticky items (tape, plastic films, etc.) are also washed off the recycled fibers to join the waste solids. The shiny finish on glossy magazine–type paper is produced using a fine kaolin clay coating, which also becomes solid waste during recycling. The chemicals used in paper manufacture, including dyes, inks, bleach, and sizing, can also be harmful to the environment when they are released into water supplies and nearby land after use. The industry has, sometimes with government prompting, cleared up a large amount of pollution, and federal requirements now demand pollution-free paper production. The cost of such clean-up efforts is passed on to the consumer.
- *Cost of equipment*—The cost of equipment can be prohibitive and can cost thou-sands of dollars to purchase and install.

    Invest the extra time, space, employee training and retraining to sort waste and recyclable materials. Employee involvement and training are key elements to the success of the recycling program. Creating a strong, well planned system at the beginning will save many headaches and additional costs for the organization at a later point in time.
- *In-processing recycling limitations*—In general, printers can not cost-effectively implement in-process recycling due to space and resource limitations. Once the paper has been torn from a large roll of raw material or marked with ink, it must be returned to pulp form. Due to the equipment needed for this, most printers and pub-lishers work with a vendor to perform recycling.

In terms of considerations, the printing and publishing sector is very much in the pub-lic eye. Conveying a green image can help to bolster customer opinions and increase sales. Printing on recycled paper and informing customers are key steps in this process.

29.4   Potential Technologies and Strategies

From an equipment standpoint, the purchase of a compactor for paper waste is usually cost-justified in this sector. The compactor will reduce the number of trips that a recycling vendor must perform to the facility. This will reduce costs and save space within the facility. Below is a list of general strategies that can be applied by area:

- Waste and emission reduction opportunities for printers
  - Practice good housekeeping.
  - Cover all solvents.
  - Limit solvent use by using pumps or squeeze bottles, rather than pails, to wet cleanup cloths.
  - Keep hazardous wastes segregated from nonhazardous and each other.
  - Do not allow personal stockpiles of hazardous materials (e.g., cleanup solvents).
- Waste accounting
  - Collect accurate data on wastes and emissions to identify key reduction opportunities.
  - Establish accountability for waste generation and provide incentives for reduction.
  - Provide feedback on waste reduction performance to employees
  - Order and manage to minimize date expiration of materials.
  - Centralize responsibility for storing and distributing solvents.
  - Use returnable containers.
  - Use returnable plastic or wood pallets.
  - Require that all potentially hazardous samples be preapproved and vendor must accept unused portion.
- Alternative materials
  - Use inks that reduce the volatile organic compounds (VOC) emissions, for example, vegetable-based; water-based; ultraviolet; and electron-beam drying.
  - For jobs still using inks with heavy metals, find alternatives.
  - Eliminate or reduce alcohol in fountain solution.
  - Consider using waterless offset printing.
  - Use nonhazardous, low- or no-VOC solution to clean equipment.
- Printing process
  - Use standard sequence on process colors to minimize color changes for presses.
  - Improve quality control to reduce rejects.
  - Improve accuracy of counting methods, reducing excess quantities printed to accommodate inaccuracy.
  - Use web break detector and automatic splicer.
  - Properly store ink to prevent skin from forming.
  - Use refrigerative cooling to reduce evaporative losses of fountain solution.
  - Run similar jobs simultaneously to reduce cleanup.

- Reuse waste paper or collect for recycling.
- Use scrap paper for press setup runs.
- Cleanup
  - Use automatic blanket washes.
  - Wring or centrifuge used cloths to recover solvent and reuse solvent in parts washer or for additional press cleaning.
  - Avoid soaking cloths in solvent; use plunger or squeeze bottle to dampen cloth.
  - Use parts washing equipment to wash press trays.
  - Use cleanup solution with lower VOC content and lower vapor pressure.
  - Clean ink fountains only when changing color; use spray skin overnight.
  - Provide marked, accessible containers for segregated collection of used solvents.
- Waste Inks
  - Consider reusing as house colors.
  - Carefully label and store special-order colors for future reuse.
  - Mix to make black ink for internal or external use.
  - Recycle after processing.
  - Donate for reuse by printing schools or others.
- Finishing
  - Use water-based adhesives.
  - Minimize coatings that hinder recycling.
- Educate customers—Customer choices and specifications affect environmental impacts of a printing process. It is important that customers receive the right information and pricing signals to encourage purchasing decisions that reduce environmental impacts.

29.5   Implementation and Approach

In implementing solid waste minimization strategies, the first step is to investigate waste reduction opportunities. This usually involves steps to cut the waste generated from input materials. For example, methods can be implemented at printing press start-up and shutdown to reduce paper waste scrap. In addition, scrap paper may be an option to test runs at start-up versus the raw input material. The next step is to identify recyclers to handle the paper wastes. Since a large amount of paper waste is generated, the organization can usually negotiate for a higher price per ton. They often serve as valuable partners in the process. Finally, the publisher or printer may consider enhancing its image by planting trees and notifying customers of its recycling efforts.

29.6   Case Study

In fall 2007, a solid waste audit was conducted by The University of Toledo Waste Minimization Team at a Toledo, Ohio–based printing company that employed 85 people and generated $1.6 million in annual revenue. The goals of the audit were to

- Increase recycling levels
- Reduce operating costs
- Improve waste materials flows and staging within the facility

The facility was rebuilt in 1974 after a major fire; in addition, printing machine and computer upgrades had rendered the layout obsolete. The team conducted a waste audit and determined that the facility produced 7500 tons of solid waste per year and paid about $52,000 in waste removal costs. Under current practices about 15 percent or 1100 tons were recycled in the form of papers and the company was not generating any revenue form the sale of these recyclable materials.

The team conducted a value stream analysis for the key processes and determined that many intersecting and inefficient process flows existed. Limited standardization was in place and few checks and balances were in place to ensure that recyclable material was transported to the correct spot on the dock. The team recommended that the facility improve and standardize the flows including dedicated space for staging and clear transportation lanes. The staging areas and lanes were clearly labeled and training was provided to facility personnel regarding the new system. The recommended flows were outlined using a computer-aided design template. In addition, an economic justification of a baler was conducted based on the amount of paper and cardboard generated. The company was able to justify the purchase of a baler with a payback period of 3.2 years and a return on investment (ROI) of 17 percent working with a commodity broker for the baled paper in Detroit, Michigan. The waste hauler contract was also renegotiated to reduce the amount of pulls by switching to an on-call system versus regularly schedule collection routes. As a result of the changes, the company increased its recycling level from 15 to 62 percent and achieved cost avoidance/additional revenue of nearly $31,000.

29.7   Exemplary Performers—Increasing Recycled Content in Publishing

Random House, the publishing company owned by Bertelsmann, the German media giant, announced in 2008 that it would increase the proportion of recycled paper it buys for its books to at least 30 percent by 2010, from 3 percent now.

The company currently buys 110,000 tons of uncoated paper to publish books each year. When it reaches its target of purchasing 33,000 tons of recycled content paper by 2010, it believes that will be equivalent to saving more than 550,000 trees a year and removing 8000 cars from the nation's roads, because of the resulting reduction in greenhouse gas emissions.

"We believe our new paper policy is the right thing for us to do and now is the right time to do it," Peter W. Olson, chairman and chief executive of Random House, said in a statement.

The move occurred near the time publishers, agents, booksellers, and authors were gathering in Washington for Book Expo America, an annual industry convention. The Green Press Initiative, a nonprofit organization that has been working with publishers to achieve recycling goals, announced a "Book Industry Treatise on Responsible Paper Use" that calls for raising the industry's average use of recycled fiber to 30 percent by 2011, from 5 percent today.

David F. Drake, director of publishing operations projects at Random House, said the company would phase in its increased use of recycled paper, starting next year when it plans for 10 percent of its paper purchases to have recycled fiber content. In addition, the company said that by 2010, 10 percent of the glossy paper it purchases for cookbooks, art volumes, and the like will come from recycled fiber.

Mr. Drake said the changes could mean a shift in suppliers. He said the company was working with its current mills to see how much recycled paper they could produce. The timing of the changes was intended to allow the mills to meet the new demands.

29.8 Additional Information

1 www.p2pays.org/ref/21/20059.pdf.

CHEMICAL AND PETROLEUM
PROCESSING APPLICATIONS

30.1   Industry Overview

NAICS codes: all 32400s through the 32500s

INDUSTRY SNAPSHOT

15,738 chemical and petroleum operations in the United States
955,133 employees
$675.7 billion in annual sales
7.1 tons of solid waste generation per employee
Major waste streams: solidified sludge and office paper

Chemical manufacturing is divided into seven segments, six of which are covered here: basic chemicals; synthetic materials, including resin, synthetic rubber, and artificial and synthetic fibers and filaments; agricultural chemicals, including pesticides, fertilizer, and other agricultural chemicals; paint, coatings, and adhesives; cleaning preparations, including soap, cleaning compounds, and toilet preparations; and other chemical products.

The basic chemicals segment produces various petrochemicals, gases, dyes, and pigments. Petrochemicals contain carbon and hydrogen and are made primarily from petroleum and natural gas. The production of both organic and inorganic chemicals occurs in this segment. Organic chemicals are used to make a wide range of products, such as dyes, plastics, and pharmaceutical products; however, the majority of these chemicals are used in the production of other chemicals. Industrial inorganic chemicals usually are made from salts, metal compounds, other minerals, and the atmosphere. In addition to producing solid and liquid chemicals, firms involved in inorganic chemical manufacturing produce industrial gases such as oxygen, nitrogen, and helium. Many inorganic chemicals serve as processing ingredients in the manufacture of chemicals, but do not appear in the final products because they are used as catalysts—chemicals that speed up or otherwise aid a reaction.

The synthetic materials segment produces a wide variety of finished products as well as raw materials, including common plastic materials such as polyethylene, polypropylene, polyvinyl chloride (PVC), and polystyrene. Among products into which these plastics can be made are loudspeakers, toys, PVC pipes, and beverage bottles. Motor vehicle manufacturers are particularly large users of such products. This industry segment also produces plastic materials used for mixing and blending resins on a custom basis.

The agricultural chemicals segment, which employs the fewest workers in the chemical industry, supplies farmers and home gardeners with fertilizers, herbicides, pesticides, and other agricultural chemicals. The segment also includes companies involved in the formulation and preparation of agricultural and household pest-control chemicals.

The paint, coating, and adhesive products segment includes firms making paints, varnishes, putties, paint removers, sealers, adhesives, glues, and caulking. The construction and furniture industries are large customers of this segment. Other customers range from individuals refurbishing their homes to businesses needing anticorrosive paints that can withstand high temperatures.

The cleaning preparations segment is the only segment in which much of the production is geared directly toward consumers. The segment includes firms making soaps, detergents, and cleaning preparations. Cosmetics and toiletries, including perfume, lotion, and toothpaste, also are produced in this segment. Households and businesses use these products in many ways, cleaning everything from babies to bridges.

The other chemical products segment includes manufacturers of explosives, printing ink, film, toners, matches, and other miscellaneous chemicals. These products are used by consumers or in the manufacture of other products.

Chemicals generally are classified into two groups: basic chemicals and specialty chemicals. Basic chemical manufacturers produce large quantities of basic and relatively inexpensive compounds in large plants, often built specifically to make one chemical. Most basic chemicals are used to make more highly refined chemicals for the production of everyday consumer goods by other industries. Conversely, specialty chemical manufacturers produce smaller quantities of more expensive chemicals that are used less frequently. Specialty chemical manufacturers often supply larger chemical companies on a contract basis. Many traditional chemical manufacturers are divided into two separate entities, one focused on basic and the other on specialty chemicals.

The diversity of products produced by the chemical industry also reflects its component establishments. For example, firms producing synthetic materials operated relatively large plants in 2006. By contrast, manufacturers of paints, coatings, and adhesive products had a greater number of establishments, each employing a much smaller number of workers.

The chemical industry segments vary in the degree to which their workers are involved in production activities, administration and management, and research and development. Industries that make products such as cosmetics or paints that are ready for sale to the final consumer employ more administrative and marketing personnel. Industries that market their products mostly to industrial customers generally employ a greater proportion of precision production workers and a lower proportion of unskilled labor.

The chemical manufacturing subsector is based on the transformation of organic and inorganic raw materials by a chemical process and the formulation of products. This subsector distinguishes the production of basic chemicals that comprise the first industry group from the production of intermediate and end-products produced by further processing of basic chemicals that make up the remaining industry groups.

This subsector does not include all industries transforming raw materials by a chemical process. It is common for some chemical processing to occur during mining operations. These beneficiating operations, such as copper concentrating, are classified in sector 21, mining. Furthermore, the refining of crude petroleum is included in subsector 324, petroleum and coal products manufacturing. In addition, the manufacturing of aluminum oxide is included in subsector 331, primary metal manufacturing; and beverage distilleries are classified in subsector 312, beverage and tobacco product manufacturing. As in the case of these two activities, the grouping of industries into subsectors may take into account the association of the activities performed with other activities in the subsector.

The petroleum and coal products manufacturing subsector is based on the transformation of crude petroleum and coal into usable products. The dominant process is petroleum refining that involves the separation of crude petroleum into component products through such techniques as cracking and distillation.

In addition, this subsector includes establishments that primarily further process refined petroleum and coal products and produce products, such as asphalt coatings and petroleum lubricating oils. However, establishments that manufacture petrochemicals from refined petroleum are classified in industry 32511, petrochemical manufacturing.

30.2 Waste Management Goals and Opportunities

The majority of solid waste generated by this sector is solidified sludge and mixed office paper. Table 30.1 displays the composition breakdown based on survey results.

As shown in the table, the recycling rate for this sector is approximately 28 percent. As derived from the solid waste evaluation model discussed in Chap. 12, the equation that estimates the annual waste generation per year per employee for this sector can be calculated from the following:

Tons of solid waste generated per year = 7.14 × number of employees − 3.49
$$\times \text{ solid waste disposal cost per ton}$$
$$- 140.4 \text{ if the company is ISO 140001 certified}$$
$$+ 371.1$$

TABLE 30.1 CHEMICAL AND RUBBER-PROCESSING INDUSTRY SOLID WASTE COMPOSITION (SURVEY RESULTS)

MATERIAL	COMPOSITION (%)	RECYCLING (%)
Plastics	24 ± 7.4	8 ± 3.2
HDPE	5 ± 1.6	8 ± 3.2
PP	4 ± 1.2	7 ± 2.8
PET	3 ± 0.9	8 ± 3.2
PVC	3 ± 0.9	7 ± 2.8
Paper	20 ± 6.2	20 ± 8.0
Mixed office paper	17 ± 4.9	19 ± 7.6
Newspaper	2 ± 0.6	14 ± 5.6
OCC (cardboard)	12 ± 3.7	46 ± 17.1
Wood	9 ± 2.8	91 ± 18.4
Metals	9 ± 2.8	89 ± 4.2
Food waste	6 ± 1.9	0 ± 0.0
Glass	2 ± 0.6	5 ± 2.0
Yard waste	2 ± 0.6	0 ± 0.0
Other	16 ± N/A	0 ± 0.0
Overall recycling level		27.7

30.3   Constraints and Considerations

The critical constraints in the chemical and rubber sector relate to the hazardous materials and by-products that are used and generated during operations. These hazardous materials can contaminate potentially recyclable materials, such as chemicals on plastic tubs or wood pallets. Once contaminated, these materials must also be treated as hazardous. The majority of the waste generated by this sector is plastic waste used to house input materials and paper streams.

30.4   Potential Technologies and Strategies

To combat the contaminate issue; several strategies are available ranging from process changes to equipment purchases. The first low-cost strategy involves separating waste to avoid contamination. Such strategies to prevent contamination include

- Clearly label the recycling bins and waste containers on site.
- Post lists of recyclable and nonrecyclable materials.
- Conduct regular site visits to verify that bins are not contaminated.

If contamination cannot be avoided, washing machines can be purchased to remove contaminates, or a recycling vendor can be hired who has the means to clean the recyclable materials. The downside of utilizing a vendor to wash parts is an additional cost. A cost comparison and break-even analysis should be performed to determine if hiring a vendor or completing the work in-house is more cost-effective.

30.5 Implementation and Approach

In implementing solid waste minimization strategies, the first step is to identify potential haulers and recyclers to handle the designated materials. They often serve as valuable partners in the process. Obtain and retain verification records (waste hauler receipts and waste management reports) to confirm that diverted materials have been recycled or diverted as intended.

30.6 Case Study

In 2002, The University of Toledo Waste Minimization Team conducted a solid waste audit at a large oil refinery located in the Midwest. The company's management was interested in expanding and improving their recycling program. The recyclable waste generated at the company included office paper, newspapers, aluminum cans, plastic containers, and cardboard. At the time of the audit, the facility had small recycling programs in place for cardboard and office paper and utilized a separate waste container for storage.

The company operates 365 days a year and there were 550 employees. Results from the audit indicated that the company produced 150 tons of solid waste that was disposed of in 12 dumpsters located within the facility. The monthly solid waste hauling cost was $13,200.

After analyzing the overall waste stream, the Waste Assessment Team determined the major waste streams by weight. These waste streams and their subsequent generation are shown in Table 30.2 and Fig. 30.1. Table 30.3 and Fig. 30.2 display the major waste streams by volume and the estimated percentage volume of recyclable waste streams.

Based upon the implementation of recycling key components in the waste stream, Table 30.4 represents an analysis of the potential savings. These savings reflect potential revenue, which may be generated by the sale of the recyclable materials.

TABLE 30.2 ESTIMATED WEIGHT OF GENERATED SOLID WASTE STREAMS			
RANK	MATERIALS	POUNDS/YEAR	PERCENTAGE
1	MOP	136,898	60.26%
2	Food waste (include packaging)	79,328	34.92%
3	Aluminum cans	5,045	2.22%
4	Cardboard	2,234	0.98%
5	ONP	2,121	0.93%
6	Plastic bottles	1,569	0.69%
	Total	213,209	100%

TABLE 30.3 ESTIMATED VOLUME OF GENERATED SOLID WASTE STREAMS			
RANK	MATERIALS	CUBIC YARDS/YEAR	PERCENTAGE
1	MOP	342	48.39%
2	Food waste (include packaging)*	227	32.05%
3	Aluminum cans	84	11.89%
4	Plastic bottles	39	5.55%
5	Cardboard	9	1.27%
6	ONP	6	0.86%
	Total	707	100%

*Nonrecyclable

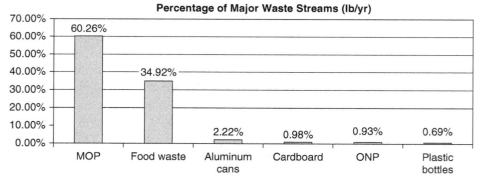

Figure 30.1   Oil refining waste audit chart for material weights.

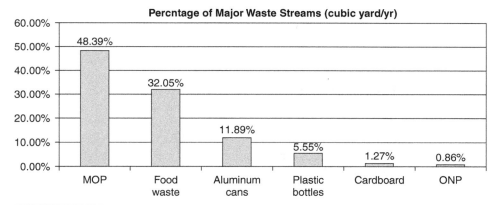

Figure 30.2   Oil refining waste audit chart for material volumes.

30.6.1   RECOMMENDATIONS

1 Implement a recycling program for office paper, aluminum cans, cardboard, plastic bottles, and news papers.

■ Potential annual saving of $900.

■ The annual revenue generated from the sale of the recyclables (paper, plastic, and aluminum cans) is approximately $900 at current market prices. Approximately 90 percent of the $900 could be generated from the sale of office paper. Additional savings will be realized by the reduction of the solid waste hauling bill (please see next recommendation).

■ Institute a desk-side container program, where the employees sort their recyclable paper into special containers beside their desks. Then, the desk-side containers are emptied into intermediate collection centers. An informational booklet is included in the attachment section of this report, which describes very effective ways to begin an office recycling program.

TABLE 30.4   OIL REFINING WASTE AUDIT SAVINGS FROM THE SALE OF RECYCLABLE MATERIALS

COMPONENT	TONS/YEAR	VALUE ($/TON)	VALUE ($/YEAR)	ANNUAL BENEFITS (%)
MOP	68.4	12.5	856	90.45
Aluminum cans	2.52	25	63	6.67
Cardboard	1.12	15	17	1.77
Plastic bottles	0.78	10	8	0.83
ONP	1.06	2.5	3	0.28
Total	73.9		947	100

2 Reduce the frequency of waste hauling.

- Potential annual saving of $2600.
- Savings are possible by reducing the frequency of times the solid waste at the company is collected by the waste hauler. Currently the waste hauler removes waste from the company three times per week. The total volume of recyclable waste streams is estimated to be 480 yd^3/year. By removing recyclables from the waste stream, the company could divert nearly 60 percent of the volume of their waste for recycling. The company, therefore, can adjust their waste hauling schedules to accommodate the reduction of the materials to be landfilled, thereby reducing the waste hauling cost. For this study, a conservative reduction in waste hauling cost is estimated to be approximately 20 percent. This will incur estimated annual savings of $2600.
- Please contact your waste hauler for more details, once the office recycling program has been implemented.

3 Set printer and copier defaults to print on both sides of the paper.

- Setting copiers to default to duplex copies would reduce paper use in the copier by 10 percent.
- Identify which copiers were changed to default duplex.
- Indicate exactly how to select single-sided copies when needed.

4 Assign a program coordinator.

- A successful recycling program requires an enthusiastic coordinator to keep the program running smoothly.
- The responsibilities of the coordinator may include selecting the recycler, developing the collection system, as well as tracking and measuring the success of the program.

5 Establish an effective education program.

- The goal is to educate employees in the importance of recycling.
- Place handouts describing what can be recycled and where to recycle it in every building and hand them to the employees so they can start recycling.
- Hang posters and flyers in the facility to remind employees to sort their waste before throwing it away.
- Be certain that every recycling bin is clearly marked and easy to use.

6 Purchase supplies made from recycled materials.

- Purchasing products made from recycled materials is critical to closing the recycling loop, and creating a demand for recycled goods.
- Develop a purchasing policy that includes the purchase of recycled content products.
- Purchase remanufactured toner cartridges and printer ribbons.

30.7   Exemplary Performers—Waste Minimization in Oil Refineries

Chevron is an exemplary company in terms of waste minimization. Protecting the earth's natural resources is as important to Chevron as providing the energy sources essential to improving quality of life. The company actively participates in all state and federal pollution prevention programs. At the Chevron Richmond Refinery, proper waste minimization and management is a high priority. Employees work to find better ways to reduce waste, reduce what is generated, and treat and dispose of what remains. The Chevron Richmond Refinery uses a variety of techniques to reduce or recycle waste and transform it into useful products. Since 1986, the company has reduced the off-site disposal of routinely generated wastes by more than 75 percent by decreasing waste sources, recycling waste, and treating waste inside the refinery.

In terms of recycling, the refinery recycles everything from laser toner cartridges, used batteries, light bulbs, and aerosol cans to newspaper. The current diversion rate is close to 50 percent. The company has been a frequent winner of the annual WRAP (Waste Reduction Awards Program) award from the California Integrated Waste Management Board.

Following are the results of a number of the Chevron Richmond Refinery's recovery and recycling initiatives:

- 1,022,000 barrels (152,000 tons) of oil products were recovered into refinery process units for conversion to salable products.
- Process by-product gas was treated to remove 99.96 percent of its hydrogen sulfide (H_2S) and used as fuel gas, replacing natural gas purchases equivalent to 30 billion standard cubic feet.
- 970 tons of spent catalyst was recycled through metals reclamation, catalyst regeneration, and metallurgical smelters to recover the metals for reuse as catalyst and in the steel and copper industry.
- 1558 tons of scrap metal from demolition activities was recycled.
- 20 percent by volume of solid waste was recycled, primarily wood and cardboard.

31

PLASTIC AND RUBBER

MANUFACTURING

31.1 Industry Overview

NAICS codes: all 32600s

INDUSTRY SNAPSHOT

15,529 plastic- and rubber-manufacturing operations in the United States
983,757 employees
$174.4 billion in annual sales
6.4 tons of solid waste generation per employee
Major waste streams: mixed plastic and office paper

Industries in the plastics and rubber products manufacturing subsector make goods by processing plastics materials and raw rubber. The core technology employed by establishments in this subsector is that of plastics or rubber-product production. Plastics and rubber are combined in the same subsector because plastics are increasingly being used as a substitute for rubber; however, the subsector is generally restricted to the production of products made of just one material, either solely plastics or rubber.

Many manufacturing activities use plastics or rubber, for example the manufacture of footwear, or furniture. Typically, the production process of these products involves more than one material. In these cases, technologies that allow disparate materials to be formed and combined are of central importance in describing the manufacturing activity. In the North American Industry Classification System (NAICS), such activities (the footwear and furniture manufacturing) are not classified in the plastics and rubber products manufacturing subsector because the core technologies for these activities are diverse and involve multiple materials.

Within the plastics and rubber products manufacturing subsector, a distinction is made between plastics and rubber products at the industry group level, although it is not a rigid distinction, as seen from the definition of industry 32622, rubber and plastics hoses and belting manufacturing. As materials technology progresses, plastics are increasingly being used as a substitute for rubber; and eventually, the distinction may disappear as a basis for establishing classification.

The simplest definition of a polymer is something made up of many units. Polymers are chains of molecules. Each link of the chain is usually made of carbon, hydrogen, oxygen, and/or silicon. To make the chain, many links, are hooked, or polymerized, together.

To create polymers, petroleum and other products are heated under controlled conditions and broken down into smaller molecules called monomers. These monomers are the building blocks for polymers. Different combinations of monomers produce plastic resins with different characteristics, such as strength or molding capability.

Plastics can be divided into two major categories: thermosets and thermoplastics. A thermoset is a polymer that solidifies or sets irreversibly when heated. They are useful for their durability and strength, and are therefore used primarily in automobiles and construction applications. Other uses are adhesives, inks, and coatings.

A thermoplastic is a polymer in which the molecules are held together by weak bonds, creating plastics that soften when exposed to heat and return to original condition at room temperature. Thermoplastics can easily be shaped and molded into products such as milk jugs, floor coverings, credit cards, and carpet fibers.

Plastic resins are processed in several ways, including extrusion, injection molding, blow molding, and rotational molding. All of these processes involve using heat and/or pressure to form plastic resin into useful products, such as containers or plastic films.

31.2   Waste Management Goals and Opportunities

The majority of solid waste generated by this sector is mixed plastics and mixed office paper. Table 31.1 displays the composition breakdown based on survey results.

As shown in the table, the recycling rate for this sector is approximately 37 percent. As derived from the solid waste evaluation model discussed in Chap. 12, the

TABLE 31.1 PLASTIC AND RUBBER-MANUFACTURING INDUSTRY SOLID WASTE COMPOSITION (SURVEY RESULTS)		
MATERIAL	COMPOSITION (%)	RECYCLING (%)
Plastics	21 ± 6.5	8 ± 3.2
HDPE	7 ± 2.2	8 ± 3.2
PP	5 ± 1.6	7 ± 2.8
PET	5 ± 1.6	8 ± 3.2
PVC	4 ± 1.2	7 ± 2.8
Paper	18 ± 5.6	20 ± 8.0
Mixed office paper	12 ± 3.5	19 ± 7.6
Newspaper	3 ± 0.9	14 ± 5.6
OCC (cardboard)	12 ± 3.7	46 ± 17.1
Solidified sludge	10 ± 3.1	69 ± 14.2
Wood	9 ± 2.8	91 ± 18.4
Metals	7 ± 2.2	89 ± 4.2
Food waste	6 ± 1.9	0 ± 0.0
Glass	2 ± 0.6	5 ± 2.0
Yard waste	2 ± 0.6	0 ± 0.0
Other	13 ± N/A	0 ± 0.0
Overall recycling level		36.7

equation that estimates the annual waste generation per year per employee for this sector can be calculated from the following:

Tons of solid waste generated per year = 6.64 × number of employees
− 4.13 × solid waste disposal cost per ton
− 153.7 if the company is ISO 140001 certified
+ 295.4

31.3   Constraints and Considerations

All plastics can be recycled; however, the extent to which they are recycled depends upon both economic and logistic factors. As a valuable and finite resource, the optimum use for most plastic after its first use is to be recycled, preferably into a product that can be recycled again. Benefits of recycling waste plastics:

■ Less plastic waste going to landfill
■ Less oil used for plastic production
■ Less energy consumed

Where the origin of the material is industrial scrap, the recovered end product is generally referred to as *reprocessed* to distinguish it from *recycled* material, which is derived from genuine postuse products. There is also some recycling of plastics products which have undergone a full service life and have then been reclaimed for further use. Material of this type is called *postuse material*. It can arise from industrial, commercial, and domestic sources. Recent years have seen a growth in postconsumer plastics recycling.

Most of the plastics recycled are from the commercial and industrial sectors, with bottles being recovered from domestic sources. This pattern is because the main requirements for effective recycling of postuse plastics are

- Resource-efficient reclamation of the postuse products
- Facilities to sort and compact the reclaimed products
- End-use applications for the recycled plastics materials and these conditions are more easily met from commercial postuse waste

In addition, heavily contaminated plastic waste requires special washing and drying facilities.

According to the American Plastics Council (APC), more than 1800 U.S. businesses handle or reclaim postconsumer plastics. Plastics from MSW are usually collected from curbside recycling bins or drop-off sites. Then, they go to a material-recovery facility, where they are sorted either mechanically or manually from other recyclables. The resulting mixed plastics are sorted by plastic type, baled, and sent to a reclamation facility. At the reclamation facility, the scrap plastic is passed across a shaker screen to remove trash and dirt, and then washed and ground into small flakes. A flotation tank then further separates contaminants, based on their different densities. Flakes are then dried, melted, filtered, and formed into pellets. The pellets are shipped to product-manufacturing plants, where they are made into new plastic products.

In 1997, APC estimated that roughly one-half of all U.S. communities—nearly 19,400—collected plastics for recycling, primarily PET and HDPE bottles, such as soda bottles. Roughly 7400 communities collected plastics at the curb, and approximately 12,000 communities collected plastics through drop-off centers.

While overall recovery of plastics for recycling is relatively small—1.4 million tons, or 3.9 percent of plastics generated in 2003—recovery of some plastic containers has reached higher levels. PET soft drink bottles were recovered at a rate of 25 percent in 2003. Recovery of HDPE milk and water bottles was estimated at about 32 percent in 2003. Significant recovery of plastics from lead-acid battery casings and from some other containers also was reported in 1999.

Plastics are recycled for both economic and environmental reasons. Recycling and reuse of plastics have the obvious benefit of decreasing the amount of used plastics that end up in landfills. With increased plastics recycling, fewer natural resources need to be extracted to produce virgin plastic.

According to APC, plastics production accounts for 4 percent of U.S. energy consumption. Though they are derived from nonrenewable natural resources, plastics' adaptable characteristics often enable manufacturers to reduce the material used, energy consumed, and waste generated in making a variety of products.

31.4   Potential Technologies, Strategies, and Approach

In all of the processes used to convert plastic raw materials into end products there is an inevitable arising of some scrap material. This results from the start-up and shutdown periods of the processing machinery, from out of specification products, and from quality-control samples. Material of this type is termed *industrial scrap*. Almost all molding companies recycle their own plastics waste, or scrap, in house. The specialist plastics recycling industry also recycles material of this type, mostly skeletal waste from thermoforming and off-specification moldings.

Mechanical recycling is where the plastics, which soften on heating, are, reformed into molding granules to make new products. The process involves collection, sorting, baling then size reduction into flake (film and sheet) or granules, which may then need washing and drying. This is then recompounded with additives and/or more virgin raw material, extruded and chopped into pellets ready for reuse. Feedstock or chemical recycling is where the plastics are broken down into their chemical constituents usually by heat and pressure. Good design for recycling should:

- Allow for easy disassembly of a product into its component parts
- Avoid the use of too many different types of plastic

Markets for some recycled plastic resins, such as PET and HDPE, are stable in most geographical areas in the United States and are expanding in many others. Currently, both the capacity to process postconsumer plastics and the market demand for recovered plastic resin exceed the amount of postconsumer plastics recovered from the waste stream. The primary market for recycled PET bottles continues to be fiber for carpet and textiles, while the primary market for recycled HDPE is bottles, according to APC.

In the future, new end uses for recycled PET bottles might include coating for corrugated paper and other natural fibers to make waterproof products like shipping containers. PET can even be recycled into clothing, such as fleece jackets. Recovered HDPE can be manufactured into recycled-content landscape and garden products, such as lawn chairs and garden edging.

The plastic lumber industry also is beginning to expand. New American Society for Testing and Materials (ASTM) standards and test methods are paving the way for use of these materials in decks, marinas, and other structural applications for both residential and commercial properties.

Source reduction is the process of reducing the amount of waste that is generated. The plastics industry has successfully been able to reduce the amount of material needed to make packaging for consumer products. Plastic milk jugs weigh 30 percent less than they did 20 years ago. Plastic packaging is generally more lightweight than its alternatives, such as glass, paper, or metal.

31.5   Case Study

STARTEX produced multilayered plastic film sheeting and used a blow-film extrusion process to make plastic packaging for commercial, medical, food, construction, and agricultural products. It manufactured about 60 million pounds of packaging materials annually.

STARTEX was disposing of approximately 16 to 20 tons (two or more 40-yd dumpsters) of solid waste weekly. This waste consisted of polyethylene film scrap (extruded plastic waste), resin pellets, and cardboard (from Gaylord boxes and paper cores). As production increased, the waste volume increased to three 40-yd dumpsters weekly, or approximately 1150 tons annually. In addition, disposal costs were increasing by approximately $3 per ton annually.

STARTEX addressed its growing waste problem by establishing a waste reduction team. The team consisted of five to eight employees who worked to identify the types, amounts, and sources of solid waste generated, and to find ways to reduce this waste.

To start the waste reduction process, the waste reduction team held early morning dumpster dives, during which team members would sort and document the types and volumes of solid wastes being disposed of in dumpsters at STARTEX.

After collecting this initial data, the team developed a strategy for reducing solid wastes, which included

- Focus the first waste reduction efforts on a waste that could be greatly and quickly reduced to show results that would stimulate employee participation.
- Review and evaluate the process(es) generating a waste to determine how it can be modified to reduce or eliminate the waste.
- Explore possibilities for reusing or recycling wastes that cannot be reused.
- A critical part of implementing the waste reduction strategy was to provide training to employees on waste reduction techniques, and encouragement and motivation to participate in the overall effort.

The most successful waste reduction technique used at STARTEX was good housekeeping. By preventing spills and raw material contamination, waste generation was greatly reduced. The four primary wastes reduced at STARTEX and the techniques used to reduce or eliminate these wastes are detailed below.

Polyfilm scrap (extruded plastic waste from the manufacturing process) contributed to over 50 percent of the total waste generated at STARTEX. Approximately 8 tons of scrap was generated weekly. Because this scrap was the largest contributor to STARTEX's waste volume, it was the focus of the first waste reduction project. By showing quick and large reductions of this waste, the waste reduction team speculated that STARTEX employees would see the results of their efforts and be more likely to accept and contribute to the overall waste reduction process.

After comparing purchasing data with waste volume data, the waste reduction team found that only 80 to 90 percent of the virgin polysheeting fed into the extrusion equipment became product. After evaluating reuse options, the team determined that polyfilm scrap could be fed directly back into the extrusion process along with virgin material to produce lower-grade packaging (not medical or food packaging).

The waste reduction team documented procedures on how to reuse scrap in the extrusion process and provided training to employees.

By reusing polyfilm scrap, STARTEX reduced the amount of scrap it disposed of by 97 percent—from approximately 16,000 lb/week to 1000 lb every 3 weeks.

Transferring resin pellets to production areas often resulted in losses before the waste reduction efforts began. STARTEX's waste reduction team found that approximately 248,500 lb of resin pellets fell onto the floor while being loaded into process equipment, and were subsequently disposed of.

The team evaluated ways to prevent pellet loss and implemented the following changes. First, the conveyor systems were improved to convey the pellets directly into the processing area. Second, after it was found that up to 5 percent of all resin pellets used remained inside the cardboard Gaylord boxes (caught under the flaps), the boxes were replaced with steel tanks. Since steel tanks had no edges or flaps, pellets could be completely removed.

Next, the team determined that spillage, however minimal, would continue to occur. The alternative to disposal was to collect the pellets and send them off site to be washed and reprocessed at a cost that was significantly lower than the purchase cost of virgin pellets. These reprocessed pellets were then used for making lower-grade packaging.

To implement the changes in handling resin pellets, the waste reduction team documented procedures for cleaning up pellets and provided training to employees.

STARTEX continues to receive virgin pellets in 1000-lb quantity cardboard Gaylord boxes, which are then emptied into steel tanks. The boxes cannot be recycled, however, because they are contaminated with resin pellets that are difficult to completely remove. After exploring alternative disposal options, STARTEX found numerous companies that would purchase the boxes for reuse, if they are properly broken down. STARTEX received approximately $36,000 annually from the sale of its used boxes.

Plastic sheeting produced at STARTEX is shipped out on solid paper cores. Cores were purchased in 12-ft lengths and were cut to fit the various widths of plastic products. STARTEX generated approximately 5000 to 6000 lb of paper core waste every 3 weeks and was anxious to find a recycling market for this material. However, the waste reduction team was unable to find a recycler who would accept the cores. STARTEX staff contacted the supplier of the paper cores and found that it shredded and reused its core waste. Since the supplier was only a few miles away, STARTEX worked out an agreement with the supplier to take back the waste cores. One requirement of the supplier was that the cores be free of plastic waste. To help employees keep the cores clean and to prevent accidental contamination from other wastes, STARTEX purchased open-wire cages for a total of $8000, which were used exclusively to collect and store the waste cores.

In terms of results and cost savings, STARTEX had disposed of approximately 2500 tons of solid waste at a cost of $90,000 ($35 per ton). After implementing the techniques to reduce or reuse wastes highlighted above, it disposed of about 290 tons at a cost of $18,000 ($62 per ton). This was a saving of $72,000 annually (not including the increase in disposal costs per ton) and an 88 percent reduction in total waste volume.

In addition to reduced disposal costs, selling the used Gaylord boxes generated approximately $36,000 annually.

Since employees saw the results of their efforts, most were committed to STARTEX's continuous process improvement efforts to reduce waste. A monthly newsletter to employees included articles from the new material conservation team (formerly the waste reduction team) that highlighted overall waste reduction successes. The articles also showed amounts and disposal costs of wastes still produced, and encouraged employees to find solutions to waste problems.

Customers reacted very positively to the clean shop appearance that resulted from the waste reduction efforts and good housekeeping practices at STARTEX. Customers from the medical industry who were required to audit STARTEX's operation were particularly pleased with the clean results.

The most significant factors contributing to the waste reduction success at STARTEX were

- Forming a team, composed of employees from throughout the operation, to focus on waste reduction.
- Training employees.
- Communicating to employees through an in-house monthly newsletter providing updates on waste reduction efforts, recognition of company-wide accomplishments in reducing waste and encouragement to employees to be ever mindful about reducing wastes, cutting costs, and keeping a clean shop.

31.6   Additional Information

1 www.epa.gov/cpg/
2 www.epa.gov/epaoswer/non-hw/recycle/jtr/comm/plastic.htm
3 www.epa.gov/opptintr/epp/
4 www.mntap.umn.edu/A-ZWastes/90-WasteReductTeam.htm

PRIMARY AND FABRICATED

METAL INDUSTRIES

32.1   Industry Overview

NAICS code: all 33100s and 33200s

INDUSTRY SNAPSHOT

67,413 metal operations in the United States
2,065,244 employees
$386.4 billion in annual sales
8.2 tons of solid waste generation per employee
Major waste streams: metals

The U.S. primary metals industry includes 5000 companies with combined annual sales of about $140 billion. Large companies include Nucor and U.S. Steel (steel); Alcoa (aluminum); and Phelps Dodge (copper). The industry is highly concentrated; the 50 largest producers hold more than 90 percent of the raw steel market. Secondary production of products from raw steel and other metals is also concentrated.

Demand comes largely from the manufacturers of durable goods like motor vehicles, machinery, containers, and construction steel. The profitability of individual companies depends largely on efficient operations, because most products are commodities sold based on price. Big companies have large economies of scale in production. Accordingly, most producers of secondary products buy raw metal from the large producers. Small companies can compete by operating efficient local minimills or producing specialty products. The industry is highly automated; average annual revenue per worker is close to $300,000.

The industry includes manufacturers and processors of steel, iron, aluminum, copper, and specialty metals like titanium, molybdenum, and beryllium. Steel products account for about 50 percent of the market. Companies are involved in three major types of activities. *Primary processing* is the separation of metal from ores in a furnace to produce slabs or ingots of metal. *Secondary processing* involves mainly the rolling or drawing of metal slabs into sheets, plates, foil, bars, and wire. *Foundry operations* produce metal shapes by pouring molten metal into casts or molds. Some producers have fully integrated operations, from mining raw materials to manufacturing finished products, but most operate in just one type of activity.

Steel production first involves converting iron ore or scrap iron into molten steel. The ore-based process uses a blast or oxygen furnace in a blast mill, and the scrap-based process uses an electric arc furnace in a minimill. Next, molten steel is poured and solidified in a continuous caster to produce semifinished products, like steel slabs, billets, and blooms. These materials are put through a mechanical and heat-treatment process known as hot rolling, and some hot-rolled sheets are rolled again at lower temperatures (cold rolling) to form finished flat products such as plates, coils, or sheets, or long products such as wire, bars, rails, or beams. These products may then be coated with protective anticorrosion material.

The production of aluminum, copper, and other metals is similar. Metal is separated from an ore by melting it. Metal alloys are produced by adding various elements to the main metal. For example, 17 percent chrome and 8 percent nickel are added to iron to create stainless steel. The different properties and characteristics of metal are produced by altering the chemical composition and the different stages of the process, such as rolling, finishing, and heat treatment.

Primary production of metals requires large amounts of ore and large amounts of energy, so producers often locate near ore deposits (copper companies); coal fields; or sources of cheap electricity (aluminum companies). To ensure a supply of raw materials, many primary producers control their own ore deposits. Transporting the finished product is typically by rail. Producers can make thousands of different products because metals can be made in many different grades of hardness or other properties. A producer of castings and forgings, such as Citation, sells 20,000 products to 2000 customers.

The technology of making metals with desired physical and chemical properties is highly complex. Modern production technology allows better control of the process and is more energy-efficient, but is also expensive to install. Many modern plants are highly automated, partly to reduce the need for expensive labor. Computerized inventory systems are used to track thousands of products at multiple locations.

Industries in the primary metal manufacturing subsector smelt and/or refine ferrous and nonferrous metals from ore, pig, or scrap, using electrometallurgical and other metallurgical process techniques. Establishments in this subsector also manufacture metal alloys and superalloys by introducing other chemical elements to pure metals. The output of smelting and refining, usually in ingot form, is used in rolling, drawing, and extruding operations to make sheet, strip, bar, rod, or wire, and in molten form to make castings and other basic metal products.

Primary manufacturing of ferrous and nonferrous metals begins with ore or concentrate as the primary input. Establishments manufacturing primary metals from ore and/or concentrate remain classified in the primary smelting, primary refining, or iron and steel mill industries regardless of the form of their output. Establishments primarily engaged in secondary smelting and/or secondary refining recover ferrous and nonferrous metals from scrap and/or dross. The output of the secondary smelting and/or secondary refining industries is limited to shapes, such as ingot or billet, that will be further processed. Recovery of metals from scrap often occurs in establishments that are primarily engaged in activities such as rolling, drawing, extruding, or similar processes.

Excluded from the primary metal manufacturing subsector are establishments primarily engaged in manufacturing ferrous and nonferrous forgings (except ferrous forgings made in steel mills) and stampings. Although forging, stamping, and casting are all methods used to make metal shapes, forging and stamping do not use molten metals and are included in subsector 332, fabricated metal product manufacturing. Establishments primarily engaged in operating coke ovens are classified in industry 32419, other petroleum and coal products manufacturing.

Industries in the fabricated metal product manufacturing subsector transform metal into intermediate or end products, other than machinery, computers and electronics, and metal furniture or treating metals and metal formed products fabricated elsewhere. Important fabricated metal processes are forging, stamping, bending, forming, and machining, used to shape individual pieces of metal; and other processes, such as welding and assembling, used to join separate parts together. Establishments in this subsector may use one of these processes or a combination of these processes.

The North American Industry Classification System (NAICS) structure for this subsector distinguishes the forging and stamping processes in a single industry. The remaining industries in the subsector group establishments based on similar combinations of processes used to make products.

The manufacturing performed in the fabricated metal product manufacturing subsector begins with manufactured metal shapes. The establishments in this sector further fabricate the purchased metal shapes into a product. For instance, the spring and wire product manufacturing industry starts with wire and fabricates such items.

Within manufacturing there are other establishments that make the same products made by this subsector; only these establishments begin production further back in the production process. These establishments have a more integrated operation. For instance, one establishment may manufacture steel, draw it into wire, and make wire products in the same establishment. Such operations are classified in the primary metal manufacturing subsector.

32.2 Waste Management Goals and Opportunities

The majority of solid waste generated by this sector is metals. Table 32.1 displays the composition breakdown based on survey results.

Industries involved in the manufacture of metals generate a variety of wastes from numerous processes and activities. All the steps of the manufacturing process including

TABLE 32.1 METAL-MANUFACTURING INDUSTRY SOLID WASTE COMPOSITION (SURVEY RESULTS)

MATERIAL	COMPOSITION (%)	RECYCLING (%)
Metals	58 ± 18.4	99 ± 4.1
Ferrous metals	39 ± 14.0	98 ± 4.2
Nonferrous metals	19 ± 6.8	99 ± 4.1
Mixed plastics	9 ± 3.3	18 ± 8.3
Paper	8 ± 3.0	12 ± 5.5
Mixed office paper	7 ± 2.5	12 ± 5.5
Newspaper	1 ± 0.4	8 ± 3.7
OCC (cardboard)	8 ± 3.0	57 ± 18.6
Wood	7 ± 2.6	34 ± 13.9
Food waste	4 ± 1.5	0 ± 0.0
Chemical	3 ± 1.1	0 ± 0.0
Yard waste	1 ± 0.4	0 ± 0.0
Glass	1 ± 0.4	12 ± 5.5
Other	1 ± N/A	0 ± 0.0
Overall recycling level		67.6

machining operations, part cleaning and stripping, surface treatment, plating, and paint applications have the potential to produce liquid, hazardous, and nonhazardous wastes. The nonhazardous solid waste must be handled properly to avoid damage to the environment or to the public and worker health. The cost of disposing of these wastes in a safe manner can be high.

As shown in the table, the recycling rate for this sector is approximately 68 percent. As derived from the solid waste evaluation model discussed in Chap. 12, the equation that estimates the annual waste generation per year per employee for this sector can be calculated from the following:

Tons of solid waste generated per year = 8.21 × number of employees
− 7.71 × solid waste disposal cost per ton
− 214.93 if the company is ISO 140001 certified
+ 756.5

32.3 Economics

Pound for pound, metals generate the highest revenue for recyclable materials and usually offer a wide variety of potential processors in most areas. Table 32.2 provides a listing of the prices for metal on the recycling commodity markets for September 2008. This price point generally creates a low threshold to economically justify metal

TABLE 32.2 WORLD CARBON STEEL TRANSACTION PRICES					
WORLD STEEL PRICES	**HOT ROLLED STEEL COIL ($/TON)**	**HOT ROLLED STEEL PLATE ($/TON)**	**COLD ROLLED STEEL COIL ($/TON)**	**STEEL WIRE ROD ($/TON)**	**MEDIUM STEEL SECTIONS ($/TON)**
Apr 2007	617	788	698	577	798
May 2007	623	800	696	606	815
Jun 2007	611	800	686	602	812
Jul 2007	599	808	681	590	819
Aug 2007	603	814	686	594	825
Sep 2007	602	810	673	580	821
Oct 2007	611	826	680	584	844
Nov 2007	615	833	688	584	853
Dec 2007	630	837	705	598	859
Jan 2008	639	847	716	621	871
Feb 2008	699	887	772	687	905
Mar 2008	800	978	890	758	970
Apr 2008	915	1065	985	852	1042
May 2008	998	1160	1080	920	1105
Jun 2008	1073	1225	1144	1005	1184

All steel prices above are dollars/metric ton. Steel price information updated September 2008.

Source: MEPS Steel Prices On-line. To obtain current steel prices including forecasts, please visit www.meps.co.uk/world-price.htm.

MEPS STEEL PRODUCT PRICE LEVELS ACROSS 2007–2008. Current & historic pricing levels.

recycling versus other materials. For example, in June 2008, 1 ton of aluminum scrap was worth over $2800 on the recycling commodity market.

32.4 Constraints and Considerations

Below is a list of the key considerations in regards to recycling for metal manufactures

- *Unit value*—The market price for various metals
- *Availability*—The amount and frequency of the various metal material generated
- *Capacity*—The available space to store or stage recyclable materials
- *Disposal cost*—The cost per unit for disposal; this is the key economic value required to determine the break-even point versus recycling
- *Environmental impact*—The result on the environment if the metal material is not disposed of properly

■ *Purity*—Measures the contamination in the pure metal; most recycling vendors establish minimum contamination levels for acceptance

■ *Contamination*—Measures the amount or level of harmful materials in contact with the metals, such as hazardous cutting fluids or cleaning agents

■ *Ease of collection*—Rates the level of difficulty required for the facility to collect the metal materials for recycling

■ *Marketability*—Measures the ability to contract with vendors to perform recycling operations

■ *Sortability*—Measures the ability to sort the various metals for recycling

■ *Public support*—Measures the level of public awareness for recycling metals; the higher the level, the greater the impact of public opinion on the organization to recycle

■ *Legislative support*—Measures the amount of government support and assistance for recycling metals; includes government grant and support programs

32.5   Potential Technologies and Strategies

One of the most cost-effective strategies to minimize waste is to reduce the amount of scrap generated by reduction of off-specification products. This can be accomplished via statistical process control and data analysis.

Consider outsourcing of nonessential sheet metal operations. Carrier Corporation generates scrap metal (copper, aluminum, and steel) from the fabrication processes throughout the Carrier Plant. Fabrication includes the following processes: heavy presses, brake presses, fin presses, cutoff shears, uncoilers and straighteners, maintenance, and tool and die operations. Their current strategy is to continue with the outsourcing of nonessential sheet metal fabrication throughout the plant. Educate the hourly and salary workforce and explain to them the adverse factors of the waste that is being generated from the fabrication process. Educate all employees of the benefits from recycling the metal products we use and ask them to participate. Urge employees for suggestions to help minimize waste generation in their work areas and all other areas of the plant. Develop vendor certification to ensure purchased scrap contains only trace amounts of lead.

Loss prevention and housekeeping are two low-cost methods to reduce solid waste. Introducing inspections methods for incoming raw material will avoid the disposal of unusable or damaged material, hence reducing scrap. Ordering materials in reusable packaging will reduce the disposal of these containers and may reduce the need for cleaning incoming materials.

32.6   Implementation and Approach

The segregation of metal dust and scrap by type is the best method to increase the value of the metals for resale. When on-site recycling is not possible due to space or economic constraints, utilize a subcontractor—most will take these materials at no cost due to the relatively high resale value.

32.7   Case Study—Aluminum Recycling

The recycling of aluminum scrap has significant economic, energy, environmental, and resource-savings implications. Compared with the primary aluminum production, aluminum recycling has a great advantage due to lower production cost (Campbell, 1996; Henstock, 1996). In order to efficiently recycle metals the industry is faced by various issues which include scrap sampling, scrap purchasing, metal recovery (based on recoverable metal in scrap) and yield (based on total mass of scrap), production cost and hence profit margins, product quality, environmental issues, and regulation. The chemical composition of the molten aluminum product is controlled not only by the process operation, but also to a large extent by proper selection of charged aluminum scrap.

As the real metal content of the scrap remains unknown, metal yield becomes a crucial factor for the recycling of aluminum scrap. Unfortunately, due to aluminum's high reactivity, metal yield of aluminum is a function of numerous parameters such as surface area–to-volume ratio (due to oxidized surface), shape of the scrap, type of alloy, scrap history, contaminants (e.g., oxides, water, oil, and paint), and amount of required flux additives in the melting process. For example, any increased level of contamination on scrap reduces metal recovery due to reaction with aluminum, and further lowers the metal yield. Figure 32.1 shows a metal scrap pile commonly found in the United States.

Figure 32.1   Metal scrap.

32.8   Additional Information

1 Sibley, Scott and Butterman, William, "Metal Recycling in the US," *Resources, Conservation, and Recycling,* vol. 15, 1995, pp. 259–267.

ELECTRONICS, SEMICONDUCTORS, AND OTHER ELECTRICAL EQUIPMENT

33.1 Industry Overview

NAICS code: all 33400s

INDUSTRY SNAPSHOT

6499 electronic manufacturing operations in the United States
494,370 employees
$102.9 billion in annual sales
3.1 tons of solid waste generation per employee
Major waste streams: papers, plastics, and metals

Industries in the computer and electronic product manufacturing subsector group include establishments that manufacture computers, computer peripherals, communications equipment, and similar electronic products, and establishments that manufacture components for such products. The computer and electronic product manufacturing industries have been combined in the hierarchy of the North American Industry Classification System (NAICS) because of the economic significance they have attained. Their rapid growth suggests that they will become even more important to the economies of all three North American countries in the future, and in addition their manufacturing processes are fundamentally different from the manufacturing processes of other machinery and equipment. The design and use of integrated circuits and the application of highly specialized miniaturization technologies are common elements in the production technologies of the computer and electronic subsector. Convergence of technology motivates this NAICS subsector. Digitalization of sound recording, for example, causes both the medium (the compact disc) and the equipment to resemble the technologies for recording, storing, transmitting, and manipulating data. Communications technology and equipment have been converging with computer technology. When technologically related components are in the same sector, it makes it easier to adjust the classification for future changes, without needing to redefine its basic structure. The creation of the computer and electronic product manufacturing subsector will assist in delineating new and emerging industries because the activities that will serve as the probable sources of new industries, such as computer manufacturing and communications-equipment manufacturing, or computers and audio equipment, are brought together. As new activities emerge, they are less likely therefore, to cross the subsector boundaries of the classification.

Semiconductors are electrical devices that perform functions such as processing and display, power handling, and conversion between light and electrical energy. Semiconductor production processes involve over 100 proprietary solutions and 200 generic materials to achieve the final product. This industry generates significant quantities of spent solvents and liquids containing metals. The economic and liability incentives caused by increasing disposal costs and more stringent regulations are resulting in efforts to minimize such wastes at the source.

Semiconductors are unique substances, which, under different conditions, can act as either conductors or insulators of electricity. Semiconductor processors turn one of these substances—silicon—into microchips, also known as integrated circuits. These microchips contain millions of tiny electronic components and are used in a wide range of products, from personal computers and cellular telephones to airplanes and missile-guidance systems.

To manufacture microchips, semiconductor processors start with cylinders of silicon called ingots. First, the ingots are sliced into thin wafers. Using automated equipment, workers or robots polish the wafers, imprint precise microscopic patterns of the circuitry onto them using photolithography, etch out patterns with acids, and replace the patterns with conductors, such as aluminum or copper. The wafers then receive a chemical bath to make them smooth, and the imprint process begins again on a new layer with the next pattern. A complex chip may contain more than 20 layers of circuitry. Once the process is complete, wafers are then cut into individual chips, which are enclosed in a casing and shipped to equipment manufacturers.

The manufacturing and slicing of wafers to create semiconductors takes place in clean rooms—production areas that are kept free of all airborne matter because the circuitry on a chip is so small that even microscopic particles can make it unusable. All semiconductor processors working in clean rooms must wear special lightweight outer garments known as "bunny suits." These garments fit over clothing to prevent lint and other particles from contaminating the clean room.

There are two types of semiconductor processors: operators and technicians. Operators start and monitor the equipment that performs the various production tasks. They spend the majority of their time at computer terminals, monitoring the operation of equipment to ensure that each of the tasks in the production of the wafer is performed correctly. Operators may also transfer wafer carriers from one station to the next, though the lifting of heavy wafer carriers is done by robots in most new fabricating plants.

Technicians are generally more experienced workers who troubleshoot production problems and make equipment adjustments and repairs. They take the lead in assuring quality control and in maintaining equipment. They also test completed chips to make sure they work properly. To keep equipment repairs to a minimum, technicians perform diagnostic analyses and run computations. For example, technicians may determine if a flaw in a chip is due to contamination and peculiar to that wafer, or if the flaw is inherent in the manufacturing process.

33.2   Waste Management Goals and Opportunities

The majority of solid waste generated by this sector is paper, plastics, and metals. Table 33.1 displays the composition breakdown based on survey results.

TABLE 33.1 ELECTRONIC- AND SEMICONDUCTOR-MANUFACTURING INDUSTRY SOLID WASTE COMPOSITION (SURVEY RESULTS)		
MATERIAL	COMPOSITION (%)	RECYCLING (%)
Paper	28 ± 2.6	16 ± 5.4
Mixed office paper	23 ± 2.4	17 ± 5.8
Newspaper	3 ± 0.4	6 ± 2.0
Paper (other)	2 ± 0.2	0 ± 0.0
Plastics	21 ± 3.5	41 ± 15.2
LDPE	7 ± 2.2	42 ± 15.2
PP	4 ± 1.2	39 ± 15.6
PS	4 ± 1.2	38 ± 14.9
HDPE	3 ± 0.9	42 ± 14.2
PET	3 ± 0.9	41 ± 17.2
Metals	17 ± 4.1	99 ± 13.1
Ferrous metals	11 ± 3.9	99 ± 13.1
Nonferrous metals	4 ± 2.1	99 ± 13.1
Aluminum cans	2 ± 0.8	41 ± 11.0
OCC (cardboard)	8 ± 1.4	71 ± 24.1
Wood	7 ± 1.3	47 ± 16.0
Food waste	6 ± 1.7	0 ± 0.0
Glass	3 ± 0.9	7 ± 2.4
Chemical	1 ± 0.8	14 ± 4.8
Yard waste	1 ± 0.4	0 ± 0.0
Other	8 ± N/A	0 ± 0.0
Overall recycling level		39.2

As shown in the table, the recycling rate for this sector is approximately 39 percent. As derived from the solid waste evaluation model discussed in Chap. 12, the equation that estimates the annual waste generation per year per employee for this sector can be calculated from the following:

Tons of solid waste generated per year = 2.29 × number of employees − 0.83
$$\times \text{ solid waste disposal cost per ton}$$
$$- 56.31 \text{ if the company is ISO 140001 certified}$$
$$+ 126.6$$

33.3   Constraints and Considerations

Due to their hazardous material contents, electronic and electric equipment waste may cause environmental problems during the waste management phase if it is not properly pretreated. Many countries have drafted legislation to improve the reuse, recycling, and other forms of recovery of such wastes so as to reduce disposal. The waste streams generated from electric and electronic manufacturing, being a mixture of various materials, can be regarded as a resource of metals, such as copper, aluminum, and gold. Effective separation of these materials based on the differences on their physical characteristics is the key for developing a mechanical recycling system. Therefore, an in-depth characterization of this specific material stream is imperative. Major hazardous components in waste electric and electronic equipments and their description:

- *Batteries*—Heavy metals such as lead, mercury, and cadmium are present in batteries.
- *Cathode ray tubes (CRTs)*—Lead in the cone glass and fluorescent coating cover the inside of panel glass.
- *Mercury-containing components, such as switches*—Mercury is used in thermostats, sensors, relays, and switches (e.g., on printed circuit boards and in measuring equipment and discharge lamps); it is also used in medical equipment, data transmission, telecommunication, and mobile phones.
- *Asbestos waste*—Asbestos waste has to be treated selectively.
- *Toner cartridges, liquid and pasty, as well as color toner*—Toner and toner cartridges have to be removed from any separately collected waste electrical and electronic equipment (WEEE).
- *Printed circuit boards*—In printed circuit boards, cadmium occurs in certain components, such as surface mounted device (SMD), chip resistors, infrared detectors, and semiconductors.
- *Polychlorinated biphenyl (PCB)–containing capacitors*—PCB-containing capacitors have to be removed for safe destruction.
- *Liquid crystal displays (LCDs)*—LCDs of a surface greater them 100 cm^2 have to be removed from WEEE.

■ *Plastics containing halogenated flame retardants*—During incineration/combustion of the plastics halogenated flame retardants can produce toxic components.

■ *Equipment containing Chlorofluorocarbons (CFCs)*—Present in the foam and the refrigerating circuit must be properly extracted and destroyed.

■ *Gas discharge lamps*—Mercury has to be removed.

The manufacturing process for semiconductor devices is a complex operation that involves physical and chemical processes such as: oxidation, photolithography, etching, doping, and metallization. These processes generated solid wastes, usually in the form of contaminated metals.

33.4   Potential Technologies and Strategies

The State of California has been a large supporter of environmental efforts in the semiconductor field. California's Department of Toxic Substance Control (DTSC); the Office of Pollution Prevention and Technology Development (OPPTD); and the Semiconductor Environmental, Safety, and Health Association (SESHA) co-sponsored a mini-conference bringing DTSC and semiconductor facilities together to share pollution-prevention strategies and provide the industry with the latest regulatory updates.

The conference provided companies with opportunities to further reduce hazardous waste generation by sharing various waste minimization approaches. Conference topics included

■ *Industrial Ecology: Promises and Challenges*—Waste of one facility can be a raw material for another. A Kalundborg, Sweden case study was presented, noting how five industrial businesses collaborated for mutual economic and environmental benefit.

■ *Waste Minimization: Is it Cost Effective during Decommissioning and Decontamination?*—Facilities should identify the tasks and decisions to be made at each step of the decommissioning process to identify savings from waste minimization.

■ *Chemical Management Services in the Silicon Valley*—Chemical management services provide a strategic, long-term benefit that contract with a service provider to supply and manage the customer's chemicals and related services. The provider's compensation is tied primarily to quantity and quality of services delivered, not chemical volume. These chemical services are often performed more effectively and at a lower cost than companies can do themselves. This approach provides an excellent source-reduction opportunity.

■ *Senate Bill 14 Update*—Source-reduction measures implemented by semiconductor facilities as reported in the latest Senate Bill 14 documents were discussed with the intent that other facilities could learn and apply the successful pollution-prevention measures that others had implemented.

- *Common Violations during CUPA Inspections*—This was a practical presentation focusing on the violations Certified Unified Program Agencies (CUPA) inspectors commonly find, and the appropriate avoidance techniques.
- *Universal Waste and Senate Bill 20*—This discussion about California's universal waste rule included what wastes were designated as universal wastes, and the standards for their management. Additionally, an update was provided on Senate Bill 20, which establishes a program for collection and recycling of covered electronic devices.
- *Pollution Prevention in the European Union (WEEE and Restriction of Hazardous Substances [RoHS] Directives)*—The European Union set various dates at which electronic companies must comply with the EU's directives on electronic- and electrical-equipment manufacturers to take back end-of-life equipment (WEEE), and restrictions of the use of certain hazardous substances in electrical or electronic components (RoHS).

In addition to these strategies, several other viable source-reduction options were identified at the conference. Waste minimization in this sector involves the use of processes, practices, or products that reduce or eliminate the generation of pollutants and wastes. The techniques that can be considered for waste minimization in the industry are

- *Product changes*—Product substitution, product conservation, and change in product composition
- *Input-material changes*—Material substitution and fewer contaminants in materials used
- *Technology changes*—Change in production processes and equipment, piping, and waste-separation
- *Operating practices and process changes*—Prevention of material or product losses, waste-stream segregation, production scheduling, and overflow controls
- *Production-process changes*—Changes in temperature, pressure, automation, equipment
- *Product reformulations*—Changes in design, composition, or specification of final product
- *Recycling and reuse*—Reuse in same or another process and usable by-product production
- *Administrative steps*—Inventory control, employee programs

The evaluation of source-reduction criteria includes

- Amount of waste reduced
- Technical feasibility
- Economic feasibility
- Effects on product quality

■ Employee health and safety
■ Regulatory compliance
■ Releases and discharges to other media

33.5   Implementation and Approach

The implementation of waste minimization in the electronics and semiconductor field is a four-step process based on the hierarchy of solid waste management:

1 Minimize the generation of wastes as much as feasible.
2 Minimize the hazard of the waste generated as much as feasible.
3 Manage the waste on-site where possible (treatment).
4 Select the off-site disposal technology which reduces the long-term liabilities as much as possible, first examining recycling options, followed by treatment, incineration, and finally landfill methods as a last resort.

Hazardous wastes are one of the major waste streams generated in this sector. It is important to ensure that the wastes are handled and disposed of in accordance with all laws and regulations and that off-site disposers are handling wastes properly and they present no unnecessary liabilities.

In terms of off-site disposal facilities it is very important that the facility has the ability, commitment, financial resources, and high level of compliance to handle a company's wastes. Steps to accomplish this include

■ Always audit a new facility before sending any wastes to them.
■ Periodically audit existing facilities—changes in ownership.
■ Use an audit check sheet.
■ There are services available that provide audit reports.
■ Comes down to a gut-check level of comfort.
■ Sometimes Toxic Substance Disposal Facilities (TSDF) facilities are not readily available in local areas or have limited capabilities.
■ Don't spread out the liability too much. Try to limit to only a few good facilities with multiple capabilities.
■ Superfund, Comprehensive Environmental Response, Compensation, and Liability Act (CERCLA), clean-up costs, liability—good key words when discussing options with management.

In regards to source reduction Senate Bill 14 (The Hazardous Waste Source Reduction and Management Review Act of 1989) requires a source-reduction evaluation review and plan every 4 years. This focuses on wastes that represent 5 percent or more of the total, including aqueous wastes and manifested wastes. The process requires a hazardous waste management performance report covering the previous 4 years as well and it must be made available to the public on request.

33.6   Case Study

The operations of high-tech companies are facing severe challenges from rapid changing of outside environment. The corporate vision should include the environmental aspects and policies to ensure continuous growth of the operation. Besides, in order to realize business globalization, it is required not only to develop high-quality products, but also to implement environment management for the operation. Especially for most of semiconductor companies, a strong environmental protection policy is inevitable as a key factor for success of the manufacturing operation.

In 2003, a solid waste minimization study was conducted at a large semiconductor manufacturing facility in California. The global company employs over 3500 individuals and generated over $24 billion in 2007. At the time of the audit, the company was pursuing new product and technology development as well as promotion and implementation of a waste minimization project. Figure 33.1 shows a semiconductor wafer manufactured at the facility.

The largest waste minimization efforts in the semiconductor industry are in source reduction. An effective source-reduction method is the use of acid spray processors for cleaning and etching wafers instead of the standard batch processing in an acid bath. Acid spray processing can reduce chemical consumption by using the cleaning chemical in an on-demand mode. In batch processing, once the chemicals are mixed they have a specific lifetime after which the cleaning ability of the solution is substantially

Figure 33.1   Semiconductor wafer.

reduced (Briones, 2006). The advantage of using cleaning chemicals only when they are needed, as in spray processors, is that it eliminates the need to replenish spent chemicals that result from batch processing.

Another substantial reduction of process chemicals can be accomplished through the use of gas- and vapor-phase processing, as compared to liquid-phase processing in the wafer-cleaning sequence. An advantage of cleaning with chemicals in a gas phase is that large volumes of wafers can be processed relatively quickly with smaller volumes of chemicals. For example, one available technique can clean 90,000 wafers using a 2.2-kg (5-lb) bottle of anhydrous hydrogen fluoride [HF] gas (Hara and Peterson, 1988). This compares to a conservative estimate of 9360 L of 12:1 HF deionized water mixture to clean the same number of wafers.

The waste minimization process began by establishing a task team and conducting a solid waste audit. Results from the audit indicated that the 310 employee facility generated 850 tons of solid waste per year and paid \$50,000 to dispose of solid waste. As a result of the analysis, the total cost reduction and additional revenue generation was about \$35,000 per year. Finally, industrial waste minimization activities will be combined with an environmental management system (EMS) for implementing the environment policies of the company for continuous improvement. The following examples document source reduction in the semiconductor industry for this case study.

33.6.1 EXAMPLE 1

The company involved in this study concentrated on waste elimination through changes in product design and manufacturing technology (Briones, 2006). At this plant, during a period when manufacturing output increased by 10 percent, chromic acid use dropped by 96 percent due to manufacturing-technology changes. The reduction was accomplished by developing a copper nitrate–based etchant that contained no chromium and generated no hazardous waste (Briones, 2006). This etchant has a long lifetime, no sludge buildup with repeated use, no unusual gases generated, and is slow to heat under normal usage. As a result, the copper-based etchant was substituted for chromium-based etchants on all but one silicon product. The chromium-treatment sludge at the facility was reduced drastically. Between 1988 and 1991, the plant reduced the amount of chromium sludge generated by 81 percent. The chromium content of the sludge also was reduced from 47,000 ppm (parts per million) by weight to 6200 ppm. A number of other waste minimization activities were instituted which saved the plant more than \$380,000 annually (Briones, 2006).

33.6.2 EXAMPLE 2

This study involved the use of an automated photoresist dispensing system (Briones, 2006). Photoresist is one of the most expensive chemicals used in the semiconductor-manufacturing process. Typical costs range from \$53 to \$106 per liter (\$200–\$400 per gallon). Photoresist waste is not one of the larger waste streams, but the cost of the chemicals motivated this facility to install an automated dispensing system. Prior to automation, the 1-gallon bottles storing the photoresist chemicals were visually monitored and replaced by the operators. Typically, operators replaced used bottles containing approximately 5 cm (2 in) of photoresist in the bottom to prevent the system from running dry. Eventually the waste from each bottle was mixed and disposed of off-site. The time and labor required for monitoring, the high cost of new chemical, and the disposal costs involved with the waste made these activities of concern.

An automated system was installed to provide maximum usage of photoresist from the supply bottles and continuous operation with minimal operator assistance. The automated system reduced photoresist waste by 50 percent.

33.6.3 EXAMPLE 3

A California semiconductor manufacturer studied approaches to reduce solvent wastes (Briones, 2006). The conclusions indicated that solvent segregation, solvent substitution, direct solvent reuse, distillation, and evaporation could accomplish a 70 percent solvent reduction.

Segregation of chlorinated solvents from nonchlorinated solvents was recommended to facilitate waste reduction by increasing recyclability. Minimizing water mixing, separating chlorinated from nonchlorinated solvent wastes, separating aliphatic from aromatic solvent wastes, and separating Freon from methylene chloride facilitated recycling.

33.7   Additional Information

1 www.dtsc.ca.gov/PollutionPrevention/P2_Semiconductor-Ind-Conf.cfm.
2 Cui, Jirang and Forssberg, Eric, "Mechanical Recycling of Waste Electric and Electronic Equipment: A Review," *Journal of Hazardous Materials B 99,* 2003, pp 243-263.

34

INDUSTRIAL AND COMMERCIAL MACHINERY INCLUDING AUTOMOTIVE MANUFACTURING

34.1 Industry Overview

NAICS code: all 33300s

INDUSTRY SNAPSHOT

28,306 machinery manufacturing operations in the United States
1,172,889 employees
$252.5 billion in annual sales
3.0 tons of solid waste generation per employee
Major waste streams: paper, plastic, wood, and metals

Industries in the machinery manufacturing subsector create end products that apply mechanical force, for example, the application of gears and levers, to perform work. Some important processes for the manufacture of machinery like forging, stamping, bending, forming, and machining are used to shape individual pieces of metal. Processes, such as welding and assembling are used to join separate parts together. Although these processes are similar to those used in metal-fabricating establishments, machinery manufacturing is different because it typically employs multiple metal-forming processes in manufacturing the various parts of the machine. Moreover, complex assembly operations are an inherent part of the production process.

In general, design considerations are very important in machinery production. Establishments specialize in making machinery designed for particular applications. Thus, design is considered to be part of the production process for the purpose of implementing the North American Industry Classification System (NAICS). The NAICS structure reflects this by defining industries and industry groups that make machinery for different applications. A broad distinction exists between machinery that is generally used in a variety of industrial applications (i.e., general-purpose machinery) and machinery that is designed to be used in a particular industry (i.e., special-purpose machinery). Three industry groups consist of special-purpose machinery—agricultural, construction, and mining-machinery manufacturing; industrial-machinery manufacturing; and commercial- and service industry–machinery manufacturing. The other industry groups make general-purpose machinery: ventilation, heating, air-conditioning, and commercial refrigeration–equipment manufacturing; metalworking-machinery manufacturing; engine, turbine, and power transmission–equipment manufacturing; and other general-purpose machinery manufacturing.

The global automotive industry is a highly diversified sector that comprises manufacturers, suppliers, dealers, retailers, original equipment manufacturers, aftermarket parts manufacturers, automotive engineers, motor mechanics, auto electricians, spray painters or body repairers, fuel producers, environmental and transport safety groups, and trade unions. The United States, Japan, China, Germany, and South Korea are the top five automobile-manufacturing nations throughout the world. The United States is the world's largest producer and consumer of motor vehicles and automobiles accounting for 6.6 million direct and spin-off jobs and represents nearly 10 percent of the $10 trillion U.S. economy. The automobile is one of the most important industries in the world, which provides employment to 25 million people in the world.

The automobile and automotive-parts manufacturers constitute a major chunk of the automotive industry throughout the world. The automotive-manufacturing sector consists of automobile and light-truck manufacturers, motor vehicle–body manufacturers, and motor vehicle–parts and supplies manufacturers. This sector is engaged in manufacturing of automotive and light-duty motor vehicles, motor vehicle bodies, chassis, cabs, trucks, automobile and utility trailers, buses, military vehicles, and motor vehicle gasoline engines.

The worldwide automobile industry is largely dominated by five leading automobile-manufacturing corporations, namely, Toyota, General Motors, Ford Motor Company, Volkswagen AG, and Daimler Chrysler. These corporations have their presence in almost every country and they continue to invest into production facilities in emerging markets, namely, Latin America, Middle East, Eastern Europe, China, Malaysia, and other markets in Southeast Asia with the main aim of reducing their production costs.

The automotive industry occupies a leading position in the global economy, accounting for 9.5 percent of world merchandise trade and 12.9 percent of world export of manufacturers. This industry manufactures self-powered vehicles, including passenger cars, motorcycles, buses, trucks, farm equipment, other commercial vehicles, and automotive components and parts.

The United States is the major revenue source for the global automotive industry with a market share of $432.1 billion, occupying 37.2 percent of the world's marketplace. In the year 2006, America sold around 16.5 million vehicles. The automobile industry's revenue was $85.6 billion in the year 2006. There were 200 countries that conducted foreign trade with the United States in 2006, exactly the same as year 2005. The leading countries that did trade with the United States in 2006 were: Canada, $49,844,026,000 (28.67 percent); Japan, $45,906,127,000 (26.40 percent); Germany, $25,749,794,000 (14.81 percent); Mexico, $17,745,627,000 (10.21 percent); and Korea, $9,525,028,000 (5.48 percent).

Northeastern United States and Southern Great Lakes Region, Northwestern Europe, Western Russia and the Ukraine, and Japan are the major manufacturing regions of automotive products in the world.

In North America, the prominent automotive manufacturing regions are New England, New York and the Mid-Atlantic, Central New York, Pittsburgh/Cleveland, Western Great Lakes, St. Lawrence Valley, Ohio and Eastern Indiana, Kanawha and middle Ohio Valley, St. Louis, the Southeastern region, Gulf Coast, Central Florida, and the West Coast.

The European Union has the largest automotive-production regions in the world. The key automobile-manufacturing regions are United Kingdom, Rhine-Ruhr River Valley, Upper Rhine–Alsace-Lorraine region, and the Po Valley in Italy.

In the Western Russian and Ukraine region, the leading industrial regions are Moscow, the Ukraine region, the Volga region, the Urals regions, and the Kuznetsh Basin region.

The highlighting features of the global automotive industry are

- It offers support to other industries such as iron, steel, rubber, glass, plastic, petroleum, textiles, oil and gas, paints and coatings, transportation industries.
- Rising foreign investments have led to the rapid growth in terms of automobile production and exports. Overseas companies are making huge investments and are installing extensive production capacities in developing countries.
- Continuous investment in research and development has resulted in increased productivity and better quality automobiles and automotive accessories and parts.
- Increases in standards of living and purchasing-power parity have resulted in the increased demand for automobiles especially four-wheelers in developing nations, mostly in the South Asian region.

This sector provides employment to a major chunk of the human population in the world, that is, 25 million. This industry not only provides millions of jobs to the people, but also produces billions of dollars in terms of worldwide revenues. Adequate infrastructural facilities in the form of power supply, machinery, capital, and ready availability of raw materials and labor help in the tremendous growth of this industry.

The automotive industry is witnessing tremendous and unprecedented changes these days. This industry is slowly and gradually shifting toward Asian countries, mainly because of saturation of the automobile industry in the Western world. The principal driving markets for the Asian automotive industry are China, India, and Association of Southeast Asian Nations (ASEAN) nations. Low-cost vehicles, namely, scooters, motorcycles, mopeds, and bicycles have led to the massive growth of some of the fastest-developing economies like China and India. The future of the automotive industry in the Asian countries such as Thailand, Philippines, Indonesia, and Malaysia is bright and promising because of the ASEAN free-trade area under which the export tariffs are very low.

On a global scale, the assets of the top 10 automotive corporations accounts for 28 percent of the assets of the world's top 50 companies, 29 percent of their employment, and 30 percent of their total sales. In the year 2006, the United States sold around 16 million new automobiles; Western Europe sold around 15 million, while China and India sold 4 million and 1 million, respectively. Latin America, Middle East, Eastern Europe, China, Malaysia, and other South-Asian nations are now emerging as the dominant markets of the automotive industry. Most of the major automotive players are shifting their production facilities in these emerging markets with the main purpose of gaining better access and reduction in their production costs. There is an estimation that the automotive markets in South America and Asia will witness a boom in the near future. The various factors such as cheap financing and price discounts, rising income levels, and infrastructure developments will assist in the growth and development of the automotive sector in the majority of Asian nations.

Due to rising pressures on cost and quality, computer-aided designing and computer-aided manufacturing tools are increasingly adopted by the automotive companies so as to save months of time in designing and improving the quality of automobiles. The other technologies being used by the automotive industry are rapid prototyping, virtual reality, on-board systems, global positioning systems, and display maps. Most of the automotive manufacturers are now resorting to environment-friendly–fuel vehicles like electric, fuel cell, and hybrid cars.

The future seems encouraging for this industry in terms of the expected surge in global demand and upsurge in investments. Several trends such as overcapacity in developed markets, globalization, technology advances, regulation and environmental consideration, and market fragmentation and product proliferation will result in the rapid growth of this sector.

34.2   Waste Management Goals and Opportunities

The majority of solid waste generated by this sector are paper and plastics. Table 34.1 displays the composition breakdown based on survey results.

As shown in the table, the recycling rate for this sector is approximately 30 percent. As derived from the solid waste evaluation model discussed in Chap. 12, the equation that estimates the annual waste generation per year per employee for this sector can be calculated from the following:

Tons of solid waste generated per year = 2.97 × number of employees − 1.73
$$\times \text{ solid waste disposal cost per ton}$$
$$- 56.22 \text{ if the company is ISO 140001 certified}$$
$$+ 412.8$$

TABLE 34.1   MACHINERY AND TRANSPORTATION-EQUIPMENT MANUFACTURING INDUSTRY SOLID WASTE COMPOSITION (SURVEY RESULTS)

MATERIAL	COMPOSITION (%)	RECYCLING (%)
Paper	23 ± 6.7	16 ± 4.3
Mixed office paper	21 ± 6.5	16 ± 4.3
Newspaper	2 ± 0.6	14 ± 3.8
Plastics	17 ± 4.9	11 ± 3.0
LDPE	6 ± 1.7	13 ± 3.5
HDPE	4 ± 1.1	12 ± 3.2
PET	3 ± 0.8	11 ± 3.0
PVC	2 ± 0.6	10 ± 2.7
PP	2 ± 0.6	11 ± 3.0
Wood	15 ± 4.4	42 ± 11.3
Metals	11 ± 3.2	81 ± 8.1
Ferrous metals	7 ± 1.7	80 ± 8.2
Nonferrous metals	2 ± 0.5	79 ± 8.0
Composite metals	2 ± 0.5	80 ± 12.0
OCC (cardboard)	10 ± 2.9	51 ± 13.8
Food Waste	7 ± 2.0	0 ± 0.0
Rubber (tires)	3 ± 0.9	45 ± 12.2
Chemical/oils	2 ± 0.6	91 ± 8.1
Glass	2 ± 0.6	42 ± 11.3
Fabrics/textiles	2 ± 0.6	4 ± 1.1
Yard waste	1 ± 0.3	0 ± 0.0
Other	7 ± N/A	0 ± 0.0
Overall recycling level		30.0

34.3   Economics

The economic downturn, higher gas prices, and cyclical patterns of the automotive market in the United States have negatively impacted the demand for automobiles. To compound the issue, the United States is shifting from large-vehicle production to smaller more fuel-efficient vehicles. Not only are U.S. auto manufacturers generating less revenue, they are investing in the equipment and processes to move toward smaller-vehicle production. This reduces and in some cases eliminates any free-capital funds to invest in waste minimization activities, such as equipment purchases.

34.4   Constraints and Considerations

The machinery and automotive-manufacturing sector faces several unique challenges and considerations. As mentioned in the previous section, the economy of this sector has taken a downturn, reducing potential funding for environmental projects. This creates a situation where only low-cost process changes may be justified and larger capital-investment projects may have to be put on hold.

A key consideration in this sector is public image. Competition can be fierce and the perception of a brand can significantly impact sales and customer loyalty. Many of these organizations are using advertising to demonstrate their levels of environmental awareness in terms of waste reduction and lower vehicle admissions. This also trends to the suppliers in this sector, creating pressure for these organizations to become green as well. This impacts the supply chain and purchasing decisions throughout the industry.

Another constraint that is gaining support from headquarter offices is to implement waste reduction or any other similar projects. Particularly in the automotive sector, the decision process to justify capital expenditures extends well beyond the facility. In some cases, even if a proposed project is justified economically it can be rejected at the headquarter level due to lack of capital funds or the lack of perceived integration into corporate strategy.

From a facility level, constraints and considerations include contamination issues and the use of available space to stage and store recyclables. Cutting fluids, oils, and paints are used frequently in machinery manufacturing. If not controlled properly, these fluids can contaminate normally recyclable materials, rendering them hazardous, for example, if lubricating oil from a crankshaft contaminates wood pallets. From a space standpoint, many manufacturing facilities are utilizing all available space, creating difficulties when an area must be identified to stage recyclables. This applies to both the interior and exterior of the building. Loading dock congestion and determining an outside space to stage a roll-off recycling container can be challenging.

34.5   Potential Technologies and Strategies

Several low-cost strategies are available to reduce or eliminate solid waste generation. Once the waste audit has been conducted and annual generation amounts and compositions have been determined, the hierarchy of solid waste reduction can be applied to reduce, reuse, and recycle components of the waste stream. Several very common waste minimization applications in the sector, in order of preference, include

- *Waste elimination*—Several process changes or packaging can be implemented to eliminate waste. This range from reducing packaging materials, scrap and defect rates, paperless faxes, and setting copier defaults to print on both sides of the paper.
- *Waste reuse*—Several relatively simple options exist to reuse materials that have been traditionally disposed of at landfills. Two very common examples are pallets

and component bindings for in-house transportation. Many organizations that currently utilize wood pallets have been very successful in implementing use of returnable plastic pallets. In terms of in-house bindings, many companies use rubber bands to secure subcomponents for transfer. Upon inspection and evaluation, many of these companies dispose of this as general waste. A simple process of collecting and reusing rubber bands saved one organization $3500 per year. Part washers may be required to clean various caps and plugs that could be reused.

- *Recycle*—If waste elimination or reduction options are not available, recycling of scrap, paper, plastic, and metal waste should be implemented, in most cases working with an outsourced supplier.
- *Energy recovery*—Finally, if recycling options are not available, energy recovery may be utilized to convert the waste into electricity at specialized sites. Internet searches will generate several organizations in local areas.

34.6 Implementation and Approach

To implement the proposed approaches listed earlier, several approaches are recommended. In unionized environments, buy-in and support from the local union is critical. This will ensure that employees support and understand the program. This will significantly increase the success rate of the program. An employee suggestion box and rewards program can also generate great ideas. Many companies have taken an approach where the originator of a process change is entitled to a percentage of the costs savings usually in the magnitude of 1 to 5 percent over the first year.

Second, many manufacturing organizations, specifically the automotive industry, have been successful in tracking waste generation by work unit or production unit. This may create healthy competition between employees and build teamwork. In addition, awards can be given to top performing units on a quarterly or annual basis.

Supply-chain modifications and "the greening" of suppliers may also generate excellent waste minimization results. This may be necessary when considering returnable containers and the reuse of certain items. In line with this, is the concept of zero landfill facilities. A zero landfill facility applies the aggressive philosophy of implementing waste minimization working toward the goal of zero landfill. This is accomplished through numerous organization-wide methods, including the focused use of composting and energy recovery.

34.7 Case Study

In 2002 The University of Toledo waste minimization team conducted a solid waste audit for an automobile-stamping plant in Northwest Ohio. The objectives of this project were to gain an overall understanding of the plant, identify all major solid waste streams, and then to suggest some areas for reducing, reusing, and recycling materials in the waste stream. At the time of the audit the facility had a small recycling program. The plant collected several materials including mixed office paper, cardboard, aluminum cans, metal-stamping scrap, plastic-stamping scrap, and oily wastewater. The facility was 790,000 ft^2 and operated 7 days per week. The annual waste hauling costs were $25,080, which included special wastes.

The waste materials currently recycled at the plant are shown in Table 34.2. The table also displays the total annual weight (in terms of pounds) of the currently

TABLE 34.2   TOTAL ANNUAL WASTE STREAM FOR AUTOMOTIVE CASE STUDY

MATERIAL	YEAR 2000 (TONS)	YEAR 2001 (TONS)	AVERAGE (TONS/YEAR)	AVERAGE (LB/YEAR)	TOTAL (%)
Metal-stamping scrap	46,836.66	50,954.51	48,895.58	97,791,165	90.769
Oily wastewater	5,294.51	4,432.70	4,863.61	9,727,210	9.029
Diesel and water*		37.21	37.21	74,420	0.069
Paper, uncontaminated	11.48	13.23	12.36	24,710	0.023
Steam booth sludge	6.14	17.88	12.01	24,020	0.022
Gasoline and water	12.01	4.92	8.47	16,930	0.016
OCC	7.51	5.53	6.52	13,040	0.012
Used oil	2.30	8.64	5.47	10,940	0.010
Parts washer	5.25	5.42	5.34	10,670	0.010
Pallets	4.58	5.48	5.03	10,060	0.009
Empty drums	3.66	5.46	4.56	9,120	0.008
Sump/dike water*		4.17	4.17	8,340	0.008
Waste paint	4.34	1.03	2.69	5,370	0.005
Ethylene glycol	0.69	4.59	2.64	5,280	0.005
Aerosols	0.81	0.70	0.76	1,510	0.001
Lamp ballasts	0.39	0.50	0.45	890	0.001
Surfactants	0.57	0.16	0.37	730	0.001
CFC (R-11) *		0.35	0.35	700	0.001
Alkaline/lithium	0.20	0.23	0.22	430	0.000
Lead acid batteries	0.01	0.33	0.17	340	0.000
HID lamps	0.20	0.10	0.15	300	0.000
Medical waste	0.14	0.11	0.13	250	0.000
Cylinder lab pack*		0.03	0.03	60	0.000
NiCd batteries*		0.02	0.02	40	0.000
Chlorinated solvent lab pack*		0.01	0.01	20	0.000
Total	52,191	55,503	53,868	107,736,545	100

*Data only available for 2001

recycled materials. The plant recycled over 107 million pounds of solid waste each year. Of the total, metal-stamping scrap comprises over 90 percent.

Table 34.3 displays the major solid wastes not currently recycled by the plant, segmented by the percentage of total waste (in pounds). As shown in the table, the plant disposed of an estimated 1,463,280 lb (732 tons) of solid waste at the landfill each year. The table also shows that general trash (including food waste), which is nonrecyclable, comprises about 52 percent of the total landfilled waste. Floor block (15 percent) and polypropylene (11 percent) were the next two largest nonrecycled waste components.

Table 34.4 displays the major annual waste streams at the plant that are potentially recyclable in terms of annual weight (pounds). This information is useful because recycling vendors usually pay for recyclables based on weight. If these materials can

TABLE 34.3   ANNUAL WASTE STREAM NONRECYCLED MATERIALS FOR AUTOMOTIVE CASE STUDY

MATERIAL	YEAR 2000 (TONS)	YEAR 2001 (TONS)	AVERAGE (TONS/YEAR)	AVERAGE (LB/YEAR)	TOTAL (%)
Trash (including food waste)	449.44	318.64	384.04	768,080	52.490
Floor block	155.96	66.39	111.18	222,350	15.195
Polypropylene	82.03	82.47	82.25	164,500	11.242
Yard trash*		59.26	59.26	118,520	8.100
Debris*		42.99	42.99	85,980	5.876
Diesel soil and absorbents*		28.05	28.05	56,100	3.834
Oily rags	23.96	3.76	13.86	27,720	1.894
Oily basement sump sludge*		7.00	7.00	14,000	0.957
Grease†	1.51		1.51	3,020	0.206
Oil filters	0.88	0.83	0.86	1,710	0.117
Crushed lamps	0.30	0.60	0.45	900	0.062
Other/special waste†	0.20		0.20	400	0.027
Total				1,463,280	100

*Handled by specialized waste hauler
†Hazardous waste

TABLE 34.4   REVENUE FROM THE SALE OF RECYCLABLES (AUTOMOTIVE CASE STUDY)

COMPONENT	POUNDS PER YEAR	CURRENT MARKET VALUE PER TON	REVENUE (FROM SALE OF RECYCLABLES)
Polypropylene	164,500	$100	$8,225
White ledger	9,884	$127	$6,276
Newspaper	4,942	$36	$89
Mixed office paper	9,884	$14	$69
Total	189,210		$14,659

be removed from the waste stream and sent to a recycler instead of the landfill, revenue could be generated and waste hauling costs could be reduced.

The waste assessment team found the national average price paid for polypropylene plastic scrap was $0.07 per pound. The waste minimization team used $0.05 per pound to be conservative. Local recyclers contacted by the waste assessment team would pay between $0.03 and $0.08 per pound depending on the type, grade, form, and contamination of the polypropylene plastic scrap.

Based on the 24,710 lb of commingled paper (white ledger, newspaper, and mixed office paper) collected each year and estimates from the on-site survey, a total of over $6400 could be generated from the sale of these materials. These materials are currently collected but are not separated. The plant received no revenue from these materials because they are commingled.

The audit indicated that the plant could potentially generate a total of $14,659 annually from the sale of recyclables if a local vendor could be found to purchase and recondition the material. The amounts paid per ton of recyclables were gathered from current market prices.

Additional cost savings could also be realized from a reduction in hauling costs of polypropylene plastic. Waste hauling charges can be reduced by approximately 10 percent by reducing the volume of polypropylene plastic waste taken by the hauler. The plant could potentially save approximately $2508 annually (calculated from a 10 percent reduction of the annual waste hauling cost of $25,080) by removing polypropylene plastic from the wastes. The cost savings will be achieved by reducing the frequency of pickup by waste management at the plant. Should a plastic-washing system be purchased, additional savings could be realized if the polypropylene plastic could be used in place of virgin material after the washing process.

The plant could potentially save a total of $17,167 annually ($14,659 from additional recycling and $2508 from waste hauling reduction). Specific information regarding procedures to achieve these cost reductions is in Table 34.5.

TABLE 34.5   RECOMMENDATIONS FOR AUTOMOTIVE WASTE MINIMIZATION CASE STUDY

ITEM	RECOMMENDATION
Improve separation of paper waste streams (white ledger and newspaper)	Improved separation of paper waste could generate approximately $6434 annually Please contact waste management for more details, once the improved recycling program has been implemented
Purchase a plastic-washing system to clean contaminated scrap polypropylene plastic Or Sell contaminated scrap polypropylene plastic to local recycler	As mentioned previously, the facility could potentially generate approximately $10,733 annually ($8225 from sale of polypropylene plastic and $2508 from waste hauling reduction). The feasibility of purchasing a plastic-washing system should be explored. The cleaned scrap plastic could also be used in place of virgin material for additional savings Local recyclers contacted by the waste assessment team would pay between $0.03 and $0.08 per pound depending on the type, grade, form, and contamination of the polypropylene plastic scrap
Improve labeling on recycling containers	Improved labeling will avoid mixing and contaminating of recycling containers
Use monthly in-plant newsletter to increase environmental awareness among employees	Use the monthly plant newsletter to promote recycling and increase awareness. Discuss issues regarding the recycling procedure at the plant, the current recycling rates, drop off locations, and recycling providers

34.8   Exemplary Performers and Industry Leaders—Zero Landfill Plants

Subaru is celebrating its 3-year anniversary of zero landfill status at its manufacturing plant, Subaru of Indiana Automotive (SIA) and the sale of its 100,000th PZEV (partial zero emissions) vehicle.

Both milestones reflect the long-standing commitment Subaru has to safeguard the environment that so many of its customers avidly enjoy.

"We are pleased to mark these milestones," said Tomohiko Ikeda, chairman, president, and CEO, Subaru of America, Inc. "At Subaru, we are committed to not only maintain an effective environmental management system, but also to integrate sound environmental practices throughout our business."

The Subaru plant in Indiana sends nothing to a landfill. That means, if this book was printed and not recycled, it would result in more trash sent to landfills than the entire Subaru manufacturing operations. The Subaru plant was the first U.S. auto plant to achieve zero landfill status, with all its waste being either reused, recycled, or converted to electricity for the city of Indianapolis.

The Subaru plant is also a designated backyard wildlife habitat. Deer, coyotes, beavers, blue herons, geese, rabbits, squirrels, meadowlarks, ducks, and other animals live there in peaceful coexistence with the Subaru plant.

Just last year, Subaru was awarded the U.S. EPA's Gold Achievement Award as a top achiever in the agency's WasteWise program to reduce waste and improve recycling.

Subaru also manufactures PZEV (partial zero emission vehicles) certified Legacy, Outback, and Forester models, available for sale anywhere in the United States. The PZEV-certified, normally aspirated, Subaru Legacy, Outback, and Forester models are SmartWay certified by the EPA. To date, Subaru has sold 100,000 Subaru PZEV vehicles, which have 90 percent cleaner emissions than the average new vehicle.

PZEV vehicles are the cleanest gasoline vehicles available today and they meet emissions standards that are sometimes even cleaner than some hybrid or alternative fuel vehicles. These vehicles have such tight pollution controls, with the burning of fuel so complete, that in very smoggy urban areas, exhaust out of the tailpipe can actually be cleaner than the air outside. For more information visit www.epa.gov.

35

TRANSPORTATION, LOGISTICAL,
AND WAREHOUSING APPLICATIONS

35.1 Industry Overview

NAICS code: all 48000s through 49000s

INDUSTRY SNAPSHOT

199,618 transportation and logistic operations in the United States
3,650,859 employees
$383.2 billion in annual sales
7.3 tons of solid waste generation per employee
Major waste streams: paper and wood

The transportation and warehousing sector includes industries providing transportation of passengers and cargo, warehousing and storage for goods, scenic and sight-seeing transportation, and support activities related to modes of transportation. Establishments in these industries use transportation equipment or transportation-related facilities as a productive asset. The type of equipment depends on the mode of transportation. The modes of transportation are air, rail, water, road, and pipeline.

Logistics services involve a range of related activities that encompass the process of planning, storing, and controlling the flow of goods, services, and related information from point of origin to point of consumption. Logistics services traditionally involve transportation services (i.e., air, land, and maritime); retailing, distribution, and warehousing services; and services involving information technology applications.

Currently, $3 trillion per year is spent globally on supply-chain logistics services, and that spending is increasing at a rate of 10 percent annually. Approximately 10 percent of U.S. GDP is related to transportation activity. In 2002, 19 billion tons of freight valued at $13 trillion was carried within the United States. When analyzing the U.S. logistics market, however, one must examine inefficiencies in logistics services resulting from deficiencies in U.S. infrastructure that could create broad economic disruptions. Figure 35.1 displays the variety of methods used to transport materials. Statistics and performance of specific logistics sectors are outlined below:

- *Air transportation*—Air transportation suffers from outdated infrastructure and analysts estimate that at the current level of infrastructure investment, the system will face critical operational failures by 2015. Currently, air transportation accounts for roughly 1 percent of U.S. GDP and is a significant GDP multiplier.
- *Maritime*—Limited port capacity (i.e., on-dock and intermodal infrastructure) is causing recurring congestion, and U.S. ports are projected to reach maximum capacity by 2010. The reliability of U.S. inland waterways is deteriorating due to outdated infrastructure. Movement of goods through more than 360 ports, 1000 harbor channels, and 25,000 miles of domestic waterways accounts for approximately $750 billion of U.S. GDP. Nearly 80 percent of all trade by volume enters the United States by sea.
- *Rail*—The current capacity of the railway system is constrained; nevertheless, forecasters predict a 55 percent increase in rail traffic by 2020. The majority of U.S. railroad infrastructure is privately owned and future necessary capital investments

Figure 35.1   **Transportation industry.**

are uncertain due to the high-capital-cost, low-capital-return nature of the industry. Railways provide a key link between ports and trucking and carry 16 percent of freight moved in the United States.

■ *Trucking*—Trucking capacity is tight due to a shortage of 20,000 drivers (which is expected to rise to 111,000 by 2014. Along with high fuel prices and hazardous materials transportation costs, the capacity shortages are driving up trucking prices. Trucking dominates the U.S. domestic freight industry, moving 75 percent of freight by value and two-thirds of freight by weight.

The impact of any one of the mentioned factors within the transport sector could exacerbate capacity pressures throughout the United States and drastically affect the market conditions in the sector, and to some significant extent, the entire U.S. economy.

The transportation and warehousing sector distinguishes three basic types of activities: subsectors for each mode of transportation, a subsector for warehousing and storage, and a subsector for establishments providing support activities for transportation. In addition, there are subsectors for establishments that provide passenger transportation for scenic and sightseeing purposes, postal services, and courier services.

A separate subsector for support activities is established in the sector because, first, support activities for transportation are inherently multimodal, such as freight-transportation arrangement, or have multimodal aspects. Secondly, there are production process similarities among the support activity industries.

One of the support activities identified in the support activity subsector is the routine repair and maintenance of transportation equipment (e.g., aircraft at an airport, railroad rolling stock at a railroad terminal, or ships at a harbor or port facility). Such establishments do not perform complete overhauling or rebuilding of transportation equipment (i.e., periodic restoration of transportation equipment to original design specifications) or transportation equipment conversion (i.e., major modification to systems). An establishment that primarily performs factory (or shipyard) overhauls, rebuilding, or conversions of aircraft, railroad rolling stock, or a ship is classified in subsector 336, transportation equipment manufacturing according to the type of equipment.

Many of the establishments in this sector often operate on networks, with physical facilities, labor forces, and equipment spread over an extensive geographic area.

Warehousing establishments in this sector are distinguished from merchant wholesaling in that the warehouse establishments do not sell the goods.

Excluded from this sector are establishments primarily engaged in providing travel agent services that support transportation and other establishments, such as hotels, businesses, and government agencies. These establishments are classified in sector 56, administrative and support, waste management, and remediation services. Also, establishments primarily engaged in providing rental and leasing of transportation equipment without operators are classified in subsector 532, rental and leasing services.

35.2 Waste Management Goals and Opportunities

The majority of solid waste generated by this sector are paper and wood. Table 35.1 displays the composition breakdown based on survey results.

As shown in the table, the recycling rate for this sector is approximately 37 percent. As derived from the solid waste evaluation model discussed in Chap. 12, the equation that estimates the annual waste generation per year per employee for this sector can be calculated from the following:

Tons of solid waste generated per year $= 7.32 \times$ number of employees $+ 153.2$

35.3 Economics

From an economic standpoint, rising fuel costs have had a signification impact on the logistics and transportation section. These additional costs have reduced available capital for improvement projects, including environmental initiatives. In such conditions,

TABLE 35.1 LOGISTICS, TRANSPORTATION, AND WAREHOUSING INDUSTRY SOLID WASTE COMPOSITION (SURVEY RESULTS)		
MATERIAL	**COMPOSITION (%)**	**RECYCLING (%)**
Paper	31 ± 9.0	21 ± 9.5
Mixed office paper	26 ± 7.3	22 ± 9.9
Newspaper	3 ± 0.8	12 ± 5.4
Magazines	2 ± 0.6	4 ± 1.8
Wood	14 ± 4.1	51 ± 18.2
Metals	12 ± 3.5	81 ± 8.1
Ferrous metals	8 ± 2.2	81 ± 8.2
Aluminum cans	4 ± 1.1	82 ± 8.1
OCC (cardboard)	10 ± 2.9	61 ± 17.5
Mixed plastics	8 ± 2.3	14 ± 6.3
Used oil	6 ± 1.7	95 ± 4.1
Food waste	4 ± 1.2	0 ± 0.0
Yard waste	3 ± 0.9	0 ± 0.0
Glass	2 ± 0.6	12 ± 5.4
Other	10 ± N/A	0 ± 0.0
Overall recycling level		36.5

processing and purchasing modifications may only be feasible under these budgetary conditions.

35.4   Constraints and Considerations

The key constraints and considerations in this sector are primarily concerned with the available space within facilities (including docks) and vehicles. Space limitations, especially during peak service periods may present challenges. To remedy this, recycling container pulls can be made before peak operating periods by contractors.

Another challenge is the transporting of recyclables to a central location if many satellite offices are used. The questions of when and how to accomplish these tasks may require additional coordination for scheduling various routes.

A key consideration in this sector, as in many others, is public image. Many of these organizations are using advertising to demonstrate their levels of environmental awareness in terms of waste reduction, lower vehicle admissions, and the use of green vehicles.

35.5   Potential Technologies, Strategies, and Implementation Approach

Several strategies and technologies exist for major components of the waste stream for this sector. Specifically, options to minimize and recycle tires, wood (pallets), and paper products. The process should begin with a formal waste audit to determine annual volumes and compositions for each material. Once these have been identified, the hierarchy of solid waste minimization can be applied, first with an emphasis on waste reduction, then reuse, and finally recycling. In terms of waste reduction, the use of paperless logs can significantly reduce waste. This can be accomplished with the use of an electronic device. Along these same lines, setting copier defaults to double sided and using paperless fax machines also reduces waste.

In terms of reuse, returnable containers generally have the largest impact on operations. Traditionally, many organizations use wood pallets that are scrapped upon use. By switching to reusable plastic pallets, the waste hauling bill will be reduced. In regards to recycling, used tires often creates disposal issues. One of the most common uses for recycled tires is as an alternative fuel source for certain industries. Concrete manufacturers, for example, must use kilns to dry their products before shipping. These massive kilns are large enough to accommodate whole recycled tires in their furnaces.

Other industries, such as steel and glass production, use shredded recycled tires to augment their usual coal or natural gas fuel sources. The recycled tires must be shredded in order to fit through the feeder grates of the furnaces. One drawback is the presence of steel belts in many tires, which can build up over time and block the feeder chutes.

In order to meet strict environmental guidelines, many landfills must provide a safe covering over each day's deliveries. Instead of using a layer of fill dirt, some landfill operations are now using a layer of shredded recycled tires as a daily cover. Instead of piling whole discarded tires in a hazardous tire pile, landfill operators can receive shredded recycled tires from a local recycling center or invest in their own tire-shredding machinery.

Recycled tires are also used as a cushioning material in playgrounds and other public areas popular with children. Sometimes, the shredded recycled tires are spread over the area like mulch, which can cushion the fall of a child or reduce the impact of playground equipment. Recently, the rubber from recycled tires has been combined with other binders and foam to produce a solid safety mat for playgrounds and schoolyards.

Even certain clothing manufacturers have discovered the benefits of using recycled tires. Material composed of recycled tires is now used to form the rubber sole of some athletic shoes and work boots. Recycled tires may one day be turned into other rubber-based clothing and accessories, such as raincoats, boots, umbrellas, and hats.

One recent use of recycled tires may become a trend in larger cities. Traditional concrete sidewalks can now be replaced with similar-sized panels constructed from recycled tires and other materials. Proponents of these new sidewalk panels claim they are more resistant to the damage caused by tree roots, and they provide more stability for pedestrians. While the current cost per panel is higher than traditional concrete forms, the new rubberized panels should require far less maintenance throughout their lifespan.

Since recycled tires contain oil and carbon black, two very useful substances, scientists are still seeking ways to retrieve these materials from discarded or recycled tires. If these researchers are successful in their quest, the huge piles of scrap tires we see today will ultimately become nothing more than a memory. Recycled tires may provide enough reclaimed oil to make them worth salvaging, instead of merely discarding.

35.6   Case Study

In 1999, The University of Toledo Solid Waste Minimization Team conducted a solid waste audit at a large worldwide shipping company located in Northwest Ohio. In 1998, the company's corporate revenues exceeded $2.3 billion dollars. The facility moved between 4 and 5 millions pounds of freight across its dock every day. This facility was a break-bulk facility with 205 dock doors. The terminal received all freight collected from 20 regional satellite terminals.

The waste assessment team has identified nine major waste streams produced by the dock operations at the facility. Of these nine streams, six of the streams are potentially recyclable. By separating the potentially recyclable material from the waste stream, the company could significantly reduce both the amount of material going to the landfill as well as disposal costs. The following sections provide alternatives for dealing with the recyclable materials not currently recycled. The nine major dock waste streams are mentioned in Table 35.2.

Of the major waste streams, wood pallets constituted the largest opportunity to increase recycling rates. The company generated over 420, 000 lb of wood from their operations on an annual basis. Figure 35.2 is a picture of these pallets. This wood waste results from damaged and unnecessary pallets as well as damaged or carelessly disposed of wood blocking, which is used to secure freight in trailer-loading operations. Wood is both reusable and recyclable and should be handled in such a way that it is beneficial to both the company and the environment. In order to reduce the frequency of disposal of the wood hopper, the company examined the possibility of a precrusher grinding unit to reduce the wasted volume in the disposal container. A cost analysis of this machine indicated that it would save the company $14,000 per year with a payback period of 1.3 years.

The plastic wrap used in the freight-loading and unloading process is potentially recyclable. The company disposed of approximately 33 tons of plastic wrap per year. By instituting a collection program for plastic material, the company would enjoy a twofold benefit. First, the plastic material would be removed from the disposal stream, therefore reducing compactor volumes, frequency of disposal, and hence, disposal costs. The second benefit involves the collection of revenue for the film plastic. By recycling the film plastic generated in the dock area, the company had the potential to generate revenues up to $1500.

Over 56,000 lb of corrugated cardboard were generated yearly. The corrugated cardboard serves the same function as the boxboard throughout dock operations,

TABLE 35.2 WASTE HANDLING OPTIONS FOR THE NINE MAJOR DOCK WASTE STREAMS

WASTE STREAM	AMOUNT GENERATED (LB/YEAR)	WASTE HANDLING OPTIONS
Boxboard	25,165	Recycling
Film plastic	57,752	Recycling
Metal banding	65,773	Recycling
OCC	57,909	Recycling
Paper	26,767	Recycling/source reduction
Paper cups	27,453	Source reduction
Tape	14,298	Source reduction
Wood	420,375	Recycling
Miscellaneous	22,306	Source reduction

Figure 35.2   Wood pallets.

acting as a trailer-packing aid as well as a repair mechanism for freight. As with boxboard, the OCC is highly recyclable and segregated collection of the material would divert over 14 tons of material from the landfill.

35.7   Exemplary Performers—UPS Solid Waste Reduction Initiatives

UPS has demonstrated a strong commitment to the environment. In 2006, UPS produced 104,550 tons of solid waste. Solid waste is paper, cardboard, and plastic that would generally be taken to a landfill. Only 38,600 tons were recycled. The reduction of solid waste reduces the amount of methane that contributes to the greenhouse gas effect. Figure 35.3 is a breakdown of the products that UPS recycled in 2006.

Steps followed by UPS to increase recycling and reduce solid waste generation

- To help reduce the greenhouse gas effect, UPS is increasing its purchases of recycled materials. They purchased 36,300 tons of recycled materials in 2006 to help this initiative.
- UPS recycled 2.7 million pounds of electronic equipment from their facilities in 2006 and 19.7 million pounds since the beginning of this project to help reduce the toxic chemical wastes created by this equipment.

KPI **2006 Solid Waste Recycling**
U.S. package and supply chain and freight

///. = 1,857 pallets & wood waste

■ = 102 plastics

= 9,190 metals

▓ = 480 office paper

■ = 25,184 corrugated containers

Figure 35.3   UPS recycling levels.

■ UPS is helping its customers recycle electronics through the Asset Recovery and Recycling Management Service. This program allows companies to give UPS the equipment for recycling or proper disposal.

■ All bags used to sort small packages purchased are reusable. UPS has purchased 5.42 million reusable bags, eliminating the need for 600 plastic bags per reusable bag. This has reduced 41,400 tons of plastic waste.

■ The packaging used by UPS has been revamped to eliminate bleached paper and to use recycled materials.

■ UPS recycled 18,525 lb of batteries in 2006 with its participation with the Rechargeable Battery Recycling Corporation.

■ Sixty percent of waste produced in the corporate offices is now being recycled.

HEALTH SYSTEM APPLICATIONS

36.1   Industry Overview

NAICS code: all 62000s

INDUSTRY SNAPSHOT

704,526 health-care operations in the United States
15,052,255 employees
$1,207.3 billion in annual sales
1.1 tons of solid waste generation per employee
Major waste streams: paper, biohazard, food waste, and plastics

---

Combining medical technology and the human touch, the health-care industry administers care around the clock, responding to the needs of millions of people—from newborns to the critically ill. About 580,000 establishments make up the health-care industry; they vary greatly in terms of size, staffing patterns, and organizational structures. Nearly 77 percent of health-care establishments are offices of physicians, dentists, or other health practitioners. Although hospitals constitute only 1 percent of all health-care establishments, they employ 35 percent of all workers.

The health-care industry consists of the following nine segments:

1 *Hospitals*—Hospitals provide complete medical care, ranging from diagnostic services, to surgery, to continuous nursing care. Some hospitals specialize in treatment of the mentally ill, cancer patients, or children. Hospital-based care may be provided on an inpatient (overnight) or outpatient basis. The mix of workers needed varies, depending on the size, geographic location, goals, philosophy, funding, organization, and management style of the institution. As hospitals work to improve efficiency, care continues to shift from an inpatient to outpatient basis whenever possible. Many hospitals have expanded into long-term and home health-care services, providing a wide range of care for the communities they serve.

2 *Nursing and residential care facilities*—Nursing care facilities provide inpatient nursing, rehabilitation, and health-related personal care to those who need continuous nursing care, but do not require hospital services. Nursing aides provide the vast majority of direct care. Other facilities, such as convalescent homes, help patients who need less assistance. Residential-care facilities provide around-the-clock social and personal care to children, the elderly, and others who have limited ability to care for themselves. Workers care for residents of assisted-living facilities, alcohol- and drug-rehabilitation centers, group homes, and halfway houses. Nursing and medical care, however, are not the main functions of establishments providing residential care, as they are in nursing-care facilities.

3 *Offices of physicians*—About 37 percent of all health-care establishments fall into this industry segment. Physicians and surgeons practice privately or in groups of practitioners who have the same or different specialties. Many physicians and surgeons prefer to join group practices because they afford backup coverage, reduce overhead expenses, and facilitate consultation with peers. Physicians and surgeons are increasingly working as salaried employees of group medical practices, clinics, or integrated health systems.

4 *Offices of dentists*—About one out of every five health-care establishments is a dentist's office. Most employ only a few workers, who provide preventative, cosmetic, or emergency care. Some offices specialize in a single field of dentistry such as orthodontics or periodontics.

5 *Home health-care services*—Skilled nursing or medical care is sometimes provided in the home, under a physician's supervision. Home health-care services are provided mainly to the elderly. The development of in-home medical technologies, substantial cost savings, and patients' preference for care in the home have helped change this once small segment of the industry into one of the fastest growing parts of the economy.

6 *Offices of other health practitioners*—This segment of the industry includes the offices of chiropractors, optometrists, podiatrists, occupational and physical therapists, psychologists, audiologists, speech-language pathologists, dietitians, and other health practitioners. Demand for the services of this segment is related to the ability of patients to pay, either directly or through health insurance. Hospitals and nursing facilities may contract out for these services. This segment also includes the offices of practitioners of alternative medicine, such as acupuncturists, homeopaths, hypnotherapists, and naturopaths.

7 *Outpatient-care centers*—The diverse establishments in this group include kidney dialysis centers, outpatient-mental-health and substance-abuse centers, health-maintenance organization medical centers, and freestanding ambulatory surgical and emergency centers.

8 *Other ambulatory health-care services*—This relatively small industry segment includes ambulance and helicopter transport services, blood and organ banks, and other ambulatory health-care services, such as pacemaker-monitoring services and smoking-cessation programs.

9 *Medical and diagnostic laboratories*—Medical and diagnostic laboratories provide analytic or diagnostic services to the medical profession or directly to patients following a physician's prescription. Workers may analyze blood, take x rays and computerized tomography scans, or perform other clinical tests. Medical and diagnostic laboratories provide the fewest number of jobs in the health-care industry.

In the rapidly changing health-care industry, technological advances have made many new procedures and methods of diagnosis and treatment possible. Clinical developments, such as infection control, less invasive surgical techniques, advances in reproductive technology, and gene therapy for cancer treatment, continue to increase the longevity and improve the quality of life of many Americans. Advances in medical technology also have improved the survival rates of trauma victims and the severely ill, who need extensive care from therapists and social workers as well as other support personnel.

In addition, advances in information technology continue to improve patient care and worker efficiency with devices such as hand-held computers that record notes about each patient. Information on vital signs and orders for tests are transferred electronically to a main database; this process eliminates the need for paper and reduces record-keeping errors.

Cost containment also is shaping the health-care industry, as shown by the growing emphasis on providing services on an outpatient, ambulatory basis; limiting unnecessary or low-priority services; and stressing preventive care, which reduces the potential cost of undiagnosed, untreated medical conditions. Enrollment in managed care programs—predominantly preferred provider organizations, health-maintenance organizations, and hybrid plans such as point-of-service programs—continues to grow. These prepaid plans provide comprehensive coverage to members and control health-insurance costs by emphasizing preventive care. Cost-effectiveness also is improved with the increased use of integrated delivery systems, which combine two or more segments of the industry to increase efficiency through the streamlining of functions, primarily financial and managerial. These changes will continue to reshape not only the nature of the health-care workforce, but also the manner in which health care is provided.

The health-care and social assistance sector comprises establishments providing health care and social assistance for individuals. The sector includes both health care and social assistance because it is sometimes difficult to distinguish between the boundaries of these two activities. The industries in this sector are arranged on a continuum starting with those establishments providing medical care exclusively, continuing with those providing health care and social assistance, and finally finishing with those providing only social assistance. The services provided by establishments in this sector are delivered by trained professionals. All industries in the sector share this commonality of process, namely, labor inputs of health practitioners or social workers with the requisite expertise. Many of the industries in the sector are defined based on the educational degree held by the practitioners included in the industry.

36.2   Waste Management Goals and Opportunities

The majority of solid waste generated by this sector is paper, biohazards, food, plastics, and cardboard. Table 36.1 displays the composition breakdown based on survey results.

As shown in the table, the recycling rate for this sector is approximately 19 percent. As derived from the solid waste evaluation model discussed in Chap. 12, the equation that estimates the annual waste generation per year per employee for this sector can be calculated from the following:

$$\text{Tons of solid waste generated per year} = 0.98 \times \text{number of employees} + 21.43$$

36.3   Constraints and Considerations

In terms of solid waste management and minimization for health-care facilities, the five key constraints and considerations are

1 *Available space to store and stage materials (interior and exterior)*—This includes space within the facility and dock space for storage. The numerous deliveries that

TABLE 36.1 HEALTH SYSTEMS SOLID WASTE COMPOSITION (SURVEY RESULTS)		
MATERIAL	**COMPOSITION (%)**	**RECYCLING (%)**
Paper	36 ± 8.6	36 ± 9.7
Mixed office paper	33 ± 8.6	33 ± 8.9
Newspaper	3 ± 0.8	45 ± 12.2
Biological/hazardous	22 ± 5.3	0 ± 0.0
Food waste	17 ± 4.1	0 ± 0.0
Mixed plastics	12 ± 2.9	14 ± 3.8
OCC (cardboard)	4 ± 1.0	71 ± 19.2
Metals	3 ± 0.7	24 ± 6.5
Yard waste	3 ± 0.7	0 ± 0.0
Glass	2 ± 0.5	12 ± 3.2
Wood	1 ± 0.4	0 ± 0.0
Overall recycling level		18.9

many of these facilities receive each day of the week can also present a logistic and scheduling problem for coordinating dock usage during peak periods.

2 *The generation of biohazard waste*—Biohazard waste presents infection and disease issues that must be addressed. These waste streams must be kept separate and clearly identified as required by law.

3 *The separation of waste and contamination*—Separating recyclable waste from nonrecyclable waste creates a new process; in conjunction with this, informing patients and visitors of the separation process also creates an additional management concern.

4 *Finding appropriate suppliers for waste removal and recycling.* Well-established and reliable waste removal and recycling providers are critical for a successful program; when meeting with prospective companies, request a list of references.

5 *Delivery systems for waste and recyclable materials*—Many older facilities utilize trash shoots on upper level floors to transport waste to the lower level for removal. Creating separate processes for recycling requires coordination of processes to minimize contamination and labor costs.

36.4 Potential Technologies and Strategies

Identifying wastes with potential to be recycled or reused is an important aspect of sustainable development. Regular monitoring of recycling data enables financial benefits to be assessed and also identifies other wastes that could be recycled or reused. The reuse of medical equipment can be high risk, particularly if the procedures applied are not strictly controlled.

Materials present in health-care waste that may be suitable for recycling include paper (newspapers, general mixed paper, office paper, and data security paper), glass (green, brown, and clear glass), plastic (PET and other plastics), metals (ferrous, aluminum, mixed, and silver), biodegradable food waste, biodegradable ground-maintenance waste, wood, bricks and concrete, soil, oils and greases, residual ashes, electrical and electronic waste, and textiles and printer cartridges. As shown in Table 36.1 the overall recycling rate for healthcare facilities is 18.9 percent. By using best practice 43 percent could be achieved, showing enormous potential for diverting waste away from disposal and using waste as a resource. It is also important to identify some specific wastes for reuse such as gas bottles, which if not reused and disposed of appropriately are potentially very dangerous. In the case of unsustainable health-care facilities, potentially recyclable materials are not separated. Unsuitable items such as needles and syringes may be sterilized using an autoclave process. This will eliminate the biohazard issues associated with the waste and allow the waste to be recycled or disposed of normally which in turn will generate cost savings. The Lab Depot, Inc. (www. labdepotinc.com) offers several sterilizers that may be used for these purposes. High-level facilities collect all potentially recyclable materials for reprocessing locally into new product.

36.5   Implementation and Approach

An important component of waste minimization in the health-care field is communication, not only internal, but with patients and visitors. Specifically, the separation process and location of clearly designated containers must be emphasized. Health-care facilities also present the challenge of available space. Designating dock space and scheduling deliveries becomes crucial. The clear separation of biohazard waste is also necessary and required by law. This must be clearly labeled and the containers must remain closed. Figure 36.1 displays commonly used medical waste separation containers. The World Health Organization (1991–1993) recommended the following colors codes for waste containers:

- *Black*—Standard refuse
- *Yellow*—Any kind of waste to be disposed of in incinerators
- *Yellow with black margin*—Waste for incineration though another type of disposal is permissible
- *Light blue or transparent*—Waste for autoclaving
- *Red or white with red border*—Infectious waste for steam sterilization
- *White*—Soiled linen

36.6   Case Study

More than just a full-service hospital, South Bay Medical Center (South Bay) has become a multiservice recycling center and buy-recycled innovator under the enthusiastic direction of Christine Vandoren, director of Materials Management in Central

Figure 36.1   Medical waste separation.

Supply. Known around Redondo Beach as "the recycling cheerleader," Christine's upbeat, can-do attitude is the prescription for a successful program. Few people could imagine what just one person, who is committed to recycling can do. Christine's story is an example worth following.

South Bay's recycling program was her idea. The catalyst was the birth of her first nephew. When her physician sister had her first baby, Christine's concern for the next generation hit home. She could see the hospital's trash bins from her office. Thanks to these visual and family reminders, Christine has helped South Bay reduce its trash pickup to once a week in favor of recycling trucks that now visit the facility twice as often.

"Recycling has to be a passion," said Christine. "It just won't happen if you have to force someone or tell them to do it." The hospital's motto is: "Planting the seeds of a healthy community." Located a mile from the beach, South Bay is expanding into beach cleanups and "heal the bay" programs, extending the huge array of integrated waste management practices on-site. Christine often advises other facilities and businesses in the community, bearing recycling bins to give away, or reusable lunch bags for the kids she turns onto recycling in their classrooms. "Look at your trash," she tells them. "Ninety percent of what's in there is recyclable."

Purchasing recycled products is an integral part of South Bay's full-range of environmentally oriented policies. They researched many products now used and asked their regular vendors to find them. South Bay purchases these recycled-content products [recycled (%)/postconsumer (%)]: napkins (100/80), sharps containers (100/3–25), copy paper (100/75), sterile wrap (100), patient tray covers (100/100), and pulse oximeter probes (100/80). Recycled-content products account for approximately 40 percent of their overall purchases.

What's new in recycled products? Sage, a company based in Chicago, makes the sharps containers purchased by South Bay. This product has a 30 percent recycled content, the maximum allowed by the FDA. (These containers are used to dispose of needles and greater recycled content might compromise the thickness and protection provided by the containers.) Both needles and containers used to be handled as infectious wastes; they were autoclaved, then landfilled. Now, a company called SteriCycle microwaves the sharps containers and contents, grinds and washes the containers, and then turns them back into plastic pellets. These are sold back to Sage for making new sharps containers. This is one of the few existing closed-loop recycling systems. Sage ultimately lowered the price of its recycled container to match those of a similar non-recycled product. According to Christine, another health-care company is spending millions of dollars in recycled-product development. "They're trying to find safe ways to make new products out of hospital trash and infectious waste."

Christine also reports that in another new development in recycled products for the hospital industry, BioNova now manufactures recycled-content examination table paper, capes, gowns, and drapes under the private label, General Medical, at a 22 percent savings over virgin products. All these medical paper products have 100 percent recycled content except the table paper, which has 20 percent.

Kimberly Clark has a recycling program for blue sterile wrap, made of cotton and plastic. Special trash containers are placed in South Bay, diverting 5200 lb per year from the landfill. Clark picks up the receptacles, then recycles the sterile wrap into blue pellets that become blue handicapped parking stops and lining for electricity and telephone poles that repel lightning.

The South Bay Medical Center recycling program began in April 1993 on Earth Day. Through the program, the hospital recycles all standard commodities and toner cartridges, sharps containers, sterile wraps, pulse oximeter probes, telephone directories, steel, and wood. They eliminated the waste of mattress overlays (a foam product) by purchasing a new type of patient mattress. South Bay uses voice- and e-mail and has a formal environmental policy, a free barbecue on Earth Day for employees to celebrate their successful program, and donates furniture or medical equipment to schools or hospitals in other countries.

"The OB unit is our best," according to Christine. "They recycle everything. I think it's because a lot of them have kids who go to school and learn about recycling, then come home and educate them not to throw things away!"

Staff suggestions for recycling are encouraged. A flyer posted throughout the hospital reads, "Recycle your ideas and we'll put them to use." It includes an invitation to call the Recycling/Waste Management Committee. South Bay recycled 2000 Christmas trees from the community on their property—a program they ran for 2 weeks instead of the traditional single day pickup. They now plan to erect an oil-recycling station. All these efforts have paid off. Disposal and related costs are down 50 percent. The hospital has saved over $100,000 since the program began.

COMMERCIAL, RETAIL, FINANCIAL,

AND GOVERNMENT OFFICE

APPLICATIONS

37.1   Industry Overview

NAICS code: all 44000s through 45000s and 52000s

INDUSTRY SNAPSHOT

1,554,905 commercial and government offices operations in United States
21,226,492 employees
$5,860.3 billion in annual sales
0.4 tons of solid waste generation per employee
Major waste streams: paper and food waste

The retail industry is a sector of the economy that is comprised of individuals and companies engaged in the selling of finished products to end users in the general public. The retail trade sector comprises establishments engaged in retailing merchandise, generally without transformation, and rendering services incidental to the sale of merchandise.

The retailing process is the final step in the distribution of merchandise; retailers are, therefore, organized to sell merchandise in small quantities to the general public. There are two types of retailers in the industry:

1 Store retailers—Those engaged in the sale of products from physical locations, which warehouse and display merchandise with the intent of attracting customers to make purchases on site.

2 Nonstore retailers—Those engaged in the sale of products using marketing methods that do not include a physical location. Examples of nonstore retailing include

- Infomercials
- Direct-response television advertising
- Catalogue sales
- In-home demonstrations
- Vending machines
- Home delivery
- E-commerce
- Multilevel marketing

According to the U.S. Census Bureau, the total sales for the U.S. Retail Industry in 2007 (including food service and automotive) was $4.48 trillion. Generally, any business that sells finished merchandise to an end user is considered to be part of the retail industry. Sales figures and economic data are sometimes reported separately for restaurants and automotive-related businesses, but by definition they are considered to be members of the retail industry as well.

According to the U.S. Department of Labor, nearly 15.5 million people worked in the U.S. retail industry in the fourth quarter of 2007. That number dropped to 15.3 million in May, 2008. From a global standpoint, of the world's 10 largest retailers, 6 of the companies are from the United States and 4 are from European countries. These top 10 retailers had a combined sales of $978.5 billion in 2007, according to international consulting group, Deloitte.

Store retailers operate fixed point-of-sale locations, located and designed to attract a high volume of walk-in customers. In general, retail stores have extensive displays of merchandise and use mass-media advertising to attract customers. They typically sell merchandise to the general public for personal or household consumption, but some also serve business and institutional clients. These include establishments such as office supply stores, computer and software stores, building materials dealers, plumbing supply stores, and electrical supply stores. Catalog showrooms, gasoline service stations, automotive dealers, and mobile home dealers are treated as store retailers.

In addition to retailing merchandise, some types of store retailers are also engaged in the provision of after-sales services, such as repair and installation. For example, new automobile dealers, electronics and appliance stores, and musical instrument and supplies stores often provide repair services. As a general rule, establishments engaged in retailing merchandise and providing after-sales services are classified in this sector.

The first 11 subsectors of retail trade are store retailers. The establishments are grouped into industries and industry groups typically based on one or more of the following criteria:

- The merchandise line or lines carried by the store; for example, specialty stores are distinguished from general-line stores.
- The usual trade designation of the establishments. This criterion applies in cases where a store type is well recognized by the industry and the public, but difficult to

define strictly in terms of merchandise lines carried; for example, pharmacies, hardware stores, and department stores.

■ Capital requirements in terms of display equipment; for example, food stores have equipment requirements not found in other retail industries.

■ Human resource requirements in terms of expertise; for example, the staff of an automobile dealer requires knowledge in financing, registering, and licensing issues that are not necessary in other retail industries.

Nonstore retailers, like store retailers, are organized to serve the general public, but their retailing methods differ. The establishments of this subsector reach customers and market merchandise with methods such as the broadcasting of infomercials, the broadcasting and publishing of direct-response advertising, the publishing of paper and electronic catalogs, door-to-door solicitation, in-home demonstration, selling from portable stalls (street vendors, except food), and distribution through vending machines. Establishments engaged in the direct sale (nonstore) of products, such as home-heating-oil dealers and home-delivery-newspaper routes are included here.

The buying of goods for resale is a characteristic of retail trade establishments that particularly distinguishes them from establishments in the agriculture, manufacturing, and construction industries. For example, farms that sell their products at or from the point of production are not classified in retail, but rather in agriculture. Similarly, establishments that both manufacture and sell their products to the general public are not classified in retail, but rather in manufacturing. However, establishments that engage in processing activities incidental to retailing are classified in retail. This includes establishments such as optical goods stores that do in-store grinding of lenses, and meat and seafood markets.

Wholesalers also engage in the buying of goods for resale, but they are not usually organized to serve the general public. They typically operate from a warehouse or office and neither the design nor the location of these premises is intended to solicit a high volume of walk-in traffic. Wholesalers supply institutional, industrial, wholesale, and retail clients; their operations are, therefore, generally organized to purchase, sell, and deliver merchandise in larger quantities. However, dealers of durable nonconsumer goods, such as farm machinery and heavy-duty trucks, are included in wholesale trade even if they often sell these products in single units.

The finance and insurance sector comprises establishments primarily engaged in financial transactions (transactions involving the creation, liquidation, or change in ownership of financial assets) and/or in facilitating financial transactions. Three principal types of activities are identified:

1 Raising funds by taking deposits and/or issuing securities and, in the process, incurring liabilities. Establishments engaged in this activity use raised funds to acquire financial assets by making loans and/or purchasing securities. Putting themselves at risk, they channel funds from lenders to borrowers and transform or repackage the funds with respect to maturity, scale, and risk. This activity is known as financial intermediation.

2 Pooling of risk by underwriting insurance and annuities. Establishments engaged in this activity collect fees, insurance premiums, or annuity considerations; build up reserves; invest those reserves; and make contractual payments. Fees are based on the expected incidence of the insured risk and the expected return on investment.

3 Providing specialized services facilitating or supporting financial intermediation, insurance, and employee benefit programs. In addition, monetary authorities charged with monetary control are included in this sector.

The subsectors, industry groups, and industries within the NAICS finance and insurance sector are defined on the basis of their unique production processes. As with all industries, the production processes are distinguished by their use of specialized human resources and specialized physical capital. In addition, the way in which these establishments acquire and allocate financial capital, their source of funds, and the use of those funds provides a third basis for distinguishing characteristics of the production process. For instance, the production process in raising funds through deposit-taking is different from the process of raising funds in bond or money markets. The process of making loans to individuals also requires different production processes than does the creation of investment pools or the underwriting of securities.

Most of the finance and insurance subsectors contain one or more industry groups of (1) intermediaries with similar patterns of raising and using funds and (2) establishments engaged in activities that facilitate, or are otherwise related to that type of financial or insurance intermediation. Industries within this sector are defined in terms of activities for which a production process can be specified, and many of these activities are not exclusive to a particular type of financial institution. To deal with the varied activities taking place within existing financial institutions, the approach is to split these institutions into components performing specialized services. This requires defining the units engaged in providing those services and developing procedures that allow for their delineation. These units are the equivalents for finance and insurance of the establishments defined for other industries.

The output of many financial services, as well as the inputs and the processes by which they are combined, cannot be observed at a single location and can only be defined at a higher level of the organizational structure of the enterprise. Additionally, a number of independent activities that represent separate and distinct production processes may take place at a single location belonging to a multilocation financial firm. Activities are more likely to be homogeneous with respect to production characteristics than are locations, at least in financial services. The classification defines activities broadly enough that it can be used both by those classifying by location and by those employing a more top-down approach to the delineation of the establishment.

Establishments engaged in activities that facilitate, or are otherwise related to, the various types of intermediation have been included in individual subsectors, rather than in a separate subsector dedicated to services alone because these services are performed by intermediaries, as well as by specialist establishments, and the extent to which the activity of the intermediaries can be separately identified is not clear.

The finance and insurance sector has been defined to encompass establishments primarily engaged in financial transactions; that is, transactions involving the creation, liquidation, or change in ownership of financial assets or in facilitating financial transactions. Financial industries are extensive users of electronic means for facilitating the verification of financial balances, authorizing transactions, transferring funds to and from transactors' accounts, notifying banks (or credit card issuers) of the individual transactions, and providing daily summaries. Since these transaction-processing activities are integral to the production of finance and insurance services, establishments that principally provide a financial transaction-processing service are classified to this sector, rather than to the data-processing industry in the information sector

Legal entities that hold portfolios of assets on behalf of others are significant and data on them are required for a variety of purposes. Thus for NAICS, these funds, trusts, and other financial vehicles are the fifth subsector of the finance and insurance sector. These entities earn interest, dividends, and other property income, but have little or no employment and no revenue from the sale of services. Separate establishments and employees devoted to the management of funds are classified in industry group 5239, other financial investment activities.

37.2   Waste Management Goals and Opportunities

The majority of solid waste generated by this sector is paper and food waste. Table 37.1 displays the composition breakdown based on survey results.

TABLE 37.1 COMMERCIAL, RETAIL, FINANCIAL, AND GOVERNMENT OFFICE SOLID WASTE COMPOSITION (SURVEY RESULTS)		
MATERIAL	**COMPOSITION (%)**	**RECYCLING (%)**
Paper	45 ± 11.3	19 ± 3.1
Mixed office paper	34 ± 8.5	23 ± 3.4
Newspaper	6 ± 1.5	12 ± 0.7
Magazines	2 ± 0.5	0 ± 0.7
Other	3 ± N/A	0 ± 0
Food waste	18 ± 4.5	0 ± 0.0
OCC (cardboard)	6 ± 1.5	14 ± 1.9
Yard waste	5 ± 1.3	0 ± 0.0
Mixed plastics	4 ± 1.0	0 ± 0.0
Furniture/office eqpt.	3 ± 0.8	0 ± 0.0
Wood	2 ± 0.5	0 ± 0.0
Aluminum cans	2 ± 0.5	45 ± 4.8
Glass	1 ± 0.3	10 ± 1.2
Other	14 ± N/A	0 ± 0
Overall recycling rate		10.4

As shown in the table, the recycling rate for this sector is approximately 10 percent. As derived from the solid waste evaluation model discussed in Chap. 12, the equation that estimates the annual waste generation per year per employee for this sector can be calculated from the following:

Tons of solid waste generated per year $= 0.43 \times$ number of employees $+ 0.94$

37.3   Constraints and Considerations

In general, establishing waste minimization programs in office settings is relatively low-cost and numerous vendors are available in most areas to assist with collection and processing at no charge. In light of the relative ease and low cost, several constraints and considerations should be evaluated before finalizing the program, these include

- *Lack of available space*—Many offices are very tight for space, in terms of outside space for vendor containers, internal space for collection cans in common areas, and desk-side space for individual recycle bins.
- *Fostering employee buy-in*—Creating awareness of recycling programs and encouraging employee involvement in the program go hand in hand. Understanding the psychology of recycling and honing in on employee motivation is key.
- *Fluctuations in the local commodity market for paper*—Drops in the commodity markets will affect the revenue generated from the sale of office paper and low markets may reduce the number of available vendors.
- *Contracting with suitable vendors and relationship management*—Developing strong relationships with organizations and vendors that will support and strengthen facility recycling programs is critical. Many of these companies offer value added advice and additional networking opportunities.
- *Security concerns for confidential documents*—Many organizations and consumers are concerned about private or proprietary information getting into the wrong hands. Safeguards, contracts, and standard processes usually remedy these issues.
- *Separation of recyclables*—Sorting recyclable materials and removing them from the waste stream can present challenges in office settings set; employees may be inclined to do what is easiest. If the general waste bin is closer than the recycling bin and requires less effort, employees may not have the motivation to recycle.

37.4   Potential Technologies and Strategies

Several proven strategies exist to implement and sustain office-oriented-recycling programs. Most are very low cost, but require a moderate to high degree of communication and coordination. Following is a brief list of such methods:

■ *Establish waste and recycling agreements with nearby businesses*—Businesses that operate in business centers or industrial parks can achieve significant cost savings by consolidating waste hauling and recycling contracts. Business tenant meetings or board meetings are appropriate venues to discuss these options.

■ *Create incentive programs for employees*—Motivating employees to recycle is one of the most important components of any waste minimization program. Offering awards to top-performing units or individuals can help achieve this goal. The awards may include money, corporate logo gifts, paid time off, or public recognition. A company-wide luncheon or breakfast may also be provided if the organizational waste minimization goals are met.

■ *Set defaults to double sided for copiers*—Studies have shown that setting copier defaults to double sided may reduce paper waste by 40 percent. Most copiers are set to single sided and users must select double sided.

■ *Use paperless fax machines*—Send files electronically to fax machines and set defaults to not print a fax confirmation.

■ *Provide employees with desk-side recycling bins*—Providing each employee in the office with a desk-side recycling container will enhance motivation to recycle. The container will serve as a visual reminder to recycle and reduce travel time to a consolidated recycle bin. The advantages of desk-side recycling include the maximum potential for collecting the most amount of paper for recycling and provide the most convenient method for employees to dispose of both recyclable and nonrecyclable materials. Disadvantages include the potential for contamination of recyclable papers if trash is disposed of into the wrong bin and the initial cost incurred for extra recycle bins.

■ *Provide employees with periodic newsletters*—These letters may contain information on recycling performance trends for the company, and announce awards and new programs.

■ *Appoint a "green cop" to monitor individual employee performance*—Designating an employee to police the material entering the waste stream and recycling bins can motivate employees to follow proper procedures. The individual should be instructed to report all findings (conformance and nonconformance) to managers and other employees.

■ *Shredding of confidential documents*—Establish a relationship with a vendor to recycle shredded paper. Many waste removal and recycling companies offer confidential shredding services. Ensure that the company is certified by the National Association of Information Destruction (NAID). NAID offers a certification program on a voluntary basis to all NAID member companies providing information-destruction services. Through the program, NAID members may seek annual certification audits for both mobile and plant-based operations in paper or printed media, micromedia, or computer hard drive destruction. The NAID certification program establishes standards for a secure destruction process including such areas as operational security, employee hiring and screening, the destruction process, responsible disposal and insurance.

■ *Implement reverse recycling*—An emerging concept involves eliminating desk-side waste bins and only providing employees with desk-side recycling bins. The custodial service is instructed to service only recycling bins at employees' desks and the employee is responsible for transporting nonrecyclable waste to centralized containers in employee break rooms. The purpose of the concept is motivating employees to recycle by creating barriers to generating nonrecyclable waste. By employing this concept, a Toledo, Ohio–based company was able to reduce waste disposal costs by 40 percent and consolidate general waste removal from 5 days to 2 days per week.

37.5   Implementation and Approach

Below are general procedures for implementing a solid waste reduction program in an office setting:

- Describe the direction, objectives, and proposed goal(s) of the recycling program.
- Define the requirements necessary to accomplish this.
- Identify the waste categories and types.
- Review and evaluate
  - Current collection and disposal practice(s)
  - Current paper collection concept
  - Current waste disposal contract(s)
- Assess composition of general cafeteria trash.
- Identify new wastes not currently identified or recovered.
- Develop campaign concept and requirements.
- Develop facility departments' interface and support requirements.
- Schedule for program scope work.
- Develop capital and expense budget outline.
- Search out and contact local reclaiming and recycling vendors.
- Use services of local consultant(s) as needed.

Office paper constitutes 45 percent of the waste generated for these businesses and marks the starting point for recycling programs in this sector. Office paper recycling has a high level of visibility, helps to foster employee buy-in, and can generate early wins for the recycling program. The following are step-by-step procedures for instituting an office paper–recycling program. The steps include

- Determining what papers need to be recycled
- Locating a vendor who meets your requirements
- Procedures for setting up a program
- Monitoring your recycling progress throughout the year
- Keeping the program alive

The success of any office recycling program depends on the support and cooperation from every employee, from the highest levels of management to the employees who carry out the actual collection procedures. In addition, effective relationship management with recycling vendors is critical. Below is a short list of vendor requirements for office paper–recycling programs:

- Reliability of service in terms of on-time pickups, accurate tonnage reimbursements and monthly reporting, and timely reimbursements
- Ability to handle documents in accordance with security guidelines
- Willingness to pick up all kinds of papers including high grade, mixed grade, file stock, and corrugated
- Maximum prices offered for mixed grades and file stock which comprise most of recyclable tonnage

Monitoring recycling program performance is an important step to ensure the long-term success of the program. Below is a list of activities that may be applied to accomplish this goal:

- Keep accurate and up-to-date statistics:
  - Include tonnage figures
  - Include dollars received from recycling vendor
  - Include estimated cost-avoidance figures
- Periodically inspect office areas to determine if employees are properly disposing of recyclable materials.
- Periodically arrange to have outside garbage unit analyzed by following unit to transfer station to weigh contents and observe contents of unit:
  - Determine what percent of contents consists of recyclable materials
  - Determine if container is being pulled prematurely
- Post recycling statistics in your building so that employees can see your progress.
- Publish updated statistics and articles in company bulletins and newsletters.
- Display recycling posters throughout the year.

37.6   Case Study

In August 1999 an action plan was developed to improve the office paper–recycling program at One Government Center, in Toledo, Ohio. Representatives from the Ohio Department of Natural Resources (ODNR), the Reuben Company, Keep Toledo/Lucas County Beautiful, and the Lucas County Solid Waste Management District (the District) attended an introductory meeting to develop the action plan. The goals of the action plan were

- Improve the material-handling system.
- Develop solutions for storage constraints.
- Develop a short-term material-handling strategy during building reconstruction.
- Conduct a volume assessment of the current recycling system.
- Interview several contacts on the recycling program's effectiveness.
- Access the cost-effectiveness and service arrangement for the compactor.

The waste minimization team utilized various on-site data-collection methods at One Government Center and the landfill that received solid waste from the building.

Numerous trash sorts were conducted at both locations to estimate volume and content of One Government Center's waste and recycling streams. Employee interviews were also conducted and questionnaires were distributed to collect specific data.

The methodology used to conduct the office paper–volume assessment involved two primary steps. First, the recycling containers on each floor were emptied and the contents placed on the loading dock. The second step began exactly 1 week after emptying the recycling containers on each floor. For this step, the contents of each recycling container on every floor were weighed and recorded. This weekly estimate was then converted to a daily and an annual estimate. At the time of the assessment, only paper was collected for recycling. Figure 37.1 summarizes the data.

From the on-site data-collection phase, it was estimated that 3975 lb of mixed office paper are collected through the system per month (47,700 lb annually). A local vendor collects all recyclables (mixed office paper and aluminum cans) at no-cost/no-profit to One Government Center, but does not provide the volumes collected to the management.

Table 37.2 summarizes the amount of material entering the waste stream at One Government Center. This information was calculated from the waste assessment at One Government Center and the trash sort at the landfill. On average 34,800 lb of waste per month is disposed of at the landfill by One Government Center (based upon a 6-month sampling) and 3975 lb of mixed office paper are recycled per month. From a trash sort conducted at the landfill, approximately 15 percent was cardboard, 25 percent was mixed office paper, 5 percent was metal, aluminum cans, and wood, and 55 percent was nonrecyclable waste.

At the time of the audit, One Government Center recycled 31 percent of the mixed office paper that entered their waste stream and approximately 69 percent entered the landfill. Aluminum cans are also recycled now after the mixed office paper–recycling program has been reestablished. Currently One Government Center recycles 10 percent of all material entering their waste stream and 22 percent of all recyclable materials in their waste stream are recyclable. Table 37.3 summarizes the level of recycling for each material type.

The current collection system utilized by One Government Center is not fully standardized in regards to the collection process and varies from floor to floor. Major floor-to-floor variations observed included

- An estimated 20 percent of employees did not have a desk-side recycling box.
- On some floors the custodial staff emptied the desk-side containers on a routine basis (or when the boxes were full) and on other floors the employees were required to empty their own recycling boxes at a central location.
- Some employees requested that their containers not be emptied in order to keep the paper for reference.
- Some employees used their desk-side containers for other purposes (such as extra storage or as a filing box).
- Some floors utilized more central-collection totes than other floors.
- Some floors utilized dilapidated cardboard central totes, while other floors utilized sturdy plastic totes.

One Government Center

Mixed Office Paper Recycle Audit

Floor Number	Audit Date	Recycling Period	Notes	Total Weight (lb)	Tare Weight (lb)	Paper Weight (lb)	Total Floor Weight (lb)	Working Days	Working Day Average Weight (lb)	Annual Average Weight (lb)	Percentage of Total	Number of Employees	Working Day Average Weight (lb)/per Employee
22	11/8/99	11/2/99-11/8/99		54.4	22.7	31.7	31.7	5	6.34	1585	3%	47	0.13
21	11/8/99	11/2/99-11/8/99		88.2	54.4	33.8	33.8	5	6.76	1690	4%	65	0.10
20	11/8/99	11/2/99-11/8/99		91.2	22.7	68.5	68.5	5	13.70	3425	7%	80	0.17
19	11/8/99	11/2/99-11/8/99		45.2	22.7	22.5	22.5	5	4.50	1125	2%	40	0.11
18	11/8/99	11/2/99-11/8/99	Bin by elevator	42.0	22.7	19.3		5					
18	11/8/99	11/2/99-11/8/99		66.6	22.7	43.9	63.2	5	12.64	3160	7%	140	0.09
17	11/8/99	11/2/99-11/8/99		60.2	22.7	37.5	37.5	5	7.50	1875	4%	80	0.09
16	11/8/99	11/2/99-11/8/99		41.2	22.7	18.5	18.5	5	3.70	925	2%	45	0.08
15	11/22/99	11/16/99-11/22/99		74.8	31.0	43.8	43.8	5	8.76	2190	5%	5	1.75
14	11/22/99	11/16/99-11/22/99	State tax	31.0	22.7	8.3	8.3	5	1.66	415	1%	30	0.06
13	11/22/99	11/16/99-11/22/99		96.8	22.7	74.1	74.1	5	14.82	3705	8%	62	0.24
12	11/22/99	11/16/99-11/22/99		103.4	22.7	80.7	80.7	5	16.14	4035	8%	95	0.17
11	11/22/99	11/16/99-11/22/99		123.0	22.7	100.3	100.3	5	20.06	5015	11%	35	0.57
10	11/22/99	11/16/99-11/22/99		74.4	52.8	21.6	21.6	5	4.32	1080	2%	34	0.13
9	11/22/99	11/16/99-11/22/99		52.8	22.7	30.1	30.1	5	6.02	1505	3%	51	0.12
8	11/15/99	11/9/99-11/15/99		76.2	25.2	51.0	51.0	4	10.20	2550	5%	67	0.15
7	11/15/99	11/9/99-11/15/99		25.2	22.7	2.5		4					
7	11/15/99	11/9/99-11/15/99		100.0	81.4	18.6	21.1	4	4.22	1055	2%	40	0.11
6	11/15/99	11/9/99-11/15/99		81.4	54.0	27.4	27.4	4	5.48	1370	3%	86	0.06
5	11/15/99	11/9/99-11/15/99		54.0	22.7	31.3	31.3	4	6.26	1565	3%	19	0.33
4	11/15/99	11/9/99-11/15/99	Much CPO	98.2	22.7	75.5		4					
4	11/15/99	11/9/99-11/15/99		135.0	54.1	80.9	156.4	4	31.28	7820	16%	116	0.27
3	11/15/99	11/9/99-11/15/99		54.1	22.7	31.4	31.4	4	6.28	1570	3%	12	0.52
Totals									191	47,660		1149	0.17

* = employees number estimate

Figure 37.1 Data collection sheet at One Government Center.

TABLE 37.2	MONTHLY WASTE GENERATION AT ONE GOVERNMENT CENTER	
MATERIAL TYPE	**POUND PER MONTH**	**TOTAL (% OF)**
Mixed office paper	12,675	25
Cardboard	5,200	15
Metal, wood, and aluminum cans	1,740	5
Nonrecyclable waste	19,140	55
Total waste generation	38,755	100

Each employee was issued a cardboard box to place under his or her desk to collect waste mixed office paper. During on-site data collection at the facility, it was observed that approximately 80 percent of employees still use the cardboard box, and an estimated 20 percent of employees did not have or use the collection box (the 20 percent stated they either do not recycle, use neighbor's boxes, or directly carry all waste mixed office paper to a central-collection point). Typically the process utilized to collect the waste mixed office paper consists of four primary steps:

1 Employees deposit waste paper in designated desk-side containers.
2 The desk-side containers are emptied into centrally located totes on each floor by either the employee directly or by the custodial staff.
3 When the totes are full, the custodial staff empties them on the dock in designated gaylords and then returns the totes to their original location.
4 On a weekly basis, Lake Erie Recycling collects the mixed office paper in the gaylords.

Standardization of the process is the most critical improvement to the recycling system. All employees should be encouraged to utilize a desk-side-recycling container (unless reasonable conditions exist to prevent or deter its use). The custodial staff should use a standardized process to collect the mixed office paper.

TABLE 37.3	MONTHLY RECYCLING LEVELS AT ONE GOVERNMENT CENTER		
MATERIAL TYPE	**AMOUNT GENERATED (LB/MONTH)**	**AMOUNT RECYCLED (LB/MONTH)**	**RECYCLED (%)**
Mixed office paper	12,675	3,975	31
Cardboard	5,200	0	0
Aluminum cans	N/A	N/A	N/A
Total recyclable generation	17,875	3,975	22

In, assessing the current material sorts, several key findings were made:

■ Office paper is not currently sorted; all recyclable paper is placed into the same containers.
■ Contamination exists in the mixed office paper containers, such as improperly disposed of nonrecyclable materials. This includes food waste, newspaper, plastic, aluminum cans, Styrofoam cups, and plastic wrappers.

Proposed modifications include

■ *Training of employees*—Focuse on what materials may be deposited in the mixed office paper–recycling bins.
■ *Development and utilization of signage*—Describe what may be deposited in the recycling bins.
■ *Recycling collector modifications*—In particular require that the current recycling collector (Lake Erie Recycling) provide monthly reports on the amount of materials collected. Also, require the recycling collector to purchase the materials at the current fair market price. The additional revenue could be used to fund future environmental programs.
■ *Waste hauling modifications*—Require that waste will only be collected on-call when the 20-yd^3 compactor is full. The general waste disposed of at the landfill is serviced by BFI Inc., and is collected from One Government Center on an on-call basis. Past records indicate that the waste has consistently been collected on every Monday (Tuesday is observed as a holiday), and that the total weight has varied from 2.9 tons to 5.0 tons. One Government Center currently uses a 20-yd^3 packer for the general waste, but management has limited ability to determine when it is full, and the waste hauler should be contacted to collect it. As a result, the waste is generally collected every Monday, and may or may not be a full load.

During the facility visits to One Government Center, a select number of random interviews were conducted with employees. Key finding from those interviews include

■ Limited room to store recycling bins under some desks
■ Perceived lack of management support
■ Perceived lack of understanding of what is recyclable
■ More lack of interest/action from top management (poor example to staff); approximately 85 percent of staff utilized the desk-side bin, whereas compared with 40 percent of the top management
■ Strong interest by staff in recycling and improving environmental quality

Upon the random interviewing of the select number of employees at One Government Center, the assessment team discovered that many of the employees (including management) did not fully understand the recycling system or the standard procedures. Primarily the employees expressed concerns regarding when and where to dump their desk-side containers, what to do if they lose or damage the container, and what is building management's role in emptying the containers. Another important concern from the employees involved the guidelines regarding which materials may be placed in the recycling bins (for example newspapers). Overall most employees at One Government Center have a positive attitude on recycling and are willing to make necessary changes to improve the recycling system.

The custodial crew was very helpful in conducting the waste assessment. The custodial crew is responsible for emptying the large centrally located totes on each floor, when they become full. This is usually done on an on-call basis or when the custodian visually notices the tote is full. The custodial crew is contracted from an outside company (not government employees). Confidential paper must be shredded in some departments, but this is usually collected by an outside contractor as well. A small amount of mixed office-paper shred is placed into the centrally located totes to be recycled. Some key findings and areas for improvement determined from interviews with the custodians were

■ All emptying of waste baskets and recycling totes are conducted during second shift, when few government employees are present.
■ Floor-by-floor variation on procedures exists due to lack of communication or from specific floor requests.
■ One service elevator is available at the facility, and limits the transportation of waste and recyclables to the loading dock located in the basement.
■ Office elevators are used by the custodial staff, but only for limited floor-to-floor movement of staff or supplies.
■ A gauge for the 20-yd^3 compactor (used for general waste) may help to better determine when the compactor is full. This may aid management in determining when the compactor is full and when to call and have it hauled. This will help to reduce costs.

The staff of One Government Center has expressed concerns regarding the small dock that is used to store waste and recyclables. Figure 37.2 shows a diagram of the dock. The dock has limited storage space, and the following items have been developed to increase space utilization:

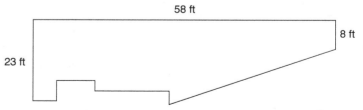

The dock is irregular in shape, and has approximately 600 ft^2 of useable surface.

Figure 37.2   Diagram of the dock at One Government Center.

■ Recycling collection modifications, in particular, require the collector to provide a wider range of services and collect more materials. Investigate the use of a cardboard baler to conserve space; the current recycling collector may provide one at no cost or for a small fee. A payback period may be developed once the fee is known.
■ Utilize stackable gaylords for mixed office paper. This would increase space utilization and allow for more mixed office paper to be stored on the dock. A small forklift may be required to transport the gaylords.

Three primary issues are recommended in regards to material handling. These are

1 Train all One Government Center employees on the recycling system and clearly define their role.
2 Eliminate excessive handling that may be taking place due to the different collection procedures that vary floor to floor and reduce multiple handling of mixed office paper by the custodial staff. Standardization of the procedure will help to achieve this goal.
3 Develop a systematic collection procedure for mixed office paper and document it (use the recommendations mentioned earlier as a guide).

With the mentioned training and improved collection system it is possible for One Government Center to increase the recycling rate from its present 10 percent to a maximum of 46 percent and become a leader in office recycling.

37.7 Exemplary Performers—Office Paper–Recycling Programs

The American Forest & Paper Association (AF&PA) announced in 2006 that Mr. Joel Ostroff of Macon County, North Carolina had received the years Ed Hurley Memorial Paper Recycling Award. The award recognizes an individual who has had a significant and positive influence in advocating paper recycling.

In 1990, with the assistance of a part-time instructor, Mr. Ostroff organized and established the Terraphile Society at Brevard Community College. The student-based environmental organization was so successful that by 1993 the college was recycling 10,000 lb of newspapers and nearly 5000 lb of office paper each month. The Terraphile Society also became actively engaged in public education and outreach. As a result of its college-wide efforts, the Terraphile Society became the most successful collegiate environmental group in Florida, winning awards from the State of Florida Education Department, numerous environmental groups, and the U.S. Environmental Protection Agency's WasteWise program.

After his retirement from Brevard Community College, Mr. Ostroff began his career as Macon County's recycling coordinator. In the 5 years that he served in this position the paper-recycling rate has increased by approximately 35 percent. He had enrolled the five largest schools in the county in a paper-recycling program, and now plans

478   COMMERCIAL, RETAIL, FINANCIAL, AND GOVERNMENT OFFICE APPLICATIONS

to incorporate this for the remaining schools in the near future. Working with local volunteer agencies and the Job Corps, he annually holds an educational fair and a poster contest to emphasize the importance of paper recycling.

The AF&PA Recycling Awards were created to recognize outstanding individual, business, community, and school paper-recycling efforts. In 2005, a record high of 51.5 percent of the paper consumed in the United States was recovered for recycling. AF&PA has set an ambitious goal of 55 percent recovery by 2012. It is only through the continued efforts of the millions of Americans who recycle at home, work, and school that this goal will be achieved.

Through his work at Smurfit-Stone Container Corporation, and his tireless service to AF&PA, Ed Hurley was actively involved in the paperboard, containerboard, and recycling industries for 35 years. Mr. Hurley was a staunch advocate for paper recycling, and the award in his name is given to an individual similarly committed to promoting paper recycling.

ACCOMMODATIONS AND FOOD

SERVICE APPLICATIONS

38.1   Industry Overview

NAICS code: all 72000s

INDUSTRY SNAPSHOT

565,590 hospitality operations in the United States
10,120,951 employees
$449.5 billion in annual sales
2.0 tons of solid waste generation per employee
Major waste streams: paper, food waste, and cardboard

The accommodation and food services sector comprises establishments providing customers with lodging and/or preparing meals, snacks, and beverages for immediate consumption. The sector includes both accommodation and food services establishments because the two activities are often combined at the same establishment.

Industries in the accommodation subsector provide lodging or short-term accommodations for travelers, vacationers, and others. There is a wide range of establishments in these industries. Some provide lodging only; while others provide meals, laundry services, and recreational facilities, as well as lodging. Lodging establishments are classified in this subsector even if the provision of complementary services generates more revenue. The types of complementary services provided vary from establishment to establishment.

Pretax income for the U.S. hotel industry in 2007 increased 5.3 percent to a record $28 billion, according to the recently released Hotel Operating Statistics (HOST) Study 2008 issued by Smith Travel Research (STR).

The industry posted an all-time best $139.4 billion in revenue in 2007—which is over $6 billion more than it generated in 2006 ($133.4 billion)—and for the second consecutive year, gross operating profit as a percentage of revenue came in at 41.3 percent.

"2007 was another excellent year for the U.S. hotel industry with record revenues and record profits," said Mark Lomanno, STR's president. "However, as the American economy slows, we are expecting a tougher operating climate for U.S. hotels in the near future."

The HOST Study is the most extensive and definitive database on U.S. hotel industry revenues and expenses. The 2008 version is derived from the operating statements of more than 5200 hotels.

People travel for a variety of reasons, including for vacations, business, and visits to friends and relatives. For many of these travelers, hotels and other accommodations will be where they stay while out of town. For others, hotels may be more than just a place to stay, but destinations in themselves. Resort hotels and casino hotels, for example, offer a variety of activities to keep travelers and families occupied for much of their stay.

Hotels and other accommodations are as different as the many family and business travelers they accommodate. The industry includes all types of lodging, from luxurious five-star hotels to youth hostels and recreational vehicle (RV) parks. While many provide simply a place to spend the night, others cater to longer stays by providing food service, recreational activities, and meeting rooms. In 2006, approximately 62,000 establishments provided overnight accommodations to suit many different needs and budgets.

Hotels and motels comprise the majority of establishments in this industry and are generally classified as offering either full service or limited service. Full-service properties offer a variety of services for their guests, but they almost always include at least one or more restaurant and beverage service options—from coffee bars and lunch counters to cocktail lounges and formal restaurants. They also usually provide room service. Larger full-service properties usually have a variety of retail shops on the premises, such as gift boutiques, newsstands, and drug and cosmetics counters, some of which may be geared to an exclusive clientele. Additionally, a number of full-service hotels offer guests access to laundry and valet services, swimming pools, beauty salons, and fitness centers or health spas. A small, but growing, number of luxury-hotel chains also manage condominium units in combination with their transient rooms, providing both hotel guests and condominium owners with access to the same services and amenities.

The largest hotels often have banquet rooms, exhibit halls, and spacious ballrooms to accommodate conventions, business meetings, wedding receptions, and other social gatherings. Conventions and business meetings are major sources of revenue for these properties. Some commercial hotels are known as conference hotels—fully self-contained entities specifically designed for large-scale meetings. They provide physical fitness and recreational facilities for meeting attendees, in addition to state-of-the-art audiovisual and technical equipment, a business center, and banquet services.

Limited-service hotels are free-standing properties that do not have on-site restaurants or most other amenities that must be provided by a staff other than the front desk or housekeeping. They usually offer continental breakfasts, vending machines or small packaged items, Internet access, and sometimes unattended game rooms or swimming pools in addition to daily housekeeping services. The numbers of limited-service properties have been growing. These properties are not as costly to build and maintain. They appeal to budget-conscious family vacationers and travelers who are willing to sacrifice amenities for lower room prices.

Hotels can also be categorized based on a distinguishing feature or service provided by the hotel. Conference hotels provide meeting and banquet rooms, and usually food service, to large groups of people. Resort hotels offer luxurious surroundings with a variety of recreational facilities, such as swimming pools, golf courses, tennis courts, game rooms, and health spas, as well as planned social activities and entertainment. Resorts typically are located in vacation destinations or near natural settings, such as mountains, seashores, theme parks, or other attractions. As a result, the business of many resorts fluctuates with the season. Some resort hotels and motels provide additional convention and conference facilities to encourage customers to combine business with pleasure. During the off season, many of these establishments solicit conventions, sales meetings, and incentive tours to fill their otherwise empty rooms; some resorts even close for the off-season.

Extended-stay hotels typically provide rooms or suites with fully equipped kitchens, entertainment systems, office space with computer and telephone lines, fitness centers, and other amenities. Typically, guests use these hotels for a minimum of 5 consecutive nights often while on an extended work assignment or lengthy vacation or family visit. All-suite hotels offer a living room or sitting room in addition to a bedroom.

Casino hotels combine both lodging and legalized gaming on the same premises. Along with the typical services provided by most full-service hotels, casino hotels also contain casinos where patrons can wager at table games, play slot machines, and make other bets. Some casino hotels also contain conference and convention facilities.

In addition to hotels, bed-and-breakfast inns, RV parks, campgrounds, and rooming and boarding houses provide lodging for overnight guests and are included in this industry. Bed-and-breakfast inns provide short-term lodging in private homes or small buildings converted for this purpose and are characterized by highly personalized service and inclusion of breakfast in the room rate. Their appeal is quaintness, with unusual service and decor.

RV parks and campgrounds cater to people who enjoy recreational camping at moderate prices. Some parks and campgrounds provide service stations, general stores, shower and toilet facilities, and coin-operated laundries. While some are designed for overnight travelers only, others are for vacationers who stay longer. Some camps provide accommodations, such as cabins and fixed campsites, and other amenities, such as food services, recreational facilities and equipment, and organized recreational activities. Examples of these overnight camps include children's camps, family vacation camps, hunting and fishing camps, and outdoor adventure retreats that offer trail riding, white-water rafting, hiking, fishing, game hunting, and similar activities.

Other short-term lodging facilities in this industry include guesthouses, or small cottages located on the same property as a main residence, and youth hostels—dormitory-style hotels with few frills, occupied mainly by students traveling on limited budgets. Also included are rooming and boarding houses, such as fraternity houses, sorority houses, off-campus dormitories, and workers' camps. These establishments provide temporary or longer-term accommodations that may serve as a principal residence for the period of occupancy. These establishments also may provide services such as housekeeping, meals, and laundry services.

The subsector is organized into three industry groups: traveler accommodation, recreational accommodation, and rooming and boarding houses. The traveler accommodation industry group includes establishments that primarily provide traditional types of lodging services. This group includes hotels, motels, and bed-and-breakfast inns. In addition to lodging, these establishments may provide a range of other services to their guests. The RV (recreational vehicle) parks and recreational camps industry group includes establishments that operate lodging facilities primarily designed to accommodate outdoor enthusiasts. Included are travel trailer campsites, RV parks, and outdoor adventure retreats. The rooming and boarding houses industry group includes establishments providing temporary or longer-term accommodations that for the period of occupancy may serve as a principal residence. Board (i.e., meals) may be provided but is not essential.

Establishments that manage short-stay accommodation establishments (e.g., hotels and motels) on a contractual basis are classified in this subsector if they both manage the operation and provide the operating staff. Such establishments are classified based on the type of facility managed and operated.

Industries in the food services and drinking places subsector prepare meals, snacks, and beverages to customer order for immediate on-premises and off-premises consumption. There is a wide range of establishments in these industries. Some provide food and drink only; while others provide various combinations of seating space, waiter/waitress services and incidental amenities, such as limited entertainment. The industries in the subsector are grouped based on the type and level of services provided. The industry groups are full-service restaurants; limited-service eating places; special food services, such as food-service contractors, caterers, and mobile food services; and drinking places.

Food services and drink activities at hotels and motels; amusement parks, theaters, casinos, country clubs, and similar recreational facilities; and civic and social organizations are included in this subsector only if these services are provided by a separate establishment primarily engaged in providing food and beverage services.

Excluded from this subsector are establishments operating dinner cruises. These establishments are classified in subsector 487, scenic and sightseeing transportation because those establishments utilize transportation equipment to provide scenic recreational entertainment.

38.2   Waste Management Goals and Opportunities

The majority of solid waste generated by this sector is paper, food waste, and cardboard. Table 38.1 displays the composition breakdown based on survey results.

TABLE 38.1 ACCOMMODATIONS AND FOOD SERVICE INDUSTRY SOLID WASTE COMPOSITION (SURVEY RESULTS)

MATERIAL	COMPOSITION (%)	RECYCLING (%)
Paper	36 ± 4.1	23 ± 5.5
Mixed office paper	21 ± 3.1	24 ± 5.8
Newspaper	13 ± 1.4	24 ± 5.8
Magazines	2 ± 0.9	13 ± 3.1
Food waste	32 ± 4.1	0 ± 0.0
OCC (cardboard)	6 ± 2.2	21 ± 5.0
Plastic bottles	5 ± 0.4	29 ± 7.0
Aluminum cans	4 ± 0.9	69 ± 16.6
Yard waste	4 ± 0.7	0 ± 0.0
Aerosol cans	2 ± 0.4	69 ± 16.6
Fabric	2 ± 0.2	0 ± 0.0
Glass bottles	2 ± 0.9	7 ± 1.7
Other	7 ± N/A	0 ± 0.0
Overall recycling level		15.3

As shown in the table, the recycling rate for this sector is approximately 15 percent. As derived from the solid waste evaluation model discussed in Chap. 12, the equation that estimates the annual waste generation per year per employee for this sector can be calculated from the following:

Tons of solid waste generated per year = 2.01 × number of employees + 4.05

38.3   Constraints and Considerations

In terms of solid waste management and minimization for hotels and restaurants, the four key constraints and considerations are

1 *Available space to store and stage materials (interior and exterior)*—This includes space within the facility and dock space for storage. The numerous deliveries that many of these facilities receive each day of the week can also present a logistic and scheduling problem for coordinating dock usage during peak periods.
2 *Cleanliness and sanitation*—A dirty restaurant or hotel will have a negative impact on guest perception; maintaining a clean site free from debris and odor is critical to success.

3 *The separation of waste and contamination*—Separating recyclable waste from nonrecyclable waste creates a new process; in conjunction with this, informing patrons of the separation process also creates an additional management concern.

4 *Finding appropriate suppliers for waste removal and recycling*—Well established and reliable waste removal and recycling providers are critical for a successful program; when meeting with prospective companies, request a list of references.

38.4   Potential Technologies and Strategies

The choice of waste minimization methods will depend on many factors. These include the quantity and type of food discards, availability of space for on-site recovery, existence of haulers and/or end users for off-site recovery, and program costs. Food-discard-recovery methods include making donations, processing this waste into animal feed, rendering, and composting. Off-site methods involve food-discard generators, haulers, and end users. The following is a list of diversion options for their food waste in preferred order of implementation:

1 *Food donations*—Nonperishable and unspoiled perishable food can be donated to local food banks, soup kitchens, and shelters. Local and national programs frequently offer free pickup and provide reusable containers to donors. Because these donations recycle food and help feed people in need of assistance, this option should be considered before looking at other alternatives. Smaller food-collection organizations are also appropriate. For a list of contact information and needs of small food-collection organizations, check yellow pages under "food pantries" or "shelters."

2 *Source reduction*—Source reduction, including reuse, can help reduce waste disposal and handling costs, because it avoids the costs of recycling, municipal composting, landfilling, and combustion. Source reduction also conserves resources and reduces pollution, including greenhouse gases that contribute to global warming. By doing a careful audit of the waste stream, a business can determine the percentage of food and organic wastes that are present in their trash. Once the potential for waste reduction is established, a business can reduce the quantity of food they buy, purchase precut foods, or explore the possibilities of portion control at restaurants.

3 *Animal feed*—Recovering food discards as animal feed is not new. In many areas hog farmers have traditionally relied on food discards to feed their livestock. Farmers may provide storage containers and free or low-cost pickup service. Coffee grounds and foods with high salt content are not usually accepted, because they can be harmful to livestock. At least one company is using technology to convert food discards into a high-quality, dry, pelletized animal feed.

4 *Rendering*—Liquid fats and solid meat products can be used as raw materials in the rendering industry, which converts them into animal food, cosmetics, soap, and other products. Many companies will provide storage barrels and free pickup service. Check the yellow pages for "rendering" or "grease trap."

5 *Composting*—Composting can be done both on- and off-site. The availability of land space, haulers, and/or end users in your area will help you decide which option is best for you. If you compost on-site, you will need to consider feed stocks, sitting, and operational issues. Composting can take many forms:

a *Un aerated static-pile composting*—Organic discards are piled and mixed with a bulking material. This method is best suited for small operations; it cannot accommodate meat or grease.

b *Aerated windrow/pile composting*—Organics are formed into rows or long piles and aerated either passively or mechanically. This method can accommodate large quantities of organics. It cannot accommodate large amounts of meat or grease.

c *In-vessel composting*—Composting that occurs in a vessel or enclosed in a building that has temperature and moisture controlled systems. They come in a variety of sizes and have some type of mechanical mixing or aerating system. In-vessel composting can process larger quantities in a relatively small area more quickly than windrow composting and can accommodate animal products.

d *Vermi composting*—Worms (usually red worms) break down organic materials into a high-value compost (worm castings). This method is faster than windrow or in-vessel composting and produces high-quality compost. Animal products or grease cannot be composted using this method.

In addition to food waste, linens may be recycled as well. Numerous companies exist to collect and recycle damaged or worn linens such as bed sheets and towels. Most of the organizations provide containers and collect the materials at no cost. These companies process the collected material back into raw material such as fabric and thread.

Depending on the amount of cardboard generated, a baler may be cost justified and operationally feasible if funds and space are available. A feasibility analysis should be conducted as discussed in the general waste audit process earlier in the book.

38.5   Implementation and Approach

Following are tips for implementing solid waste reduction in the hospitality industry:

■ Conduct a waste audit to determine the amounts, compositions, and quality of waste items generated.

■ Consult with local and state recycling coordinators. These solid waste planners may help locate a market for food discards or provide technical advice. Some agencies award grant money for innovative projects.

■ Ask the solid waste planners to provide you with contacts and information about businesses with successful food-recovery programs. By networking with other businesses you will be able to learn from their experiences. These organizations can also provide assistance in finding haulers and end users in your area.

■ Anticipate barriers to a successful program and how you will overcome them. Learn from others. Ask employees what potential problems they see. They, after all, will be responsible for running the program.

■ Train food-service workers well and well ahead of program implementation.

■ Educate suppliers and customers on recycling processes.

■ Monitor and periodically reevaluate your program.

■ Use composting diversion to reduce your waste hauling and tipping costs.

38.6   Case Study

In 2000 The University of Toledo Solid Waste Minimization Team conducted a walk-through survey of a large restaurant in Toledo, Ohio. The objective was to gain an overall understanding of the facility, identify major solid waste streams by application, and then to suggest some areas for reducing, reusing, and recycling material in the waste stream.

The restaurant is an upscale lunch and dinner provider specializing in wood-grilled steaks. As one of Toledo's most popular restaurants, the facility was in the process of creating a formal recycling program. The facility had no formal recycling program. In late 1999 the company discontinued cardboard recycling due to the need of an additional waste container, which management felt was not visually appealing and distracted from the dining experience. There were approximately 100 employees and the facility operates 363 days per year. The restaurant used about 30 reams of paper per month, which includes order sheets for the wait staff and cash register paper. The annual solid waste hauling cost was $8400 and the waste hauler collected cardboard at no cost, but this service was discontinued. The restaurant currently utilizes a 30-yd^3 compactor for all waste. Table 38.2 lists the solid waste minimization recommendations that were presented to the restaurant management team.

38.7   Exemplary Performers— LEED Certification

The Hilton Vancouver, Washington, one of America's first sustainably designed hotels, made history in 2008 as the first hotel in the world to earn both Leadership in Energy and Environmental Design (LEED) and Green Seal certifications. Green Seal, an independent nonprofit organization providing science-based environmental certification accolade, comes 3 years after the hotel became the first major U.S. hotel—and the first Hilton Hotel—to earn LEED certification.

"The Hilton Vancouver, Washington is the only hotel in the world to achieve environmental sustainability certification from two top universally accepted and independent

TABLE 38.2 RESTAURANT WASTE MINIMIZATION RECOMMENDATIONS (EXAMPLE)

RECOMMENDATIONS	COMMENTS
Implement a facility recycling program for paper, plastic, and aluminum cans (Annual savings of $80)	Instituting a facility-wide recycling program would eliminate the disposal of nearly 1800 lb of solid waste The annual revenue generated from the sale of the recyclables (paper, plastic, and aluminum cans) is approximately $80 at current market prices, but greater savings will be realized by the reduction of the solid waste-hauling bill (please see next recommendation)
Reduce the frequency of waste hauling (Annual savings of $1500)	Significant savings are possible by reducing the frequency of times the solid waste at the facility is collected. Currently the waste hauler removes waste 3 times per week. By removing recyclables from the waste stream and more efficiently scheduling waste hauling frequency, the waste hauling bill will reduce by about 20%. This will incur estimated annual savings of $1500
Install a cardboard collection container.	Since a large amount of food arrives in cardboard packages, a large amount a waste could be removed from the waste stream. Contact the waste hauler and discuss installing a visually appealing container. In most cases, the waste hauler will collect the cardboard free of charge
Contact local farmers to use food waste as animal feed.	Some farmers in the Midwest process food waste and transform it into high-nutrition, low-cost animal feed. Some farmers may be interested is some types of food wastes generated
Install collection containers for glass bottles and aluminum cans	Depending on how much of these materials are generated, a portion of the waste stream could be reduced (reducing waste hauling bill) and revenue could be generated from the sale of these materials

analysts," said Jeff Diskin, senior vice president, brand management, Hilton Hotels & Resorts. "This hotel has set a new standard for helping to minimize environmental impact while providing the very best in service and amenities to guests."

The Hilton Vancouver, Washington became Green Seal certified. Based on Green Seal criteria, the hotel implemented a comprehensive environmental management program for property operations, which includes elimination of hazardous substances, use of biodegradable cleaning products, wastewater management, energy efficiency and conservation, waste minimization, reuse and recycling, and an environmentally and socially sensitive purchasing policy. The Green Seal standards are internationally recognized as credible, transparent, and fair and provide third-party validation through science-based testing that a product works as well or better than others in its class.

Standards are established with input from representatives of various industries, governments, and academia.

Additionally as part of the Green Seal program, the hotel composts or donates excess food to reduce waste, minimize its carbon footprint, and support the local community. Over the past 12 months, more than 200,000 lb of compostable material and product, including food waste, has been diverted from landfills through the hotel's efforts.

"Because the hotel had a head start with LEED certification, adopting the Green Seal policies seemed like the next logical step," said Gerry Link, general manager of the Hilton Vancouver, Washington. "Our entire team takes great pride in the fact that the Hilton Vancouver, Washington is a leader in the hospitality industry in implementing environmentally sound sustainable green practices," said Link. "In the past seven months alone we've recycled more than 28,000 pounds of cardboard and we continue to find innovative ways to support the greening of travel and meetings and our overall operations."

Owned by the City of Vancouver, Washington, and managed by Beverly Hills-based Hilton Hotels Corporation, the Hilton Vancouver, Washington integrates some of the most advanced environmentally friendly features found in the hotel industry.

Soon after opening its doors in 2005, the Hilton Vancouver, Washington became one of the first hotels in the world to earn the coveted U.S. Green Building Council (USGBC) LEED certification, a nationally accepted benchmark for the design and construction of high-performance green buildings. The certification confirms that third-party experts have verified that the hotel satisfies criteria for sustainable site development, water savings, energy efficiency, materials selection, and indoor environmental quality. Today, the hotel is one of six LEED certified hotels in the United States.

In order to achieve LEED certification, the hotel incorporated several key energy-saving and waste-reducing strategies into its design and construction, including heat-reflecting rooftops and water-efficient landscaping. All windows within the hotel are energy efficient and all plumbing fixtures are low flow to limit water use. Some of the hotel's ecofriendly measures include the following:

- *Reducing energy use*—The hotel provides alternative fueling stations for electric cars and encourages its team members to find alternative methods of transportation to and from work. The hotel runs on 30 percent less energy than local codes require and CO_2 sensors turn off heating and cooling systems in empty rooms, meeting areas, and hallways.
- *Water-efficient landscaping*—The Hilton Vancouver, Washington's irrigation system reduces water use by 50 percent and the hotel features local native plants that need little water during the area's long, dry summer season. Storm water from the hotel is funneled to underground dry wells, which provide a natural filtering mechanism for pollutants.
- *Fighting the urban island heat effect*—A white reflective roof helps the hotel dissipate heat and reflect it back into space, rather than adding to the heat island effect that can afflict urban areas.

■ *Designing green guest rooms*—Operable windows in all guest rooms allow fresh air to enter the building and help control indoor pollutants. Additionally, the hotel was built with recyclable brick and 75 percent of the construction waste from the hotel was recycled.

■ *Supporting local vendors and sustainability*—The hotel uses materials from local and regional vendors within a 500 mile radius to support community vendors and reduce transportation impact. Building supplies are often purchased from local suppliers and Gray's At The Park, the Hilton Vancouver, Washington's restaurant, uses produce from a local farmers' market and meat, fish, cheese, wine, and bread from local farmers and producers.

39

AUTOMOTIVE SERVICE

APPLICATIONS

39.1   Industry Overview

NAICS code: all 44700s

INDUSTRY SNAPSHOT

121,446 auto service and gasoline operations in the United States
926,792 employees
$249.1 billion in annual sales
0.5 tons of solid waste generation per employee
Major waste streams: paper, plastics, and metals

Industries in the gasoline stations subsector retail automotive products and automotive oils with or without convenience-store items. These establishments have specialized equipment for the storage and dispensing of automotive fuels.

Automotive service technicians inspect, maintain, and repair automobiles and light trucks that run on gasoline, electricity, or alternative fuels such as ethanol. Automotive service technicians' and mechanics' responsibilities have evolved from simple mechanical repairs to high-level technology-related work. The increasing sophistication of automobiles requires workers who can use computerized shop equipment and work with electronic components while maintaining their skills with traditional hand tools. As a result, automotive service workers are now usually called technicians rather than mechanics. (Service technicians who work on diesel-powered trucks, buses, and equipment are diesel service technicians and mechanics. Motorcycle technicians—who repair and service motorcycles, motor scooters, mopeds, and small all-terrain vehicles—are small engine mechanics.)

Today, integrated electronic systems and complex computers regulate vehicles and their performance while on the road. Technicians must have an increasingly broad knowledge of how vehicles' complex components work and interact. They also must be able to work with electronic diagnostic equipment and digital manuals and reference materials.

When mechanical or electrical troubles occur, technicians first get a description of the problem from the owner or, in a large shop, from the repair service estimator or service advisor who wrote the repair order. To locate the problem, technicians use a diagnostic approach. First, they test to see whether components and systems are secure and working properly. Then, they isolate the components or systems that might be the cause of the problem. For example, if an air-conditioner malfunctions, the technician might check for a simple problem, such as a low coolant level, or a more complex issue, such as a bad electrical connection that has shorted out the air-conditioner. As part of their investigation, technicians may test drive the vehicle or use a variety of testing equipment, including onboard and hand-held diagnostic computers or compression gauges. These tests may indicate whether a component is salvageable or whether a new one is required.

During routine service inspections, technicians test and lubricate engines and other major components. Sometimes technicians repair or replace worn parts before they cause breakdowns or damage the vehicle. Technicians usually follow a checklist to ensure that they examine every critical part. Belts, hoses, plugs, brake and fuel systems, and other potentially troublesome items are watched closely.

Service technicians use a variety of tools in their work. They use power tools, such as pneumatic wrenches to remove bolts quickly; machine tools like lathes and grinding machines to rebuild brakes; welding and flame-cutting equipment to remove and repair exhaust systems; and jacks and hoists to lift cars and engines. They also use common hand tools, such as screwdrivers, pliers, and wrenches, to work on small parts and in hard-to-reach places. Technicians usually provide their own hand tools, and many experienced workers have thousands of dollars invested in them. Employers furnish expensive power tools, engine analyzers, and other diagnostic equipment.

Computers are also commonplace in modern repair shops. Service technicians compare the readouts from computerized diagnostic-testing devices with benchmarked standards given by the manufacturer. Deviations outside of acceptable levels tell the technician to investigate that part of the vehicle more closely. Through the Internet or from software packages, most shops receive automatic updates to technical manuals and access to manufacturers' service information, technical service bulletins, and other databases that allow technicians to keep up with common problems and learn new procedures.

High-technology tools are needed to fix the computer equipment that operates everything from the engine to the radio in many cars. In fact, today most automotive systems, such as braking, transmission, and steering systems, are controlled primarily by computers and electronic components. Additionally, luxury vehicles often have integrated global positioning systems, Internet access, and other new features with which technicians will need to become familiar. Also, as more alternate-fuel vehicles are purchased, more automotive service technicians will need to learn the science behind these automobiles and how to repair them.

Automotive-service technicians in large shops often specialize in certain types of repairs. For example, transmission technicians and rebuilders work on gear trains, couplings, hydraulic pumps, and other parts of transmissions. Extensive knowledge of computer controls, the ability to diagnose electrical and hydraulic problems, and other specialized skills are needed to work on these complex components, which employ some of the most sophisticated technology used in vehicles. Tune-up technicians adjust ignition timing and valves and adjust or replace spark plugs and other parts to ensure efficient engine performance. They often use electronic-testing equipment to isolate and adjust malfunctions in fuel, ignition, and emissions control systems.

Automotive air-conditioning repairers install and repair air-conditioners and service their components, such as compressors, condensers, and controls. These workers require special training in federal and state regulations governing the handling and disposal of refrigerants. Front-end mechanics align and balance wheels and repair steering mechanisms and suspension systems. They frequently use special alignment equipment and wheel-balancing machines. Brake repairers adjust brakes, replace brake linings and pads, and make other repairs on brake systems. Some technicians specialize in both brake and front-end work.

39.2   Waste Management Goals and Opportunities

The majority of solid waste generated by this sector is paper, plastics, and metals. Table 39.1 displays the composition breakdown based on survey results.

As shown in the table, the recycling rate for this sector is approximately 33 percent. As derived from the solid waste evaluation model discussed in Chap. 12, the equation that estimates the annual waste generation per year per employee for this sector can be calculated from the following:

Tons of solid waste generated per year = 0.51 × number of employees + 6.48

39.3   Constraints and Considerations

In terms of solid waste management and minimization for auto repair facilities, the three key constraints and considerations are

1 *Available space to store and stage materials (interior and exterior)*—This includes space within the facility and dock space for storage. The numerous deliveries that many of these facilities receive each day of the week can also present a logistic and scheduling problem for coordinating dock usage during peak periods.
2 *The separation of waste and contamination*—Separating recyclable waste from non-recyclable waste creates a new process; in conjunction with this, informing staff and visitors of the separation process also creates an additional management concern.

TABLE 39.1   AUTOMOBILE REPAIR AND SERVICE SOLID WASTE COMPOSITION (SURVEY RESULTS)

MATERIAL	COMPOSITION (%)	RECYCLING (%)
Paper	26 ± 7.3	14 ± 5.3
Mixed office paper	18 ± 5.0	14 ± 5.4
Newspaper	5 ± 1.4	14 ± 5.3
Paper (other)	3 ± 0.8	12 ± 4.6
Plastics	17 ± 4.8	7 ± 2.7
HDPE	5 ± 1.4	8 ± 3.0
PET	4 ± 1.1	7 ± 2.7
LDPE	3 ± 0.8	6 ± 2.3
PP	3 ± 0.8	8 ± 3.0
PS	2 ± 0.6	0 ± 0.0
Metals	15 ± 4.2	65 ± 14.2
Ferrous meals	9 ± 2.5	66 ± 14.3
Nonferrous metals	6 ± 1.7	67 ± 14.8
OCC (Cardboard)	12 ± 3.4	36 ± 13.7
Rubber (Tires)	8 ± 2.2	95 ± 6.1
Food Waste	7 ± 2.0	0 ± 0.0
Chemical/oils	6 ± 1.7	99 ± 4.1
Glass	3 ± 0.8	31 ± 11.8
Yard waste	2 ± 0.6	0 ± 0.0
Other	4 ± N/A	0 ± 0.0
Overall recycling level		33.4

3 *Finding appropriate suppliers for waste removal and recycling.* Well established and reliable waste removal and recycling providers are critical for a successful program; when meeting with prospective companies, request a list of references.

39.4   Potential Technologies and Strategies

Auto recycling is the business of recycling automobiles for the resale of usable parts and the wholesale of scrap material. Auto recycling differs from auto wrecking in many ways, the biggest being the care and handling of resalable parts. Working parts are carefully removed by qualified technicians then inventoried using computer databases. Most auto recyclers rely on North America–wide computer software to inventory parts. This process also allows an individual recycler to access inventories across North America via the Internet. Each part is given a specific numerical code.

Auto recyclers take great pride in the image they present to the public. The days of junkyards with heaping stacks of crushed and wrecked vehicles are long gone. Today's recyclers keep very clean and well-organized yards and many provide showrooms to display the most prominent parts available.

Store fronts are attractive and located in easily accessible industrial areas. Staff is knowledgeable and well versed in both auto parts and the Hollander Interchange allowing for exemplary customer care. Many also operate full-service auto repair and/or body shops, employing licensed mechanics, bodymen, and technicians. Though these services may be offered at a rate lower than that of dealership shops, the work is on par and staff are trained in the same manner, often at the same institutions as those employed by dealership or service-specific shops.

Most Canadian provinces and American states have their own auto-recycling associations. There is also a worldwide association, the Automotive Recyclers Association or ARA for short. These governing bodies work to ensure high standards of service and ethical recycling practices are maintained. Recyclers adhere to strict environmental policies and keep in accordance with municipal, state/provincial, and federal laws when dealing with hazardous and nonhazardous waste disposal, use, and resale. The specific areas of concentration for waste minimization in this sector include

- *Tire recycling*—Whole tires can be reused in many different ways. One way, although not recycling, is for a steel mill to burn the tires for carbon replacement in steel manufacturing. Tires are also bound together and used as different types of barriers such as collision reduction, erosion control, rainwater runoff, wave action (that protects piers and marshes), and sound barriers between roadways and residences. Entire homes can be built with whole tires by ramming them full of earth and covering them with concrete, known as *Earthships*.

  Some artificial reefs are built using tires that are bonded together in groups. There is some controversy on how effective tires are as an artificial reef system; an example is the Osborne Reef Project.

  The process of stamping and cutting tires is used in some apparel products, such as sandals and as a road subbase, by connecting together the cut sidewalls to form a flexible net.

  Chipped and shredded tires are used as tire-derived fuel (TDF). TDF helps to eliminate tires from our waste stream and produces a fuel source. Tires are also used in civil engineering applications such as: subgrade fill and embankments, backfill for walls and bridge abutments, subgrade insulation for roads, landfill projects, and septic system drain fields.

  Ground and crumb rubber, also known as size-reduced rubber, can be used in both paving-type projects and in moldable products. These types of paving are rubber modified asphalt (RMA) or rubber modified concrete, and act as a substitution for an aggregate. Examples of rubber-molded products are carpet padding or underlay, flooring materials, dock bumpers, patio decks, railroad crossing blocks, livestock mats, sidewalks, rubber tiles and bricks, moveable speed bumps, and curbing/edging. Then there is plastic and rubber blend-molded products like pallets and railroad ties. Athletic and recreational areas can also be paved with the shock absorbing rubber-molded material. Rubber from tires is sometimes ground into medium-sized chunks and used as rubber mulch.

Ground up tires even find their way back to your car in the form of automotive parts, like: exhaust hangers, brake pads and shoes, acoustic insulation, and even low percentages go into making new tires.

- *Metal recycling*—Metal recycling can offer large paybacks as it demands the highest price on commodity markets. Steel and aluminum are the most common metals recycled and numerous vendors operate in most areas that provide recycling services.
- *Used auto part recycling*—Waste and material exchanges connect businesses with reusable materials with other businesses that can use them. These exchanges allow businesses to capture the value of by-products and surplus materials and generate revenue, reduce disposal costs, and reduce purchasing costs. Many waste and material exchanges exist for used automobile parts and are easily found online. This is also a low-cost method to reduce disposal costs and enhance corporate images.

39.5   Implementation and Approach

In implementing solid waste minimization strategies, the first step is to conduct a solid waste audit to identify opportunities. Once the annual waste generation amounts are known the company can identify potential haulers and recyclers to handle the designated materials. They often serve as valuable partners in the process. Obtain and retain verification records (waste hauler receipts and waste management reports) to confirm that diverted materials have been recycled or diverted as intended. This includes the identification and screening of potential waste and material exchanges, which is most easily completed with Internet searches.

Once the amounts and vendors have been selected, collection bins can be acquired and processes can be standardized. A key component to this is communication with employees and supplier to ensure processes are followed and continuously improved.

39.6   Case Study

In 2006, The University of Toledo Waste Minimization Team conducted a solid waste audit a gas station/automobile repair center in Northwest Ohio. The facility employed 20 people and did not have a recycling program at the time of the audit. Results from the solid waste audit indicated that the facility generated nearly 17 tons of solid waste per year and had annual waste removal expenses of $1500. The majority of the waste stream consisted of mixed paper, used automobile components (metals and plastics), food waste, and cardboard.

The waste minimization team recommended that the facility establish a relationship with a local metal and paper recycler based on the volumes of materials generated. As a result, the facility entered agreements with a large metal-recovery facility and paper recycler. Both vendors provided containers and transportation for the collected materials. The metal recycler paid the facility for all metal collected, which resulted in annual revenue of $1200 and the paper vendor provided collection services at no charge. The net annual financial benefit to the company from reduced waste hauling and recycling revenue was $1700 and over 5 tons of material was diverted from land-fills per year.

39.7   Industry Leaders—The Automotive Recyclers Association

The Automotive Recyclers Association (ARA) was formed in 1943, and is a nonprofit trade association that has represented an industry dedicated to the efficient removal and reuse of automotive parts, and the safe disposal of inoperable motor vehicles.

ARA services approximately 1000 member companies through direct membership and more than 2000 other companies through affiliated chapters. Suppliers of equipment and services to this industry complete ARA's membership. ARA is the only trade association serving the automotive-recycling industry in 12 countries internationally.

ARA aims to further the automotive recycling industry through various services and programs to increase public awareness of the industry's role in conserving the future through automotive recycling and to foster awareness of the industry's value as a high-quality, low-cost alternative for the automotive consumer. ARA encourages aggressive environmental management programs to assist member facilities in maintaining proper management techniques for fluid and solid waste materials generated from the disposal of motor vehicles.

40

EDUCATION APPLICATIONS

40.1 Industry Overview

NAICS code: all 61000s

INDUSTRY SNAPSHOT

49,319 educational operations in the United States
430,164 employees
$30.7 billion in annual sales
0.25 ton of solid waste generation per employee
Major waste streams: paper, food waste, yard waste

Successful operation of an educational institution requires competent administrators. Educational administrators provide instructional leadership and manage the day-to-day activities in schools, preschools, day-care centers, and colleges and universities. They also direct the educational programs of businesses, correctional institutions, museums, and job training and community service organizations.

Educational administrators set educational standards and goals and establish the policies and procedures to achieve them. They also supervise managers, support staff, teachers, counselors, librarians, coaches, and other employees. They develop academic programs, monitor students' educational progress, train and motivate teachers and other staff, manage career counseling and other student services, administer record-keeping, prepare budgets, and perform many other duties. They also handle relations with parents, prospective and current students, employers, and the community. In an organization such as a small day-care center, one administrator may handle all these functions. In universities or large school systems, responsibilities are divided among many administrators, each with a specific function.

Educational administrators who manage elementary, middle, and secondary schools are called principals. They set the academic tone and actively work with teachers to develop and maintain high curriculum standards, develop mission statements, and set performance goals and objectives. Principals confer with staff to advise, explain, or answer procedural questions. They hire, evaluate, and help improve the skills of teachers and other staff. They visit classrooms, observe teaching methods, review instructional objectives, and examine learning materials. Principals must use clear, objective guidelines for teacher appraisals, because pay often is based on performance ratings.

Principals also meet and interact with other administrators, students, parents, and representatives of community organizations. Decision-making authority has increasingly shifted from school district central offices to individual schools. School principals have greater flexibility in setting school policies and goals, but when making administrative decisions they must pay attention to the concerns of parents, teachers, and other members of the community.

Preparing budgets and reports on various subjects, including finances and attendance, and overseeing the requisition and allocation of supplies also is an important responsibility of principals. As school budgets become tighter, many principals have become more involved in public relations and fundraising to secure financial support for their schools from local businesses and the community.

Principals must take an active role to ensure that students meet national, state, and local academic standards. Many principals develop partnerships with local businesses and create school-to-work transition programs for students. Increasingly, principals must be sensitive to the needs of the rising number of non–English speaking and culturally diverse student body. In some areas, growing enrollments also are a cause for concern because they lead to overcrowding at many schools. When addressing problems of inadequate resources, administrators serve as advocates for the building of new schools or the repair of existing ones. During summer months, principals are responsible for planning for the upcoming year, overseeing summer school, participating in workshops for teachers and administrators, supervising building repairs and improvements, and working to make sure the school has adequate staff for the school year.

The educational services sector comprises establishments that provide instruction and training in a wide variety of subjects. This instruction and training is provided by specialized establishments, such as schools, colleges, universities, and training centers. These establishments may be privately owned and operated for profit or not for profit, or they may be publicly owned and operated. They may also offer food and accommodation services to their students.

Educational services are usually delivered by teachers or instructors who explain, tell, demonstrate, supervise, and direct learning. Instruction is imparted in diverse settings, such as educational institutions, the workplace, or the home through correspondence, television, or other means. It can be adapted to the particular needs of the students, for example sign language can replace verbal language for teaching students with hearing impairments. All industries in the sector share this commonality of process, namely, labor inputs of instructors with the requisite subject matter expertise and teaching ability.

Teachers play an important role in fostering the intellectual and social development of children during their formative years. The education that teachers impart plays a key role in determining the future prospects of their students. Whether in preschools or high schools or in private or public schools, teachers provide the tools and the environment for their students to develop into responsible adults.

Teachers act as facilitators or coaches, using classroom presentations or individual instruction to help students learn and apply concepts in subjects such as science, mathematics, or English. They plan, evaluate, and assign lessons; prepare, administer, and grade tests; listen to oral presentations; and maintain classroom discipline. Teachers observe and evaluate a student's performance and potential and increasingly are asked to use new assessment methods. For example, teachers may examine a portfolio of a student's artwork or writing in order to judge the student's overall progress. They then can provide additional assistance in areas in which a student needs help. Teachers also grade papers, prepare report cards, and meet with parents and school staff to discuss a student's academic progress or personal problems.

Many teachers use a hands-on approach that uses props or manipulatives to help children understand abstract concepts, solve problems, and develop critical thought processes. For example, they teach the concepts of numbers or of addition and subtraction by playing board games. As the children get older, teachers use more sophisticated materials, such as science apparatus, cameras, or computers. They also encourage collaboration in solving problems by having students work in groups to discuss and solve problems together. To be prepared for success later in life, students must be able to interact with others, adapt to new technology, and think through problems logically.

40.2   Waste Management Goals and Opportunities

Solid wastes represent a significant loss of natural resources and school district funds as well as a potential threat to student/staff health and the environment. To be responsible stewards of environmental quality, school districts should review processes and operations, and even curriculum choices. They should evaluate the economic, educational, and environmental benefits of implementing an effective waste reduction program.

Incorporating waste reduction as part of the school district's overall way of doing business can provide a number of important benefits:

- Reduced disposal costs
- Improved worker safety
- Reduced long-term liability
- Increased efficiency of school operations
- Decreased associated purchasing costs

TABLE 40.1 EDUCATION SECTOR SOLID WASTE COMPOSITION (SURVEY RESULTS)

MATERIAL	COMPOSITION (%)	RECYCLING (%)
Paper	33 ± 5.7	31 ± 6.5
Mixed office paper	30 ± 5.7	31 ± 6.5
Newspaper	3 ± 2.1	30 ± 6.3
Food waste	31 ± 6.7	0 ± 0.0
Yard waste	10 ± 1.7	0 ± 0.0
Mixed plastics	7 ± 1.3	24 ± 5.0
OCC (cardboard)	7 ± 1.4	37 ± 7.8
Aluminum cans	4 ± 0.9	84 ± 7.1
Glass	2 ± 0.7	29 ± 6.1
Other	$6 \pm N/A$	0 ± 0.0
Overall recycling level		18.4

School district waste reduction programs also foster student achievement by transforming the school environment into a laboratory for learning and providing numerous opportunities for investigation through environment-based education.

The majority of solid waste generated by this sector is paper, food waste, and yard waste. Table 40.1 displays the composition breakdown based on survey results.

As shown in the table, the recycling rate for this sector is approximately 18 percent. As derived from the solid waste evaluation model discussed in Chap. 12, the equation that estimates the annual waste generation per year per employee for this sector can be calculated from the following:

Tons of solid waste generated per year $= 0.25 \times$ number of employees $+ 35.68$

40.3   Constraints and Considerations

In terms of solid waste management and minimization for educational facilities, the five key constraints and considerations are

1 *Available space to store and stage materials (interior and exterior)*—This includes space within the facility and dock space for storage. Many school buildings are very tight on space, both in terms of classroom usage and storage.

2 *The separation of waste and contamination*—Separating recyclable waste from nonrecyclable waste creates a new process; in conjunction with this, informing students and staff of the separation process also creates an additional management concern.

3 *Finding appropriate suppliers for waste removal and recycling.* Well established and reliable waste removal and recycling providers are critical for a successful program; when meeting with prospective companies, request a list of references.

4 *Staff constraints*—Many schools have limited staff, including custodial, to perform waste collection and recycling activities.

5 *Budgetary constraints*—Many schools operate on very lean budgets and have limited funds to invest in environmental projects, including the purchase of recycling containers and promotional materials.

40.4 Potential Technologies and Strategies

The three Rs (reduce-reuse-recycle) is a resource conservation philosophy promoting a reduction in solid waste generation through changes in the manner in which materials are purchased, used, and discarded. This approach can easily be integrated into a school district's business and educational culture to conserve resources, reduce waste, and save money. A successful school district waste reduction program depends on the coordination of administration, management, faculty, staff, students, and parents.

The following provide specific waste reduction strategies that can be implemented within each program area of a school district, and, as a result, can be incorporated into the school district's daily operations. These strategies can also serve as the basis for developing administrative procedures for each district department to support a district-wide waste reduction policy:

- *District administration*—Suggestions on how to provide administrative direction and support through district-wide integrated waste management policy and administrative procedures.
- *Business services*—Information regarding contracting for waste management services and paper waste reduction opportunities for school and district offices and printing areas.
- *Facilities and planning*—Excellent resources regarding greening school district building projects, including recycling construction and demolition wastes.
- *Food service*—Waste reduction ideas for the cafeteria and kitchen, including "offer versus serve," food donations, recycling, composting, school gardens, and more.
- *Maintenance and operations*—Waste reduction suggestions for the custodial, grounds, and maintenance operations.
- *Personnel*—Ideas for incorporating waste reduction program goals into duty statements and employee evaluations.
- *Purchasing*—Tips on how to purchase for waste reduction and how to buy recycled and other environmentally preferable products.

- *Technology services*—Various Internet sites provide information and resources regarding reuse, recycling, and proper disposal of old computer equipment or other electronic devices.
- *Transportation*—Information and resources highlighting key pollution prevention strategies to reduce the amount materials purchased and disposed of in school district transportation departments.

40.5   Implementation and Approach

For schools considering starting a recycling program at their school, there are resources available to help. The following guidelines should be followed when setting up a school recycling program:

- Organize a coordination team.
  - Involve students, parents, teachers, custodial staff, local solid waste or public works departments, and community representatives.
- Determine which recyclables are in your waste stream.
  - Perform a waste composition study and categorize the trash to determine what waste can be minimized or recycled. Use the results of the audit to help create a specific recycling program.
- Identify a local market for recyclables.
- Contact local recycling facilities to see what materials they collect and what services they provide. Be sure to find out how recyclables should be separated and what items can be commingled.
- Find local recycling facilities. Contact the school's current waste hauler to see if they provide recycling services as well. If your local government solid waste office already has a curbside or business recycling program, see if the school can be added to the pickup schedule.
- Select the type of recycling program that would be best for each school.
  - Contract with a private hauler, tap into curbside recycling within the community, or establish a mini drop-off facility at the school for the entire community. To prevent the accumulation of items that you cannot recycle, make sure to have all aspects of your program in place before collecting any recyclables.
- Work out a budget for the collection program.
  - Obtain money from the school budget, PTA fundraising, or partnerships with local businesses or civic groups. Recycling should reduce the school's waste stream, so look into reducing the frequency of trash pickups and allocating those savings toward the pickup of recyclables.
  - Apply for a grant to help fund containers for your program.
- Establish a system for collecting and storing recyclables.
  - Place bins in easily accessible areas within the school. Focus on areas that generate recyclables, such as classrooms, the cafeteria, teacher lounges, and copy rooms.

- Bins can be old copy paper boxes, plastic storage containers, or a local government curbside recycling bin.
- Have students decorate the bins with their own artwork or pictures from the Recycle Guys or RE3.org Web pages. Participating helps students feel ownership of the program. Each class could decorate their own bin or the school could have a contest to pick the most creative picture for each grade level.
- Check with the fire marshal for storage and collection requirements.
- If a private hauler will be collecting the recyclables, make sure to set aside storage space for the containers that allow truck access. Designate a publicly accessible area if establishing a drop-off facility for the community.
- Educate the school and the community about the program.
  - Inform all school personnel, students, parents, and the community how the program will work. Let everyone know what can and cannot be recycled.
  - Ready-made graphics are available on the RE3.org and Recycle Guys Web sites for easy printing. Monthly newsletters or e-mails can be an effective way to inform the community and parents of the recycling program's progress. Educate volunteers and staff on the storage and collection procedures, and the location of containers.
  - Integrate environmental lesson plans and recycling education into the curriculum.
- Set overall and individual goals.
  - Convey the goals of the project to all participants and give specific examples of how each person, class, or school can help reach these goals. Tally the totals and track progress for all to see. For example, put posters in the hallways with fun facts: "Last month the paper recycled from our school saved four trees."
  - To the extent possible, keep track of how many pounds or tons of materials are collected over time to evaluate the program's performance and to set benchmarks for improvement.
- Reward the doers.
  - Let students know that a cleaner environment is a prize they can all enjoy. Other incentives can be given to students and classes who participate, such as field trips to a material-recovery facility or a landfill and RE3.org or Recycle Guys t-shirts, stickers, or posters.

40.6   Case Study

In 2003, The University of Toledo Waste Minimization Team conducted a recycling survey at a small private college in Northwest Ohio. The college had an enrollment of 2100 students and a student-to-faculty ratio of 15:1. The objectives of the recycling research were to determine recycling opportunities, identify major solid waste streams, and suggest potential areas for reducing, reusing, and recycling materials in the waste stream.

The college campus consisted of 28 individual facilities in 14 unconnected buildings. These facilities include dormitories, an auditorium building, a college, a care center, and administrative buildings. The facility had a small recycling program in place for mixed office paper, cardboard, and aluminum cans. The college established a team to explore waste reduction, cost savings, and recycling opportunities at the facility.

The procedure used to estimate the annual solid waste streams at the college's facilities involved a 4-day sampling of the 141 waste containers. The containers were each observed to determine the total amount of waste, types of materials, and quantity of each material. From this information, volumes of solid waste were calculated for each container. Using standard packing densities, weights of recyclable materials were estimated. Volumes are required to determine waste hauling cost savings, and weights are required to determine revenue generated from the sale of recyclables. Four assessments took place: once during a light day (Friday), once during a heavy day (Wednesday), once on an extra light day during the summer term (Monday), and another for "other areas" where the amount of waste would remain the same throughout the year, (Thursday). Annual waste streams were calculated using the data collected during each of these different days. This report illustrates the minimum amounts of recyclable waste produced at this campus. Although the difference between the sample data for light and heavy days is within 5 percent, the data gathered for the light days was used because it results in a more conservative analysis. On average each container was found to be approximately 35 percent full just before emptying (mostly after 1 day of use) and produced nearly 315 lb each per year.

Table 40.2 and Fig. 40.1 display the major solid wastes disposed of annually by the college. The waste streams are listed by annual tonnage and by the percentage of total waste. As shown in Table 40.2, the facility disposed of approximately 22 tons of solid waste each year. All this material goes to a landfill. Food waste, which is nonrecyclable, comprises about 22 percent of the total waste.

TABLE 40.2 ANNUAL TONS OF MAJOR SOLID WASTE STREAMS NOT BEING RECYCLED AT THE TIME OF THE AUDIT

RANK	COMPONENT	TONS PER YEAR	PERCENTAGE OF TOTAL
1	Mixed office paper	6.27	28%
2	Magazines	5.20	23%
3	Food waste	4.79	22%
4	Newspaper	2.87	14%
5	Aluminum cans	1.42	6%
6	Cardboard	1.17	5%
7	Plastic bottles	0.53	2%
	Total	22.25	100%

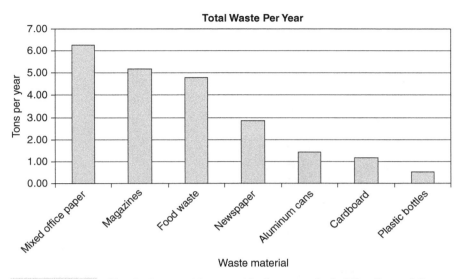

Figure 40.1   **Chart of annual tons not being recycled at the time of the audit.**

Table 40.3 displays the major potentially recyclable annual waste streams at the campus. The annual weights (tons) are shown. This information is useful because recycling vendors usually pay for recyclables based on weight. If these materials can be removed from the waste stream and sent to a recycler instead of the landfill, revenue could be generated and waste-hauling costs reduced.

TABLE 40.3 MAJOR RECYCLING OPPORTUNITIES FROM AUDIT			
COMPONENT	TONS PER YEAR	CURRENT MARKET VALUE PER TON*	REVENUE PER YEAR (FROM SALE OF RECYCLABLES)
Mixed office paper	6.27[†]	Collection by LCSWMD	$0
Magazines	5.20	Collection by LCSWMD	$0
Newspaper	2.87[†]	Collection by LCSWMD	$0
Aluminum cans	1.42	$980	$1,389
Cardboard	1.17	$65	$76
Plastic bottles	0.53	$3	$2
Totals	17.46		$1,467

*The prices in this table were retrieved from www.amm.com, and its most current information.
[†]Proposed collection for Lucas County.

Figure 40.2 Proposed recycling container for schools.

As shown in the table, the college could potentially generate over $1400 annually from the sale of mostly aluminum cans and some other recyclables to local area recyclers. Figure 40.2 shows a typical recycling container used for schools. The amount paid per ton of recyclables was gathered from current market prices. Due to the low amounts of waste generated by all components except mixed office paper and newspaper, many vendors were not likely to collect these other items. However, mixed office paper and newspaper are the recyclable items that generate enough waste for county-sponsored weekly collection. Aluminum cans may be recycled but are generated in quantities that are too small for collection by the county.

Additional cost savings could also be realized from a reduction in waste hauling costs. Based on this assessment, waste hauling charges can be reduced by approximately 75 percent by reducing the volume of waste taken by the hauler. The college could potentially save approximately $6900 annually (calculated from a 75 percent reduction of the annual waste hauling cost of $9212). This could be achieved by removing recyclables (mixed office paper, newspaper, and aluminum cans) from the waste stream. The cost savings will be achieved by reducing the frequency of waste collection at the facilities. It is also possible to recycle the other recyclable items which are not collected by the county, to further reduce the frequency of collection and subsequently lower hauling costs.

If the above recommendations are implemented, the college will save a total of $8300 ($1400 from sale of aluminum cans, $6900 from waste hauling reduction.). Specific information regarding the procedures used to achieve these cost reductions is in Table 40.4.

TABLE 40.4 RECOMMENDATIONS FOR WASTE AUDIT AT THE COLLEGE	
ITEM	**RECOMMENDATION**
Improve the school recycling program for paper, plastic, and aluminum cans	Expanding the recycling program would eliminate the disposal of over 17 tons of solid waste annually Allow Lucas County SWMD to install a paper-recycling container on the campus (if LCSWMD is able to do so). Lucas County will then collect the paper at no cost, which would reduce waste hauling costs. Arrange to sell aluminum cans to one of the local recyclers
Reduce the frequency of waste collection (Annual Savings of $6900)	Significant savings are possible by reducing the frequency of times the solid waste at the facilities is collected by Waste Management, Inc. Currently Waste Management, Inc. removes waste from the facilities either every week or every other week in various areas. The total volume of recyclable waste streams is estimated to be 515 yd^3/year. By removing recyclables from the waste stream, the campus could divert up to 75% of the volume of their waste into recycling. The organization, therefore, can adjust their waste hauling schedules to accommodate the reduction of the materials to be taken to a landfill, thereby reducing the waste hauling cost. For this study, a conservative reduction in waste hauling cost is estimated to be approximately 75%. This will generate an estimated annual savings of $6900
Use paperless fax machines	Depending on fax usage, substantial paper cost savings could be realized, as well as waste reduction
Set printer and copier defaults to print on both sides of the paper, when possible	This will aid in reducing paper waste. At $2 per ream, the college could achieve significant cost savings as well

40.7   Exemplary Performers— RecycleWorks School Recycling Award Program

San Mateo County in California has developed an awards program for schools in the community that promote recycling. RecycleWorks recognizes schools and school district recycling leaders. Awards are given annually to schools, administrators, students and/or parents who demonstrate excellence in creating or expanding a recycling program or who initiate a recycling project within the school community or district to increase awareness of resource conservation. The following schools are recipients of the award:

- Central Elementary School has won the School Award for Resource Conservation from RecycleWorks in 2003, a President's Environmental Youth award in 2004, and a Sustainable San Mateo County Award in 2005 for their "Getting Green at Central" program. They organized a school-wide recycling movement for the collection of everything from aluminum cans to tennis shoes. The money that they saved from recycling these materials paid for their new play structure.
- McKinley Institute of Technology was honored with the School Award for Resource Conservation from RecycleWorks in 2003 for their seventh grade "Bridge to English" program taught by Ms. Morgan. The students in this class took a recycling field trip to understand the processes and benefits of recycling. The class decided to improve upon their school's recycling program by conducting weekly pickups of recyclables, and educating the rest of the school about the importance of recycling. The students in the class took pride in the service they were providing to their school and to the environment.
- Students at Trinity School in Menlo Park have created a program to let the students give back to their community. In 2004, each class conducted a project focused on recycling, waste reduction, and reuse of products. The third graders sponsored a recycling program that collected paper, cans, and bottles. The second graders used leftover lunch waste to vermi compost and created rich soil for the school's garden where each grade raised a small garden of organic plants. Each month, the entire school participated in a no-waste lunch program. All the efforts of Trinity School have helped recycle, reduce, and reuse materials.

41

INTERNATIONAL MANUFACTURING
APPLICATIONS

41.1   International Waste Management Goals and Opportunities

Each year the European Union alone disposes of 1.3 billion tones of waste, some 40 million tones of it hazardous. This amounts to about 3.5 tones of solid waste for every man, woman and child.

Recycling rates increased recently in the EU. The percentage of waste being disposed of in landfills decreased as a corollary. However, landfill remains the prevailing option in many EU countries. Indeed, treatment of municipal waste in the EU in 1999 was distributed as follows:

- Landfills: 57 percent
- Incineration: 16 percent
- Recycling and composting: 20 percent
- Others 7 percent

The EU Landfill Directive (1999/31/EC) promotes a decrease in landfilling waste. Biodegradable waste counts for approximately two-thirds of total municipal waste quantities. By 2016, the EU Landfill Directive provides for a reduction of the quantity of biodegradable material to be landfilled of 35 percent of the 1995 levels.

Only a few EU member states had reached this target in 2001 whereas in accession countries the fraction of municipal waste going to landfill is generally more than 90 percent and in many cases very close to 100 percent.

The municipal waste management market is primarily driven by legislation, and here the impact of the EU Landfill Directive continues to be felt in most countries. Restrictions on the use of landfill for the disposal of municipal solid waste (MSW) are encouraging the development of alternate technologies, with nonthermal treatment proving particularly popular. With respect to opportunities in the services sector, sorting and separation services and biological treatment are both attracting substantial investment. Thermal treatment, while somewhat less popular than it was a decade ago, continues to have a role in a number of European countries.

Frost & Sullivan's current research estimates that the municipal waste management market was just over $30 billion in 2005 with a stable compound annual growth rate (CAGR) of 3 percent projected until 2012. Within the market, the role of the public sector in European MSW management remains strong. Despite moves in many countries to increase the involvement of private sector companies, public ownership of facilities such as landfill and thermal-treatment plants remains significant. As yet, no major moves toward large-scale consolidation within the market have occurred, although on a national and regional scale large waste management companies are emerging who have the potential to accelerate this process.

Overall, the Western European market for municipal waste management services has shown steady growth. This is despite a recent trend in many European countries to minimize waste volume entering the market. Waste volume has continued to show year-to-year increases in most countries, with no suggestion of any major reversal in the amount of waste being generated.

Over the medium and long terms, the European municipal waste management market is expected to continue to attract revenue as pricing factors and rising taxes raise the profile of the industry. Prices have risen in many sectors and markets, offering a boost to revenues after a period of difficult competition.

End users, though, are looking more at service quality, technical know-how, and delivery. The market has moved toward higher value pretreatment services. More stable pricing across many sectors has also had a positive impact on the market.

The nonhazardous industrial waste management services sector is generally the least technologically developed of European waste management markets, as the primary focus is on municipal and hazardous waste. Moreover, across Europe, market conditions are not uniform, and opportunities still exist for companies able to position themselves in the right countries or regions and service sectors. Revenues are estimated at $62 billion and growing relatively slowly at a CAGR of 1.5 percent due to an anticipated decline in waste volume generated.

A recent revenue spurt has been attributed to the buoyant economies of Southern Europe, due to an increase in industrial, commercial, and construction activity. Moreover, shifting patterns of waste management that are being stimulated by legislative and regulatory changes have further supported growth in this area. The market is expected to continue growing as waste is diverted toward more value-added channels, though at a slower rate as waste volume decreases.

41.2   Constraints and Considerations

The key constraints associated with solid waste minimization for international manufacturers include

■ Increased waste disposal costs
■ Increased and varying government regulations
■ Lack of recycling vendors in certain areas
■ Space constraints within and outside the building to stage materials
■ In some cities, narrow streets

41.3   Potential Technologies and Strategies

The German *Grüner Punkt* is considered the forerunner of the European scheme. It was originally introduced by *Duales System Deutschland GmbH* (DSD) in 1991 following the introduction of a packaging ordinance under the Waste Act. Since the successful introduction of the German industry–funded dual system similar Green Dot systems have been introduced in most other European countries.

The Green Dot scheme is captured under the European "Packaging and Packaging Waste Directive—94/62/EC," which is binding for all companies if their products use packaging and requires manufacturers to recover their own packaging. According to the directive, if a company does not join the Green Dot scheme they must collect recyclable packaging themselves although this is almost always impossible for mass products and only viable for low-volume producers. Regulatory authorities in individual countries are empowered to fine companies for noncompliance although enforcement varies by country. Environmentalists claim that some countries deliberately turn a blind eye to the European directive.

Since European introduction, the scheme has been rolled out to 23 European countries. In some, namely, France, Turkey, Spain, Portugal, and Bulgaria, companies joining the Green Dot scheme must use the logo. The Green Dot is used by more than 130,000 companies encompassing 460 billion packages.

The basic idea of the Green Dot is that consumers who see the logo know that the manufacturer of the product contributes to the cost of recovery and recycling. This can be with household waste collected by the authorities (e.g., in special bags—in Germany these are yellow), or in containers in public places such as car parks and outside supermarkets.

The system is financed by a Green Dot license fee paid by the producers of the products. Fees vary by country and are based on the material used in packaging (e.g., paper, plastic, metal, wood, and cardboard). Each country also has different fees for joining the scheme and ongoing fixed and variable fees. Fees also take into account the cost of collection, and sorting and recycling methods. In simple terms, the system encourages manufacturers to cut down on packaging as this saves them the cost of license fees.

In 1991, the German government passed a packaging law (*Verpackungsverordnung*) that requires manufacturers to take care of the recycling or disposal of any packaging material they sell. As a result of this law, German industry set up a dual system of waste collection, which picks up household packaging in parallel to the existing municipal waste collection systems. This industry-funded system is operated in Germany by the *Duales System Deutschland GmbH* (German for "Dual System Germany Ltd.") corporation, or short DSD.

DSD collects only packaging material from manufacturers who pay a license fee to DSD. DSD license fee payers can then add the Green Dot logo to their package labeling to indicate that this package should be placed into the separate yellow bags or yellow wheelie bins that will then be collected and emptied by DSD-operated waste collection vehicles and sorted (and where possible, recycled) in DSD facilities.

German license fees are calculated using the weight of packs, each material type used, and the volumes of product produced per annum.

41.4   Implementation and Approach

Concerns for the environment have forced many international firms to define policies that protect the environment within which they operate. To implement a waste minimization program a firm should first identify legislative and legal issues associated with waste disposal. This information is readily available on the Internet. In addition here are a few points:

- Form a task team to analyzed solid waste generation
- Conduct a solid waste audit
- Determine annual generation amounts by material type
- Identify vendors to assist in waste removal and recycling efforts
- Communicate results and expectations to all stakeholders within the firm
- Track results and continuously improve the program

The European Union has developed unique strategies for addressing waste and resources issues. Since the 1980s, a number of policies and directives have been discussed and adopted at the EU level to reduce waste generation in the European Union, with priority given to waste minimization and prevention, and reuse and recycling.

More recently, the European Union emphasized the link between resource efficiency and waste generation in two major documents: the *Sustainable Development Strategy* and the *Sixth Environmental Action Program (6EAP)*. They both set as an essential objective the decoupling of economic growth, the use resources, and the generation of waste. To achieve this objective, the European Commission has been working on a thematic strategy on the sustainable use and management of resources since 2002.

For waste that continues to be generated, the 6EAP aims at a situation where

- Most of the wastes are either reintroduced into the economic cycle, especially by recycling, or returned to the environment in a useful (e.g., composted) or harmless form.
- The quantities of waste that still need to go to final disposal are reduced to an absolute minimum and are safely destroyed or disposed of.
- Waste is treated as close as possible to where it is generated.

41.5   Case Study

In 2002, a solid waste audit was conducted for a nitrocellulose-manufacturing facility, located in Taiwan. Figure 41.1 is a picture of nitrocellulose. The task team applied the waste minimization assessment methodology developed by Environmental Protection Agency (EPA) of the United States to reduce, recycle, and recover the wastes generated at one of its satellite plants, located at Taoyuan, Taiwan. Since implementing the waste minimization program in the late 1980s, the company not only has solved waste disposal problems associated with the production of nitrocellulose and met the stringent environmental regulations, but also improved the productivity over 20 percent and maintained itself in the top five position despite harsh international competition.

Before an effective waste minimization program is developed, managerial commitment plays an essential role. In the late 1980s, a task force was formed, consisting of engineers and production managers from both research and development division and the manufacturing facility.

The task force conducted a waste assessment process using the methodology developed by the EPA. The waste assessment process contains three basic steps: (1) assessment preparation; (2) assessment; and (3) feasibility analysis. This process has been described in Chap. 8 of this book. .

After completing the feasibility analysis, the task force made a list of selected options for implementation. The primary option involved the replacement of cotton linters with wood pulp as feedstock which reduced waste acids generated in nitrocellulose production significantly. The decision was mainly based on other considerations

Figure 41.1   Nitrocellulose.

such as cost reduction of the raw materials by 30 percent, improvement of the production yield, and so on; thus it was not considered as one of the many waste minimization projects. There is another advantage for the use of wood pulp as feedstock, which had not been taken into consideration previously. Due to a threefold increase of the yield per reactor, the company was able to boost the production from 2000 to 6000 metric tons per year with minor process modification at the facility located in Taoyuan, Taiwan, and to consolidate all domestic production in Taoyuan later. The consolidation further reduced the wastes generated during the production processes.

The implementation took almost 10 years to complete at the cost of $8 million. The accumulated benefit is roughly $4 million. The task force not only continued to look for new opportunities for waste minimization at the Taoyuan facility, but also helped establish similar programs at two production sites in China as well as one in the Philippines.

The overall project was completed in 1989 at the cost of $4 million. It covered the following:

- Pusher centrifuge at a capacity of 800 kg solid throughput
- 1 ton/h acid reconcentration unit
- Facilities expansion such as cooling and chilling water systems, chillers, high-pressure boilers, electrical system, and so on
- Storage tanks, prefiltration equipment, building, and other similar items
- Piping, instrumentation, and control equipment
- Site preparation, labor, and construction cost

With the application of the structured methodology developed by EPA, the company was able to implement a successful waste minimization program at one of its production facilities located in Taoyuan, Taiwan, in the late 1980s. Since the waste minimization program was implemented, the company has generated significantly fewer amounts of wastes and improved its competitiveness in the international market by reducing over $400 per metric ton of nitrocellulose produced while the capacity was increased from 6000 to 15,000 tons/year. The effluents from the facility also meet the requirements of the stringent environmental regulations.

41.6 Exemplary Performers— European Employee Award Programs

Volkswagen has a proven strategy to promote waste minimization. The company presents an internal environmental award to creative employees. In July of 2008 Volkswagen presented its internal environmental award for the fifth time to employees who have shown exceptional commitment and creativity with regard to environmental protection.

"We were delighted with the diversity of ideas submitted for this year's environmental award: we are particularly pleased to see two award winners from the Spanish plant in Pamplona, since this international participation underscores the global significance of environmental issues. The spectrum ranged from product development to production, with the personal commitment of our employees going well beyond the normal course of duty," Günter Damme, environmental management officer at the Volkswagen brand, commented in Wolfsburg in 2008. He added that this commitment was important in order to successfully establish environmental protection in a company such as Volkswagen. "Employees in all sections of the company must bring environmental protection to life. This year's award winners and their ideas ranging from clean TDI technology through cutting down on packaging material to reworking reject piston rods set a marvelous example," Günter Damme said.

The first prize for the 2008 internal environmental award, which is presented in the production and the product categories, went to the production department at the Pamplona plant in Spain. Pablo Romero Zalba, who works in body construction and is responsible for material supplies, came up with his own idea for developing and building a device for using the residual adhesive contained in the barrels delivered to the plant. His idea not only saved resources, but also reduced special waste by 5800 kg annually.

Second prize went to Holger Jerzewski from Brunswick, who devised a method for treating blued piston rods with a solvent-free cold cleaning agent so that these rods can be used rather than going to scrap. This simple method brings annual cost savings of around 120,000 Euros.

The third prize went to Andreas Kwiecinski and Holger Eigenbrod from Kassel; they work in the hardening shop and are responsible for the repair and maintenance of the hardening furnaces. Their idea optimizes furnace insulation and uses a thicker insulating material, thus reducing the size of the furnace combustion chamber and slashing gas consumption. The idea also lowers CO_2 emissions.

In the production category, two employees were honored for their long-standing commitment to environmental protection. Since 1999, Juan–A. Ferreiro da Silva from Hanover has shown enormous personal commitment to optimizing paint applications. He helped to cut solvent consumption by approximately 2 million kg since 1999. Michael Hübscher from the Wolfsburg plant has been an expert in environmental protection since 1994 and has coordinated and monitored the use of all chemicals for hardening and galvanizing applications in prototype construction. He has shown great initiative in implementing numerous environment-related projects over the years.

First prize in the product category went to Matthias Leifheit, Jörg Worm, Uwe Behlendorf, and Christian Eigen for the development of the natural gas-powered Passat TSI-CNG. All four showed considerable initiative going well beyond their project remit in developing a power train concept featuring a turbocharged engine fuelled by natural gas. This twin-charging technology has given Volkswagen a head start in realizing a CNG engine with optimum consumption and performance characteristics. Moreover, the significantly lower weight brought by fitting thin-walled gas tanks has brought CO_2 emissions down to 129 g/km. This low value contributes to cutting fleet CO_2 emissions.

The second prize in this category went to Steffen Hunkert and his team from diesel engine development; they used intelligent engine control to optimize the interaction between performance, consumption, and emissions, thus lowering fuel consumption.

This engine application was first used in the Polo, Passat, and Audi A3, and subsequently for all other BlueMotion models. The fine tuning, combined with other vehicle measures such as lower driving resistance and longer gear ratios, culminated in a successful overall concept.

Richard Dorenkamp and his team from diesel engine development were awarded third place in the product category for their clean TDI technology Jetta BlueTDI for the U.S. market. The Jetta BlueTDI is the first diesel model in the United States to comply with the world's most stringent exhaust standard: the 90 percent reduction in NO_x emissions was achieved through engine design measures in combination with novel exhaust gas treatment.

The personal prize in this category was awarded to Dzemal Sjenar, concept definition, who was honored for his great personal commitment to designing environmentally compatible vehicles. The ideas generated by Dzemal Sjenar during the concept phase lent considerable momentum to the later development of the BlueMotion strategy and he was also involved in concept definition and prototype construction of new and alternative power train and vehicle concepts.

All Europe-based employees of the Volkswagen and Volkswagen Commercial Vehicles brands, Volkswagen Sachsen GmbH, and departments in Wolfsburg performing group functions are eligible to apply for the internal environmental award. Ideas relating to new methods, materials, and products which have already proven their worth in practice were eligible to apply for the 2008 award. Projects still at the ideas stage cannot therefore be considered.

As a result of the positive response, the internal environmental award will with effect from this year no longer be presented every 2 years but will become an annual event.

INTERNATIONAL SERVICE
APPLICATIONS

42.1   International Waste Management Goals and Opportunities in Service

For the international community, solid waste generation in service organizations is receiving increased attention. The majority of solid wastes generated by these organizations are similar to their counterparts in the United States. These wastes are mixed office papers, food waste, and yard waste. New statistics show that the paper- and board-recycling rate in Europe reached 64.5 percent in 2007, which confirms that the industry is on track to meet its voluntary target of 66 percent by 2010.

According to the European Recovered Paper Council (ERPC) the total amount of paper collected and sent to be recycled in paper mills was 60.1 million tons, an increase of 7.6 million tons (or +14.5 percent) since 2004, the base year for the target.

42.2   Constraints and Considerations

The key constraints associated with solid waste minimization for international service organizations are similar to those of manufacturing organizations. These constraints and considerations include

- Increased waste disposal costs
- Increased and varying government regulations
- Lack of recycling vendors in certain areas
- Space constraints within and outside the building to stage materials
- Separation and contamination of office papers (both the process, equipment and labor required)
- In some cities, narrow streets

42.3   Potential Technologies and Strategies

Several proven strategies exist to implement and sustain office oriented–recycling programs. Most are very low cost, but require a moderate to high degree of communication and coordination, below is a brief list:

■ *Establish waste and recycling agreements with nearby businesses*—Businesses that operate in business centers or industrial parks can achieve significant cost savings by consolidating waste hauling and recycling contracts. Business tenant meetings or board meetings are appropriate venues to discuss these options.

■ *Create incentive programs for employees*—Motivating employees to recycle is one of the most important components of any waste minimization program. Offering awards to top performing units or individuals can help achieve this goal. The awards may include money, corporate logo gifts, paid time off, or public recognition. A company-wide luncheon or breakfast may also be provided if the organizational waste minimization goals are met.

■ *Set defaults to double sided for copiers*—Studies have shown that setting copier defaults to double sided may reduce paper waste by 40 percent. Most copiers are set to single side and users must select double sided.

■ *Use paperless fax machines*—Send files electronically to fax machines and set defaults to not print a fax confirmation.

■ *Provide employees with desk side recycling bins*—Providing each employee in the office with a desk-side recycling container will enhance motivation to recycle. The container will serve as a visual reminder to recycle and reduce travel time to a consolidated recycle bin. The advantages of desk-side recycling include the maximum potential for collecting the most amount of paper for recycling and provide the most convenient method for employees to dispose of both recyclable and nonrecyclable materials. Disadvantages include the potential for contamination of recyclable papers if trash is disposed of into the wrong bin and the initial cost incurred for extra recycle bins.

■ *Provide employees with periodic newsletters*—These letters may contain information on recycling performance trends for the company, and announce awards and new programs.

■ *Appoint a green cop to monitor individual employee performance*—Designating an employee to police the material entering the waste stream and recycling bins can motivate employees to follow proper procedures. The individual should be instructed to report all findings (conformance and nonconformance) to managers and other employees.

■ *Shredding of confidential documents*—Establish a relationship with a vendor to recycle shredded paper. Many waste removal and recycling companies offer confidential shredding services. Ensure that the company is certified by the NAID. NAID offers a certification program on a voluntary basis to all member companies providing information-destruction services. Through the program, NAID members may seek annual certification audits for both mobile and plant-based operations in paper or printed media, micromedia, or computer hard drive destruction. The NAID certification program establishes standards for a secure destruction process including such areas as operational security, employee hiring and screening, the destruction process, responsible disposal, and insurance.

■ *Implement reverse recycling*—An emerging concept involves eliminating desk-side waste bins and only providing employees with desk-side recycling bins. The custodial service is instructed to service only recycling bins at employees' desks and the employee is responsible for transporting nonrecyclable waste to centralized containers in employee break rooms. The purpose of the concept is motivating employees to recycle by creating barriers to generating nonrecyclable waste. By employing this concept, a Toledo, Ohio–based company was able to reduce waste disposal costs by 40 percent and consolidate general waste removal from 5 days per week to 2.

42.4   Implementation and Approach

Below are general procedures for implementing a solid waste reduction program in an office setting:

■ Describe the direction, objectives, and proposed goal(s) of the recycling program.
■ Define the requirements necessary to accomplish this.
■ Identify the waste categories and types.
■ Review and evaluate the following:
    ■ Current collection and disposal practice(s)
    ■ Current paper collection concept
    ■ Current waste disposal contract(s)
■ Assess composition of general cafeteria trash.
■ Identify new wastes not currently identified or recovered.
■ Develop campaign concept and requirements.
■ Develop facility departments' interface and support requirements.
■ Schedule for program scope work.
■ Develop capital and expense budget outline.
■ Search out and contact local reclaiming and recycling vendors.
■ Use services of local consultant(s) as needed.

Office paper constitutes 45 percent of the waste generated for these businesses and marks the starting point for recycling programs in this sector. Office paper recycling has a high level of visibility, helps to foster employee buy-in, and can generate early wins for the recycling program. The following are step-by-step procedures for instituting an office paper–recycling program. The steps include

■ Determining what papers need to be recycled
■ Locating a vendor who meets your requirements
■ Procedures for setting up a program

■ Monitoring your recycling progress throughout the year
■ Keeping the program alive

The success of any office recycling program depends on the support and cooperation from every employee, from the highest levels of management to the employees who carry out the actual collection procedures. In addition, effective relationship management with recycling vendors is critical. Below is a short list of vendor requirements for office paper recycling programs:

■ Reliability of service in terms of on-time pickups, accurate tonnage reimbursements and monthly reporting, and timely reimbursements
■ Ability to handle documents in accordance with security guidelines
■ Willingness to pick up all kinds of papers including high grade, mixed grade, file stock, and corrugated
■ Maximum prices offered for mixed grades and file stock which comprise most of recyclable tonnage

Monitoring recycling program performance is an important step to ensure the long-term success of the program. Below is a list of activities that may be applied to accomplish this goal:

■ Keep accurate and up-to-date statistics:
　■ Include tonnage figures
　■ Include dollars received from recycling vendor
　■ Include estimated cost-avoidance figures
■ Periodically inspect office areas to determine if employees are properly disposing of recyclable materials.
■ Periodically arrange to have outside garbage unit analyzed by following unit to transfer station to weigh contents and observe contents of unit:
　■ Determine what percent of contents consists of recyclable materials
　■ Determine if container is being pulled prematurely
■ Post recycling statistics in your building so that employees can see your progress.
■ Publish updated statistics and articles in company bulletins and newsletters.
■ Display recycling posters throughout the year.

42.5   Case Study

In 2008 the Heathrow Airport in the United Kingdom undertook a large-scale sustainable development at one of the largest and most complicated building sites in Europe. BAA, the owner of seven airports in the United Kingdom, completed a major construction project, the biggest single construction project in Europe, at Terminal 5 at Heathrow. At a cost of over £4 billion, the project represents a huge program of construction works. Using traditional materials and construction techniques can have a significant environmental impact. BAA has sought to lead the U.K. construction industry in the use of sustainable construction materials:

- HCFCs and HFCs have been almost completely eliminated from the T5 project.
- Only timber that has been approved by the Forestry Stewardship Council (FSC) has been used by BAA, ensuring it's been sourced from a sustainable supply.
- Over 300,000 tones of aggregate was processed and recycled on site from demolition materials and waste concrete.
- Crushed green glass from domestic household–recycling banks was used as a base for T5 site roads.
- Around 6.5 million cubic meters of earth was moved during the project. This earth has been used to backfill excavations and landscape the terminal.
- Waste materials were segregated on site and 85 percent of the waste from the project has been recycled.

In addition, efforts have been made to reduce emissions and water usage. Heating, cooling, and powering airport terminals all require energy that contributes to CO_2 emissions. BAA is on target to reduce its own CO_2 emissions from energy by 15 percent below 1990 levels by 2010. This is in excess of the United Kingdom's targets under the Kyoto treaty and despite a predicted growth in passenger numbers of 70 percent during this period.

Water for nonpotable use at T5 is sourced from a rainwater-harvesting scheme and groundwater boreholes, reducing the demand on the public water supply by 70 percent. The rainwater-harvesting system reuses 85 percent of all the rainwater that falls on T5. All toilets, taps, and showers are water-efficient.

In terms of solid waste management, BAA has implemented a large program to recycle waste and transform it into energy. BAA has a target of recycling or composting 40 percent of its waste by 2010 and 70 percent of its waste by 2020. They also aim to send zero waste to landfill by 2020. From 2008 waste that currently goes to landfill will go to an energy-generating waste-incineration plant. BAA is also investigating turning its nonrecyclable waste into energy through an on-site anaerobic digestion plant.

Heathrow will be one of the first sites in Britain to trial generating zero-carbon energy from waste heat. The new technology will convert waste heat from Heathrow's existing boilers into zero-carbon electricity. The electricity generated will be used to power the terminals without burning more fossil fuels.

42.6   Industry Leaders—The International Solid Waste Association

The International Solid Waste Association (ISWA) is an international, independent, and nonprofit association, working in the public interest to promote and develop sustainable waste management worldwide. ISWA has members around the world and is the only worldwide association promoting sustainable and professional waste management.

The association is open to individuals and organizations from the scientific community, public institutions, and public and private companies from all over the world working in the field of and interested in waste management.

ISWA is the only worldwide association that is working on waste matters. Being part of this association gives an unparalleled access to international organizations such as UNEP, WHO, or in Europe the EU, where ISWA is an accredited association for waste questions. Through their working groups, they help these international organizations form policy by supplying technical papers and opinions regarding waste practices. It is also the only worldwide waste organization, which allows individuals to network with professionals, companies, and institutional representatives.

By being part of ISWA individuals and organizations can make a fundamental contribution to politicians and decision-makers around the world to develop waste management strategies.

REFERENCES

"A Brief History of Solid Waste Management in the U.S. During the Last 50 Years," MSW Management, Available at http://environmentalchemistry.com/yogi/environmental/wastehistory .html, Accessed March 12, 2009.

Allen, D. T., and K. S. Rosselot, *Pollution Prevention for Chemical Processes*, John Wiley and Sons, New York, 1997.

Anderberg, M., *Cluster Analysis for Applications*, Academic Press, New York, 1973.

Babarki, K., "A Theoretical Framework to Study Recycling Performance Changes Following Achievement of ISO 14001 Registration," The University of Toledo, the Department of Mechanical, Industrial, and Manufacturing Engineering, Toledo, Ohio, Dec. 2002.

Berenyi, E. B., *2001–2002 Materials Recycling and Processing in the United States*, 4th ed., Governmental Advisory Associates, Inc., Westport, Conn., 2002.

Briones, R., *California Semiconductor Industry Hazardous Waste Source Reduction Assessment Report*, EPA, doc. No. 549, California, June 2006.

California Integrated Waste Management Board, www.ciwmb.ca.gov/2000plus/ history.htm.

Chatterjee, S. and B. Price, *Regression Analysis by Example*, John Wiley and Sons, New York, 1977.

"Commercial Generation Study Palm Beach County Florida," Solid Waste Authority of Palm Beach County, West Palm Beach, Fla., 1995.

Covey, S. K., "Building Partnerships—The Ohio Materials Exchange," *Resources, Conservation and Recycling,* 2000, 28(2), pp. 265–277.

Daskalopoulos, E., "Municipal Solid Waste: A Prediction Methodology for the Generation Rate and Composition in the European Union Countries and the United States of America," *Resources, Conservation and Recycling,* Nov. 1998, 24(2), pp. 155–166.

"Decision Maker's Guide to Solid Waste Management, Volume II," United States Environmental Protection Agency, 1995, Available at http://www.epa.gov/osw/nonhaz/ municipal/dmg2/chapter5.pdf, Accessed March, 2009.

Dennis, A., and B. Wixom, *Systems Analysis and Design*, John Wiley and Sons, New York, NY, 2000.

Department of the Army (Corps of Engineers), *Solid Waste Management,* United States Army, 1990.

Diaz, L. F., G. M. Savage, L. L. Eggerth, and C. G. Golueke, *Composting and Recycling Municipal Solid Waste*, Lewis Publishers, Boca Raton, Fla., 1993.

Dielman, T. E., *Applied Regression Analysis for Business and Economics*, Wadsworth Publishing Company, Belmont, Calif., 1996.

Dillon, W., and M. Goldstein, *Multivariate Analysis*, John Wiley and Sons, New York, 1984.

"District Solid Waste Management Plan Format," Ohio EPA, 1996.

"Environment at a Glance: OECD Environmental Indicators," OECD, Paris 2006.

"Environmental Management and Monitoring for Sport Events and Facilities: A Practical Toolkit for Managers," Prepared for Department of Canadian Heritage Sport Canada by Green & Gold Inc., Mar. 1999.

Everitt, B., *Cluster Analysis*, Halsted Press, New York, 1980.

"Financial Analysis of Pollution Prevention Projects," Office of Pollution Prevention—Ohio EPA, Columbus, Ohio, Oct. 1995.

Fishbein, B., *Germany, Garbage, and the Green Dot: Challenging the Throwaway Society,* Inform Publishing, New York, 1994.

Foecke, T., "Defining Pollution Prevention and Related Terms," *Pollution Prevention Review*, 2(1), Feb. 1991/92, pp. 1003–1112.

Franchetti, M., "One Company's Trash is Another Company's Treasure: With the Rising Volumes of Recyclable Materials Entering the Waste Stream, Waste Commodity Exchanges Have Started Receiving Increased Interest from Companies Looking to Trim Their Bottom Line," *Resource Recycling: North America's Recycling and Composting Journal,* Feb. 2008, pp. 40–42.

Franchetti, M., "Reengineering the Systems Approach to Solid Waste Management and Minimization," The University of Toledo, the Department of Mechanical, Industrial, and Manufacturing Engineering, Toledo, Ohio, Dec. 1999.

Freeman, H., *Waste Minimization Opportunities Assessments Manual,* Environmental Protection Agency, Ohio, US Environmental Protection Agency, Washington, D.C., 1998.

"Full Cost Accounting for Municipal Solid Waste Management," United States Environmental Protection Agency, Available at http://www.epa.gov/epaoswer/non-hw/muncpl/fullcost/docs/fca-hanb.pdf, Accessed Mar. 2006.

"Garbage Primer," The League of Women Voters, Washington, D.C., Lyons and Burford Publishers, Dec. 1993.

"Garbage Then and Now," the Environmental Industries Associations, Washington, D.C., 2008.

Goddard, H. C., *Managing Solid Waste: Economics, Technology, and Institutions,* Praeger Publishers, New York, 1975.

Graham, R. C., *Data Analysis for the Chemical Sciences; A Guide to Statistical Techniques,* VCH Publishers, New York, 1993.

Gupta, P., C. Pexton, J. H. Harrington, B. E. Trusko, *Improving HealthCare Quality and Cost with Six Sigma*, Pearson Education, New Jersey, 2007.

Hair, J., R. Anderson, R. Tatham, and W. Black, *Multivariate Data Analysis*, Prentice Hall, Inc., Upper Saddle River, N.Y., 1998.

Haman, W., "Total Assessment Audits (TAA) in Iowa," *Resources, Conservation and Recycling*, 2000, 28, pp. 189–198.

Harris, E., *Cornell Waste Management Institute Update June 2009*. Cornell Waste Management Institute. Ithaca, N.Y., 2009.

Henry, J. G., and G. W. Heinke, *Environmental Science and Engineering*. Prentice Hall, Englewood Cliffs, N.J., 1989.

Hillier, F., and G. Lieberman, *Introduction to Operations Research*, McGraw-Hill, Inc., New York, N.Y., 1995.

"History & Background" Lucas County Solid Waste Management District, U.S. Environmental Protection Agency, Washington, D.C., Aug. 1993.

"How to Negotiate with Waste Haulers," Office of Pollution Prevention—Ohio EPA Household Hazardous Waste Management: A Manual for One-Day Community Collection Programs, Columbus, Ohio, Jan. 1994.

"Industrial Assessment Center Brochure 2001," U.S. Department of Energy, Office of Energy Efficiency and Renewable Energy, Washington D.C., 2001.

Johnson, R., and D. Wichern, *Applied Multivariate Analysis*, Prentice Hall, Inc., New York, 1988.

Kachigan, S., *Multivariate Statistical Analysis*, Radius Press, Inc., New York, 1991.

Keller, G., B. Warrack, and H. Bartel, *Statistics for Management and Economics,* Duxbury Press, Belmont, Calif., 1994.

Lorr, M., *Clustering Analysis for Social Scientists*, Jossey-Bass, Inc., San Francisco, Calif., 1983.

"Measuring the Environmental Performance of Industry, Final Report," Science and Technology Policy Research Center, University of Sussex, United Kingdom, 2001.

Miller, C., "An End to Volatility," *Waste Age*, 21 June 2005 Available at http://www.wasteage.com/mag/waste_end_volatility/index.html, Accessed Nov. 2006.

Minkley, M., "A Systems Approach to Analysis and Minimization with Emphasis on Urethane Recycling," The University of Toledo, the Department of Mechanical, Industrial, and Manufacturing Engineering, Toledo, Ohio, June, 1997.

"Mixed Waste Processing in New York City: A Pilot Test Evaluation," New York City Department of Sanitation, Oct. 1999, Available at http://www.nyc.gov/html/nycwasteless/downloads/pdf/mixed_waste.pdf, Accessed Mar. 2006.

"Mixed Waste Processing," NC Division of Pollution Prevention and Environmental Assistance, January 1997, Available at http://www.p2pays.org/ref/01/00028.htm, Accessed Feb. 2006.

Montgomery, D., *Introduction to Linear Regression Analysis*, John Wiley and Sons, New York, 2001.

Montgomery, D., *Introduction to Statistical Quality Control*, John Wiley and Sons, New York, 1997.

"Modern Marvels: History of Garbage," video produced by the History Channel, 2003.

"Municipal Solid Waste Generation, Recycling, and Disposal in the United States: 2000 Facts and Figures," Franklin and Associates for the EPA, June 2002.

"Municipal Solid Waste Generation, Recycling, and Disposal in the United States: Facts and Figures for 2003," United States Environmental Protection Agency, Washington D.C., 2003.

"Municipal Solid Waste—Basic Facts," United States Environmental Protection Agency, Available at http://www.epa.gov/epawaste/nonhaz/municipal/index.htm, Accessed March 2009.

Norusis, M., *Advanced Statistics SPSS/PC+,* SPSS, Inc., Chicago, Ill., 1986.

"OECD Environmental Data Compendium 2004," OECD, Paris 2005.

Ohio EPA. "Governor's Pollution Prevention Award Recipient: Mahoning County's Industrial Waste Minimization Project," Fact Sheet #41, 1997.

Pande, P. S., R. P. Neuman, and R. R. Cavanagh, *The Six Sigma Way*, McGraw-Hill, New York, 2000.

Parks, C. S., *Contemporary Engineering Economics*, Addison-Wesley Publishing Company, New York, 1993.

Rathje, W. and C. Murphy, *Rubbish!: The Archeology of Garbage*, University of Arizona Press, March, 2001.

PEER Consultants and CalRecovery Inc., *Material Recovery Facility Design Manual*, CRC Press, Inc., Boca Raton, Fla., 1993.

"Recycling Managers—Prices," *American Metal Market,* Available at http://www.amm.com/recman/, Accessed Nov. 2006.

"Recycling" United States Environmental Protection Agency, Available at http://www.epa.gov/osw/conserve/rrr/, Accessed March 2009.

Reitz, D. P., "Industrial Waste Analysis and Minimization Modeling with Computer Simulation," The University of Toledo, the Department of Mechanical, Industrial, and Manufacturing Engineering, Toledo, Ohio, May, 1998.

Rhyner, C. R., L. J. Schwartz, R. B. Wenger, and M. G. Kohrell, *Waste Management and Resource Recovery*, Lewis Publishers, Boca Raton, Fla., 1995.

Romesburg, C., *Cluster Analysis for Researchers*, Wadsworth Publishing Company, Inc., Belmont, Calif., 1984.

Ryding, S., "International Experiences of Environmentally Sound Product Development Based on Life Cycle Assessment," Swedish Waste Research Council, ARF Report 36, Stockholm, May 1994.

Scheaffer, R., *Elementary Survey Sampling*, Wadsworth Publishing Company, Inc., Belmont, Calif., 1996.

Shimberg, S. J., "The Hazardous and Solid Waste Amendments of 1984: What Congress Did ... and Why," *The Environmental Forum*, Mar. 1985, pp. 8–19.

Siegler, T., and N. Starr, "Sorting Out Recycling Rates," *Resource Recycling,* May 1999.

"Solid Waste Management Plan," Lucas County Solid Waste Management District, Updated 2005.

"State of Ohio Waste Characterization Study," Ohio Department of Natural Resources (Prepared by: Engineering Design & Design, Inc.), Apr. 2004.

"State Solid Waste Management Plan 2001," Ohio EPA—Division of Solid and Infectious Waste Management, Columbus, Ohio, 2001.

"Statewide Waste Characterization Study, Results and Final Report," California Integrated Waste Management Board, Sacramento, Calif., 1999.

"Strong, D. L., *Recycling in America*, 2d ed., ABC-CLIO, Inc., Santa Barbara, Calif., 1997.

Tchobanoglous, G., H. Theisen, and S. Vigil, *Integrated Solid Waste Management: Engineering Principles and Issues,* McGraw-Hill, New York, 1993.

Tryon, R., and D. Bailey, *Cluster Analysis*, McGraw Hill, Inc., New York, 1970.

2004 Annual District Report for Lucas Co. SWMD," Ohio Environmental Protection Agency—Division of Solid and Infectious Waste Management, Dec. 2005.

"2004 Municipal Recycling Report," City of Toledo—Hoffman Road Landfill, 2005.

U.S. Environmental Protection Agency, "The Nation's Hazardous Waste Management Program at a Crossroads," Report No. EPA/530-SW-90-069, EPA Washington, D.C., July 1990, p. 114.

U.S. Environmental Protection Agency, Office of Solid Waste, *RCRA Orientation Manual*, U.S. Government Printing Office, Washington, D.C., 1998, p. 160.

U.S. Environmental Protection Agency, *The Guardian: Origins of the EPA,* EPA Historical Publication-1, April 1992, (April).

United States Congress of Technology Assessment (US Congress OTA), "Green Products by Design: Choices for a Cleaner Environment," OTA-E-541, U.S. Government Printing Office, Washington, D.C., Oct. 1992.

Walpole, R. E., R. H. Myers, S. H. Myers, and, K. Ye, *Probability and Statistics for Engineers and Scientists*, Prentice Hall, New Jersey, 2002.

Whitwell, J. C., and R. K. Toner, *Conservation of Mass and Energy,* Blasidell Publishing Company, Waltham, Mass., 1969.

Wood, M. F., J. A. King, and N. P. Cheremisinoff, *Pollution Prevention Software System Handbook*, Noyes Publications, Westwood, N.J., 1997.

Yonger, M., *A First Course in Linear Regression*, PWS Publishing, Boston, Mass., 1985.

Yonger, M., *Handbook for Linear Regression*, Duxbury Press, Inc., Belmont, Calif., 1979.

Zupan, J., *Clustering of Large Data Sets*, John Wiley and Sons, Letchworth, United Kingdom, 1982.

www.apexq.com (APEXQ-Secondary is the world's largest database of secondary plastic for sale)

www.bizstats.com (Business statistics website)

www.census.gov (U.S. Census Bureau)

http://cwmi.css.cornell.edu/ (Cornell Waste Management Institute)

www.ciwmb.ca.gov/WasteChar/ (California Environmental Protection Agency)

www.ciwmb.ca.gov/WasteChar/ (California Waste Characterization Studies)

www.dep.state.fl.us/ (Florida Department of Environmental Protection)

www.dnr.state.wi.us (Wisconsin Department of Natural Resources)

www.dol.gov (U.S. Department of Labor)

www.encarta.msn.com (Encarta Reference)

www.environmental-performance.com (MEPI Project)

www.epa.gov (US Environmental Protection Agency)

www.epa.gov/epaoswer/non-hw/recycle/recmeas/ (EPA conversion factors)

www.money.cnn.com (corporate financial and operating data)

www.epa.ohio.gov (Ohio Environmental Protection Agency)

www.indstate.edu/recycle/ (Indiana Institute on Recycling)

www.oit.doe.gov/iac (US Department of Energy Industrial Technology Program)

www.recycle.net/price (Current market prices for the sale of recyclables)

www.redo.org (Reuse Development Organization)

www.superpages.com (U.S. corporation and business data warehouse)

www.surveysystem.com (Creative Research Systems)

http://www.swa.org/pdf/residential_generation_study.pdf (1995 Commercial Generation Study Palm Beach County, Florida)

www.swana.org (Solid Waste Association of North America—used for landfill disposal charges)

http://www.census2010.gov/ipc/www/idb/worldpopinfo.html.

http://www.wasteonline.org.uk/resources/InformationSheets/HistoryofWaste.htm.

INDEX

CPSIA information can be obtained
at www.ICGtesting.com
Printed in the USA
LVHW060157110123
736924LV00006B/55